Protein Design

METHODS IN MOLECULAR BIOLOGY™

John M. Walker, SERIES EDITOR

Protein Design

Methods and Applications

Edited by

Raphael Guerois

Département de Biologie Joliot–Curie
CEA Saclay
Gif-sur-Yvette, France

Manuela López de la Paz

Merz Pharmaceuticals GmbH
Frankfurt am Main
Germany

HUMANA PRESS ✳ TOTOWA, NEW JERSEY

© 2006 Humana Press Inc.
999 Riverview Drive, Suite 208
Totowa, New Jersey 07512

www.humanapress.com

This publication is printed on acid-free paper. ∞
ANSI Z39.48-1984 (American Standards Institute) Permanence of Paper for Printed Library Materials.

Cover design by Patricia F. Cleary

Cover illustration: Three-dimensional structure of CD4M33 in complex with gp120YU2, as determined by crystallography (lyyl). *See* complete caption and discussion on p. 140.

For additional copies, pricing for bulk purchases, and/or information about other Humana titles, contact Humana at the above address or at any of the following numbers: Tel.: 973-256-1699; Fax: 973-256-8341; E-mail: orders@humanapr.com; or visit our Website: www.humanapress.com

Printed in the United States of America. 10 9 8 7 6 5 4 3 2 1

eISBN 1-59745-116-9

ISSN 1064-3745

Library of Congress Cataloging-in-Publication Data

Protein design : methods and applications / edited by Raphael Guerois,
    Manuela López de la Paz.
      p. cm. — (Methods in molecular biology ; 340)
    Includes bibliographical references and index.
    ISBN 1-58829-585-0 (alk. paper)
    1. Protein engineering.   I. Guerois, Raphael. II. López de la Paz,
  Manuela. III. Series: Methods in molecular biology (Clifton, N.J.) ;
    v. 340.
  TP248.65.P76P727 2006
  660.6'3—dc22

                                                          2005055116

Preface

Proteins have evolved through selective pressure to accomplish specific functions. The functional properties of proteins depend upon their three-dimensional structures, which result from particular amino acid sequences folding into tightly packed domains. Thus, to understand and modulate protein function rationally, one definitely needs methods and algorithms to predict and decipher how amino acid sequences shape three-dimensional structures. Protein design aims precisely at providing the tools to achieve this goal.

The predictive power of rational protein design methods has dramatically increased over the past five years. A broad range of studies now illustrate how the sequence of proteins and peptides can be tuned to engineer biological tools with intended properties (*1–3*). The extensive characterization of peptides and protein mutants has enormously benefited the understanding of protein sequence-to-structure relationships. Synergies between computational and experimental approaches have also added momentum to the advancing limits of design methods. The potential applications in fundamental biochemistry and in biotechnology justify the considerable excitement that this progress has generated within the research community. The field is probably mature enough so that expert knowledge can assist researchers of diverse disciplines to rationally create or modify their favorite protein. Thus, the aim of *Protein Design: Methods and Protocols* is to account for the most up-to-date protein design and engineering strategies so that readers can undertake their own projects with maximum confidence in a successful return.

The basic concepts underlying rational design of proteins are intimately related to their three-dimensional structures. The stability of a given structure results from a complex combination of interactions that favor a specified conformation at the expense of any alternative one. Researchers have devised different strategies to extract the general principles on which protein structure is based. Proteins have been systematically mutated to address the question of how specific residues affect the stability of a given protein (*4,5*). Proteins have been also "redesigned," starting with a protein of known structure and dramatically modifying features of its construction (*6*). For the sake of simplicity, initial works in the field of design were dedicated to the elucidation of the factors contributing to the stability of elementary building blocks. Peptide model systems have been shown to be very suitable to this end. They have served to dissect the relative energetic contributions of short- and long-range interactions to a given folding motif. They have provided key insights into the relationship between sequence, folded structure, and stability *(7)*. Major accomplishments have been achieved in the design and structural characterization of helical peptides and proteins *(8,9)*. The main factors underlying α-helix stability have been largely identified, leading to advances in the rational design of helical proteins. Protein stability has been enhanced by maximizing helical propensities at specific sites, and protein structures have even been redesigned to adopt different folded topologies. The design of α-helices is surveyed in Chapter 1.

The rational understanding of β-sheet structure and stability has remained, however, more elusive, and it is only during the past five years that similar success has been achieved *(10)*. In contrast to α-helices, β-sheets are propagated by residues remote in the polypeptide backbone. As a consequence, whereas in model helical peptide structure stabilization is largely a result of interactions between neighboring residues (local interactions), nonlocal interactions make important contributions to the stability of even minimal β-sheet peptides and proteins. This fact together with the intrinsic tendency of β-sheets to aggregate can be recognized as the main impediment to a comprehensive understanding of β-sheet structures. These studies are reviewed in Chapter 2. The basic rules derived from the analysis of these model systems can be used directly by the reader to increase the stability of a given protein through the local optimization of their constitutive secondary structure blocks. The predictive power of these rules has been tested already by the design, completely *de novo*, of several peptides and miniproteins. The conjunction of rational design principles and combinatorial approaches has been very successful also at finding sequences that highly populate a desired folding motif *(11,12)* and their applicability is demonstrated in Chapter 3.

These fundamentals can be exploited also to design, modify, or improve the interaction between peptide ligands and their receptor targets. This interaction commonly involves the formation of beta structures. Yet poor bioavailability and unfavorable pharmacokinetics significantly compromise the use of peptides as drugs. An additional problem is their conformational flexibility, which results in poor binding to the target. Thus, there is a great deal of interest in designing peptidomimetics with improved structural properties as therapeutical agents by mimicking β-turn and β structures. To this end, D-amino acids have been strategically introduced in polypeptide backbones to decrease the conformational flexibility of β-hairpin and β-sheet peptides designed *de novo* *(13)*. β-peptides constitute one of the most important families of nonnatural polymers with the propensity to form well-defined secondary structures. They are attracting more and more attention because they have been found to have various applications in medicinal chemistry and biochemistry *(14)*. These topics are covered in Chapters 4 and 5.

Procedures and strategies for engineering helices or β-sheets, solvent-exposed positions, or buried ones, common folds, or rare ones differ substantially. It is often difficult to account for all these factors using simple rules or relationships. Besides, one has rapidly to face a huge combinatorial complexity while increasing the number of positions in a sequence that are to be engineered simultaneously. For that purpose, integrated computational approaches have been developed based on different strategies *(15–18)*. *Protein Design: Methods and Protocols* presents several of these algorithms, which require various degrees of computational complexity (Chapters 7–9). In addition to the basic philosophy underlying their work, the practical comments of the authors on the use of their tools will be of major interest to experimentalists selecting the strategy most adapted for their design problem.

Since protein binding is fundamentally ruled by the same laws as protein folding, the lessons learned by designing stable proteins have paved the way for important progress in the engineering of protein complex interfaces. This issue has a tremendous impact in many biological fields because it allows one to modulate the way protein-interaction networks in cells are organized. The specificity of protein–protein complex engineering is discussed through the success of three different applications (Chapters 6, 10, and 11).

A frequent pitfall hindering successful designs is the tendency of the engineered molecule to aggregate. Unspecific aggregation processes can trap most of the designed protein into amorphous aggregates. In other cases, proteins can aggregate in an organized fashion and lead to the formation of fibrillar aggregates, known as amyloid fibrils. Amyloid fibrils are also associated with a range of human disorders, such as spongiform encephalopathies, Alzheimer's disease, type II diabetes, and so forth *(19)*. Recent progress in understanding the relationship between protein sequence and protein aggregation processes have provided clues on how to escape from these conformational traps *(20)*. This knowledge may help to negatively design sequences that, while maintaining the compatibility with the template fold, either decrease or fully prevent self-association processes. Knowledge-based tools might also be applied to predict protein fragments responsible for the amyloidogenic behavior of a given pathogenic protein, and, as a further application, to design or screen for inhibitor molecules that specifically interact with these key aggregating regions, preventing aggregation or increasing clearance of the misfolded protein. Design approaches, validation methods, and application to predicting such behavior are discussed in Chapter 12. Therapeutic approaches that are currently under scrutiny for preventing or curing amyloidoses or protein misfolding diseases in general are discussed in Chapter 13.

How can I handle the design of my protein? How can I improve the binding of this peptide to my target protein? Might I avoid protein aggregation while retaining fold and stability? Which structural features should be considered with acute attention? How good are we at translating angstroms into calories? These are central questions addressed throughout *Protein Design: Methods and Protocols* with the expectation that researchers can find their way toward achieving successful designs.

Raphael Guerois
Manuela López de la Paz

REFERENCES

1. Kortemme, T. and Baker, D. (2004) Computational design of protein-protein interactions. *Curr. Opin. Chem. Biol.* **8,** 91–97.
2. Bolon, D. N., Voigt, C. A., and Mayo, S. L. (2002) De novo design of biocatalysts. *Curr. Opin. Chem. Biol.* **6,** 125–129.
3. Brannigan, J. A. and Wilkinson, A. J. (2002) Protein engineering 20 years on. *Nat. Rev. Mol. Cell Biol.* **3,** 964–970.
4. Mendes, J., Guerois, R., and Serrano, L. (2002) Energy estimation in protein design. *Curr. Opin. Struct. Biol.* **12,** 441–446.
5. Bava, K. A., Gromiha, M. M., Uedaira, H., Kitajima, K., and Sarai, A. (2004) ProTherm, version 4.0: thermodynamic database for proteins and mutants. *Nucleic Acids Res.* **32** (Database issue), D120–121.
6. Regan, L. (1999) Protein redesign. *Curr. Opin. Struct. Biol.* **9,** 494–499.
7. Serrano, L. (2000) The relationship between sequence and structure in elementary folding units. *Adv. Protein Chem.* **53,** 49–85.
8. Munoz, V. and Serrano, L. (1995) Helix design, prediction and stability. *Curr. Opin. Biotechnol.* 6, 382–386.
9. Doig, A. J. (2002) Recent advances in helix-coil theory. *Biophys. Chem.* **101–102,** 281–293.
10. Lacroix, E., Kortemme, T., Lopez de la Paz, M., and Serrano, L. (1999) The design of linear peptides that fold as monomeric beta-sheet structures. *Curr. Opin. Struct. Biol.,* 9, 487–493.
11. Pastor, M. T., Lopez de la Paz, M., Lacroix, E., Serrano, L., and Perez-Paya, E. (2002) Combinatorial approaches: a new tool to search for highly structured beta-hairpin peptides. *Proc. Natl. Acad. Sci. USA* **99,** 614–619.
12. Wei, Y., Liu, T., Sazinsky, S. L., Moffet, D. A., Pelczer, I., and Hecht, M. H. (2003) Stably folded de novo proteins from a designed combinatorial library. *Protein Sci.* **12,** 92–102.
13. Venkatraman, J., Shankaramma, S. C., and Balaram, P. (2001) Design of folded peptides. *Chem. Rev.* **101,** 3131–3152.
14. Cheng, R. P., Gellman, S. H., and DeGrado, W. F. (2001) β-Peptides: from structure to function. *Chem. Rev.* **101,** 3219–3232.

15. Kuhlman, B., Dantas, G., Ireton, G. C., Varani, G., Stoddard, B. L., and Baker, D. (2003) Design of a novel globular protein fold with atomic-level accuracy. *Science* **302,** 1364–1368.
16. Dahiyat, B. I. and Mayo, S. L. (1997) De novo protein design: fully automated sequence selection. *Science* **278,** 82–87.
17. Wernisch, L., Hery, S., and Wodak, S. J. (2000) Automatic protein design with all atom force-fields by exact and heuristic optimization. *J. Mol. Biol.* **301,** 713–736.
18. Reina, J., Lacroix, E., Hobson, S. D., Fernandez-Ballester, G., Rybin, V., Schwab, M. S., Serrano, L., and Gonzalez, C. (2002) Computer-aided design of a PDZ domain to recognize new target sequences. *Nat. Struct. Biol.* **9,** 621–627
19. Dobson, C. M. (2003) Protein folding and misfolding. *Nature* **426,** 884–890.
20. Lopez de la Paz, M. and Serrano, L. (2004) Sequence determinants of amyloid fibril formation. *Proc. Natl. Acad. Sci. USA* **101,** 87–92.

Contents

Contributors

PADMANABHAN BALARAM • *Molecular Biophysics Unit, Indian Institute of Science, Bangalore, India*

EMMANUELLE BECKER • *Département de Biologie Joliot-Curie, CEA Saclay, Gif-sur-Yvette, France*

LUKE H. BRADLEY • *Department of Chemistry, Princeton University, Princeton, NJ*

RICHARD P. CHENG • *Department of Chemistry, University at Buffalo, The State University of New York, Buffalo, NY*

AITZIBER L. CORTAJARENA • *Department of Molecular Biophysics and Biochemistry, Yale University, New Haven, CT*

ALAN R. DAVIDSON • *Department of Molecular and Medical Genetics and Department of Biochemistry, University of Toronto, Toronto, Ontario, Canada*

ANDREW J. DOIG • *Faculty of Life Sciences, The University of Manchester, Manchester, United Kingdom*

NEIL ERRINGTON • *Faculty of Life Sciences, The University of Manchester, Manchester, United Kingdom*

ALEXANDRA ESTERAS-CHOPO • *Structural Biology and Biocomputing, European Molecular Biology Laboratory, Heidelberg, Germany*

LISBELL D. ESTRADA • *Protein Misfolding Disorders Laboratory, Department of Neurology, University of Texas Medical Branch, Galveston, TX*

GREGORIO FERNANDEZ-BALLESTER • *IBMC-Universidad Miguel Hernárdez, Elche (Alicante), Spain*

KAY E. GOTTSCHALK • *Department of Applied Physics, Ludwig-Maximilians-Universität-München, Munich, Germany*

RAPHAEL GUEROIS • *Département de Biologie Joliot-Curie, CEA Saclay, Gif-sur-Yvette, France*

MICHAEL H. HECHT • *Department of Chemistry, Princeton University, Princeton, NJ*

TEUKU IQBALSYAH • *Faculty of Life Sciences, Jackson's Mill, The University of Manchester, Manchester United Kingdom*

M. ANGELES JIMÉNEZ • *Instituto de Química-Física Rocasolano, Consejo Superior de Investigaciones Científicas, Madrid, Spain*

TOMMI KAJANDER • *Department of Molecular Biophysics and Biochemistry, Yale University, New Haven, CT*

MARC J. KOYACK • *Department of Chemistry, University at Buffalo, The State University of New York, Buffalo, NY*

MANUELA LÓPEZ DE LA PAZ • *Medicinal Chemistry, Merz Pharmaceuticals GmbH, Frankfurt am Main, Germany; formerly, Structural Biology and Biocomputing, European Molecular Biology Laboratory, Heidelberg, Germany*

HOCINE MADAOUI • *Département de Biologie Joliot-Curie, CEA Saclay, Gif-sur-Yvette, France*

RADHAKRISHNAN MAHALAKSHMI • *Molecular Biophysics Unit, Indian Institute of Science, Bangalore, India*

LOÏC MARTIN • *Département d'Ingénierie et d'Etudes des Protéines, CEA Saclay, Gif-sur-Yvette, France*

DAVID PANTOJA-UCEDA • *Structural Biology Laboratory, CSAT, Valencia, Spain*

MARÍA TERESA PASTOR • *European Molecular Biology Laboratory, Heidelberg, Germany*

LYNNE REGAN • *Department of Molecular Biophysics and Biochemistry, Yale University, New Haven, CT*

CLARA M. SANTIVERI • *Instituto de Química-Fiscia Rocasolano, Consejo Superior de Investigaciones Científicas, Madrid, Spain*

GIDEON SCHREIBER • *Department of Biological Chemistry, Weizmann Institute of Science, Rehovot, Israel*

LUIS SERRANO • *Structural Biology and Biocomputing, European Molecular Biology Laboratory, Heidelberg, Germany*

YOSSI SHAUL • *Department of Biological Chemistry, Weizmann Institute of Science, Rehovot, Israel*

CLAUDIO SOTO • *Protein Misfolding Disorders Laboratory, Department of Neurology, University of Texas Medical Branch, Galveston, TX*

FRANÇOIS STRICHER • *Département d'Ingénierie et d'Etudes des Protéines, CEA Saclay, Gif-sur-Yvette, France*

PETER P. THUMFORT • *NEC Research Laboratories, Princeton, NJ*

CLAUDIO VITA • *Département d'Ingénierie et d'Etudes des Protéines, CEA Saclay, Gif-sur-Yvette, France*

JUNE YOWTAK • *Protein Misfolding Disorders Laboratory, Department of Neurology, University of Texas Medical Branch, Galveston, TX*

I

DESIGN OF STRUCTURAL ELEMENTARY MOTIFS

1

Structure and Stability of the α-Helix

Lessons for Design

Neil Errington, Teuku Iqbalsyah, and Andrew J. Doig

Summary

The α-helix is the most abundant secondary structure in proteins. We now have an excellent understanding of the rules for helix formation because of experimental studies of helices in isolated peptides and within proteins, examination of helices in crystal structures, computer modeling and simulations, and theoretical work. Here we discuss structural features that are important for designing peptide helices, including amino acid preferences for interior and terminal positions, side chain interactions, disulfide bonding, metal binding, and phosphorylation. The solubility and stability of a potential design can be predicted with helical wheels and helix/coil theory, respectively. The helical content of a peptide is most often quantified by circular dichroism, so its use is discussed in detail.

Key Words: α-Helix; circular dichroism; protein structure; protein design; protein stability.

1. Introduction

The α-helix is the most abundant secondary structure in proteins, with approx 30% of residues in this structure (*1*). Helices have been extensively studied experimentally by synthesizing monomeric helical peptides or site-directed mutagenesis in proteins. Helical peptides are readily simulated by molecular dynamics and other computational techniques. Rules for helix formation have also been discovered by surveying helices in crystal structures. In this review, we discuss structural features of the helix and their study in peptides, focusing in particular on designing stable, monomeric helical peptides and characterizing them by circular dichroism. Some earlier reviews on helix structure and stability may be found in **refs. *2–8*.**

From: *Methods in Molecular Biology, vol. 340: Protein Design: Methods and Applications*
Edited by: R. Guerois and M. López de la Paz © Humana Press Inc., Totowa, NJ

1.1. Peptide Design

1.1.1. Structure of the α-Helix

A helix combines a linear translation with an orthogonal circular rotation. In the α-helix, the linear translation is a rise of 5.4Å per turn of the helix and a circular rotation is 3.6 residues per turn. Side chains spaced $i,i+3$, $i,i+4$, and $i,i+7$ are therefore close in space and interactions between them can affect helix stability. Spacings of $i,i+2$, $i,i+5$, and $i,i+6$ place the side chain pairs on opposite faces of the helix avoiding any interaction. The helix is primarily stabilized by $i,i+4$ hydrogen bonds between backbone amide groups.

1.1.2. Terminal Positions

The residues at the N-terminus of the α-helix are called N'-N-cap-N1-N2-N3-N4 etc., where the N-cap is the residue with nonhelical ϕ, ψ angles immediately preceding the N-terminus of an α-helix and N1 is the first residue with helical ϕ, ψ angles *(9)*. The C-terminal residues are similarly called C4-C3-C2-C1-C-cap-C' etc. The N1, N2, N3, C1, C2 and C3 residues are unique because their amide groups participate in $i,i+4$ backbone-backbone hydrogen bonds using either only their CO (at the N-terminus) or NH (at the C-terminus) groups. The need for these groups to form hydrogen bonds has powerful effects on helix structure and stability *(10)*.

The amide NH groups at the helix N-terminus are satisfied predominantly by side-chain hydrogen-bond acceptors. In contrast, carbonyl CO groups at the C-terminus are satisfied primarily by backbone NH groups from the sequence after the helix *(10)*. A well-known interaction is helix capping defined as specific patterns found at or near the ends of helices *(9,11–15)*. More complex capping motifs, involving multiple side chains and changes in backbone conformation are known *(16–21)*.

The secondary amide group in a protein backbone is polarized with the oxygen negatively charged and hydrogen positively charged. In a helix, the amides are all oriented in the same direction with the positive hydrogens pointing to the N-terminus and negative oxygens pointing to the C-terminus. This can be regarded as giving a helix dipole with a positive charge at the helix N-terminus and a negative charge at the helix C-terminus *(22–24)*. In general, therefore, negatively charged groups are stabilizing at the N-terminus and positive at the C-terminus.

Measurements of the amino acid preferences for the N-cap, N1, N2, and N3 positions in the helix have been made *(25–28)*. The best N-caps are Asn, Asp, Ser, and Thr *(25)* which can accept hydrogen bonds from the N2 and N3 NH groups *(29)*. Glu has only a moderate N-cap preference despite its negative charge. In contrast, N1, N2, and N3 results suggest that helix dipole interactions are more important. The contrasting results between the different helix

N-terminal positions can be rationalized by considering the geometry of the hydrogen bonds. N-cap hydrogen bonds are close to linear *(29)* and so are strong, whereas N1 and N2 hydrogen bonds are close to 90° *(30)*, making them much weaker. **Note 1** gives the amino acid preferences for different positions in the helix.

1.1.3. Interior Positions

The residue with the highest intrinsic preference for the interior of the helix is alanine. This discovery led to the successful design of isolated, monomeric helical peptides in aqueous solution, first containing several salt bridges and a high alanine content, based on $(EAAAK)_n$ *(31,32)* and then a simple sequence with a high alanine content solubilized by several lysines *(33)*. These "AK peptides" are based on the sequence $(AAKAA)_n$, where n is typically 2–5. The Lys side chains are spaced $i,i+5$ so they are on opposite faces of the helix, giving no charge repulsion and may be substituted with Arg or Gln to give a neutral peptide. Hundreds of AK peptides have been studied, giving most of the results on helix stability in peptides. The alanines in the $(EAAAK)_n$ type peptides may be removed entirely; E_4K_4 peptides, with sequences based on $(EEEEKKKK)_n$ or EAK patterns are also helical, stabilized by large numbers of salt bridges *(5,34–36)*.

Helix formation in peptides is co-operative, with a nucleation penalty. Helix stability therefore tends to increase with length, in homopolymers at least. As the length of a homopolymer increases, the mean fraction of helix will level off below 100%, because long helices tend to break in two. In heteropolymers, observed lengths are highly sequence dependent. As helices are at best marginally stable in monomeric peptides in aqueous solution, they are readily terminated by the introduction of a strong capping residue or a residue with a low intrinsic helical preference. The length distribution of helices in proteins is very different to peptides *(1)*. Most helices are short, with 5 to 14 residues most abundant. There is a general trend for a decrease in frequency as the length increases beyond 13 residues. Helix lengths longer than 25 are rare.

1.1.4. Noncovalent Side-Chain Interactions

Side chains in the helix are spaced 3.6 residues per turn of the helix. Side chains spaced $i,i+3$, $i,i+4$, and $i,i+7$ are therefore close in space and interactions between them can affect helix stability. Spacings of $i,i+2$, $i,i+5$, and $i,i+6$ place the side-chain pairs on opposite faces of the helix avoiding any interaction. Many studies have been performed on the stabilizing effects of interactions between amino acid side chains in α-helices. These studies have identified a number of types of interaction that stabilize the helix including salt bridges *(31,34,36–42)*, hydrogen bonds *(41–44)*, hydrophobic interactions *(45–48)*, basic-aromatic interactions *(49,50)* and polar/nonpolar interactions *(51)*. The stabilizing energies of many pairs in these categories have been measured, though some have only been analyzed qualitatively. As described earlier, residue side-chains spaced *(i,i+3)* and *(i,i+4)* are on the same face of the α-helix, though it is the *(i,i+4)* spacing that receives most attention in the literature, because these are stronger. A summary of stabilizing energies for side-chain interactions is given in the table in **Note 2**. We give only those that have been measured in helical peptides with the side chain interaction energies determined by applying helix-coil theory. Almost all are attractive, with the sole exception of the Lys-Lys repulsion.

1.1.5. Examining the Design

After the helix has been designed, it is necessary to check that there are no obvious problems such as having polar or nonpolar faces, because these can lead to aggregation problems. A good tool for this is the helical wheel, examples of which are available on the internet (*see* **Note 3**). These give a graphical representation of a helix as viewed along the major axis, with the side chains indicated at the appropriate points around the wheel (i.e., from 3.6 residues per turn). Thus, it is possible to check for nonpolar faces and other unwanted side-chain interactions (**Fig. 1**).

It is also useful to check how helical the peptide may be. This is most easily done using helix-coil theory programs such as AGADIR (*see* **Note 4**) *(52–55)*, SCINT2 *(47)*, and CAP-HELIX *(47)* (*see* **Note 5**). These programs will calculate the helicity of your peptide from sequence. SCINT2 and CAP-HELIX give estimates for water as solvent at 0°C. SCINT2 is especially tailored to include contributions from side-chain interactions, and CAP-HELIX has specific terms for the terminal regions of the helix. With AGADIR, the sequence, modifications, and other parameters (e.g., temperature, ionic strength) are input as requested on the web page. The output gives a percentage helix either at the residue level or the peptide level.

1.1.6. The Helix Dipole

The α-helix has a dipole (*see* **Subheading 1.1.2.** and **refs.** *22,24,54,55* and *56–67*), which also has stability implications for charged groups in the peptide. A positively charged group at the N-terminus will be unfavorable because the N-terminus is the positive end of the dipole, likewise with negative charges at the C-terminus. Left unaltered, the N- and C-termini of a peptide have positive and negative charges, respectively, both of which are unfavorable. It is therefore common to use N-terminal acetylation or succinylation alongside amidation of the C-terminus. These will remove the respective charges and the instability associated with them. Acetylation of the N-terminus also has the

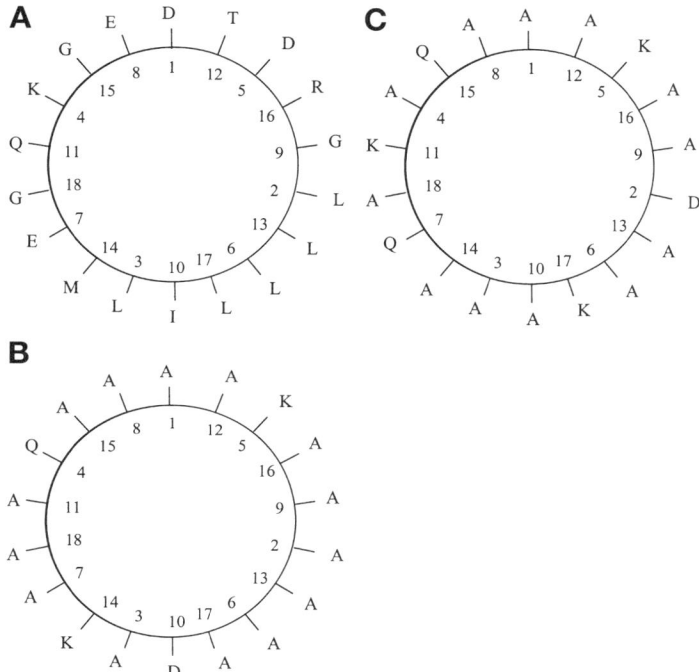

Fig. 1. Helical wheel representations of peptides using Marcel Turcotte's web application. **(A)** An example peptide, which, if helical, would have one continuous hydrophobic face (bottom right) and one charged face, which may well lead to aggregation of the peptide. **(B)** A polyalanine-based peptide with charged groups for solubility. The charged side chains are not well distributed giving a large hydrophobic face. The Lys-Asp *i,i+4* may well help stabilize the local helix conformation however. **(C)** Another polyalanine based peptide with the charged groups evenly distributed to remove exposed hydrophobic faces and thus avoid aggregation problems.

advantage that the acetyl group is a good N-cap *(13)* further promoting the stability of the N-terminus.

1.1.7. Phosphorylation

Phosphorylated amino acids have not often been studied in model α-helical peptides. Vinson and coworkers first studied phosphorylation in coiled-coil systems *(68,69)*. They found that phosphorylation of serine in the helix interior destabilized the helix *(69)*, but when interactions with the charged side chains of other amino acids (in this case, arginine) were possible the phosphorylation stabilized the coiled-coil by 1.4 kcal/mol (dimer) *(68)*. Other studies with model helices showed that serine phosphorylation is destabilizing in the helix interior

(70,71). We have shown that phosphorylation can be stabilizing by up to 2.3 kcal/mol at the helix N-terminus *(71)*, and modeling studies by Smart and McCammon *(72)* have indicated that this is due to electrostatic interactions between the phosphate group and the peptide backbone rather than helix capping. Thus, the site of phosphorylation and neighboring groups can have profound effects upon the stability of a helical peptide.

1.1.8. Disulfide Bonds

In theory, disulfide bonds are helix stabilizing given the correct side-chain spacing between the interacting residues. The effects of disulfide formation on helix stability, however, have been rarely studied. Nevertheless short peptides were stabilized by an *i, i+7* disulfide bond *(73)*. We have also studied the effect of a disulfide bond formed between Cys at Ncap and Cys at N3 (*i,i+3* spacing) of an α-helical stretch. It is found to be stabilizing by as much as 0.5 kcal/mol (manuscript in preparation). The disulfide bond formation requires oxidation of the cysteine residues, promoted by oxidizing agent such as H_2O_2. We found that bubbling air into the peptide solution is sometimes sufficient to promote disulfide bond formation. Excess oxidizing agents, however, could perturb circular dichroism (CD) spectra. The rate of disulfide formation is strongly dependent on the temperature and pH of the solution. Higher pH and temperature require less time for the reaction to complete *(74)*.

1.1.9. Metal Binding

Again, in designing a helical peptide which will bind to metal ions, the same general rules about side-chain spacing and other interactions described in **Subheading 1.1.4.** should be followed.

The first item to decide is which amino acid side chains will be used to bind the metal. A rule of thumb in metal-ligand binding is that "hard metals" prefer "hard ligands." For example, Ca and Mg prefer ligands with oxygen as the coordinating atoms (Asp, Glu) *(75)*. In contrast, soft metals, such as Cu and Zn, bind mostly to His, Cys, and Trp ligands and sometimes indirectly via water molecules *(76)*. Soft ligands are the most commonly used in published studies of helical peptide models. In the presence of Cd ions, a synthetic peptide containing Cys-His ligands *i,i+4* apart at the C-terminal region promoted helicity from 54 to 90%. The helicity of a similar peptide containing His-His ligands increased by up to 90% as a result of Cu and Zn binding *(77)*. The addition of a cis-Ru(III) ion to a 6-mer peptide, Ac-AHAAAHA-NH$_2$, changed the peptide conformation from random coil to 37% helix *(78)*. An 11-residue peptide was converted from random coil to 80% helix content by the addition of Cd ions, although the ligands used were not natural amino acids but aminodiacetic acids *(79)*. As(III) stabilizes helices when bound to Cys side chains spaced *i,i+4* by –0.7 to –1.0 kcal/mol *(80)*.

The side chains that will bind to the metal should be spaced appropriately, e.g., *i,i+4* or *i,i+3* spacing so that metals can be bound in the helical conformation. Many common biological metal ligands such as Zn and Cu have a tetrahedral binding geometry and therefore there is a strong possibility that they can induce helical peptides to dimerize when present, which will cause problems with CD measurements of helicity when trying to derive energetic parameters. This has been overcome in a short peptide with Pd ligands *(81)* by using ethylenediamine such that only two of the coordination sites of the metal are available for binding.

1.1.10. Helix Templates

A major penalty to helix formation is the loss of entropy arising from the requirement to fix three consecutive residues to form the first hydrogen bond of the helix. After this nucleation, propagation is much more favoured as only a single residue need be restricted to form each additional hydrogen bond. A way to avoid this barrier is to synthesize a template molecule that facilitates helix initiation, by fixing hydrogen bond acceptors or donors in the correct orientation for a peptide to bond in a helical geometry. The ideal template nucleates a helix with an identical geometry to a real helix. Kemp's group applied this strategy and synthesized a proline-like template that nucleated helices when a peptide chain was covalently attached to a carboxyl group *(82–86)*. Bartlett et al. reported on a hexahydroindol-4-one template *(87)* that induced 49 to 77% helicity at 0°C, depending on the method of determination, in an appended hexameric peptide. Several other templates were less successful and could only induce helicity in organic solvents *(88–90)*. Their syntheses are often lengthy and difficult, partly because of the challenging requirement of orienting several dipoles to act as hydrogen bond acceptors or donors.

2. Materials

2.1. Circular Dichroism

A spectropolarimeter with Peltier temperature control unit was used for all circular dichroism measurements.

Buffers are of various compositions, avoiding heavily absorbing components in the region of interest. Water used is the highest quality available (usually from MilliQ system).

2.2. Peptide Synthesis

Peptides are synthesized using a peptide synthesizer, using standard solid state Fmoc chemistry.

Purification is performed using reversed phase HPLC and detection at 214 nm wavelength.

3. Methods

3.1. Circular Dichroism

The following is a protocol for measuring the helix content of a peptide using circular dichroism. It is based on a Jasco spectropolarimeter; therefore, for some other models, some names of variables and output units are different. This will be covered in the notes for this section. This section also assumes that the spectropolarimeter is properly calibrated (*see* **Note 6**).

1. Start up a flow of dry nitrogen gas into the spectropolarimeter. The flow should be at least 3–5 L/min (*see* **Note 7**).
2. Start up the spectropolarimeter and software plus any temperature control equipment.
3. Leave for at least 5 min for Nitrogen purging and another 20–30 min for the light source to warm up.
4. A quartz cuvette of appropriate path length, filled with the appropriate solvent should be placed into the sample holder of the instrument (*see* **Note 8**). Ideally, solvent components should not absorb significantly in the wavelength range used because this may lead to excessive noise in or loss of CD signal.
5. Choose the wavelength range and other parameters necessary for scanning (*see* **Notes 9–14**).
6. Scan the solvent.
7. Repeat this for a solution of the peptide in the same solvent and the same cell (after cleaning). This should be done during the same working session as the solvent measurement and with the same parameters. The concentration of the sample should be known ACCURATELY (*see* **Note 13**).
8. Check that the spectrum looks typical for a helical peptide; this should have two minima at approximately 208 nm and 222 nm and a maximum at 190–195 nm (**Fig. 2A**).
9. Subtract the solvent signal from that of the peptide solution to correct for any signal from solvent.

3.2. Tests for Aggregation

It is important to have peptides that are soluble for CD measurement (*see* **Note 14**). To check peptide solubility of α-helical peptides, the following simple method can be used.

1. Record CD spectra of the buffer at 222 nm. Typical buffer used is phosphate buffer 5–10 mM containing NaCl 5–10 mM.
2. Record CD spectra of peptide dissolved in the buffer of different concentrations (normally 5 μM–200 μM) at 222 nm.
3. Subtract the average buffer reading from the average peptides readings.
4. Convert the net reading from unit of ellipticity (millidegrees) to molar ellipticity $[\theta]_{222}$ or appropriate units.

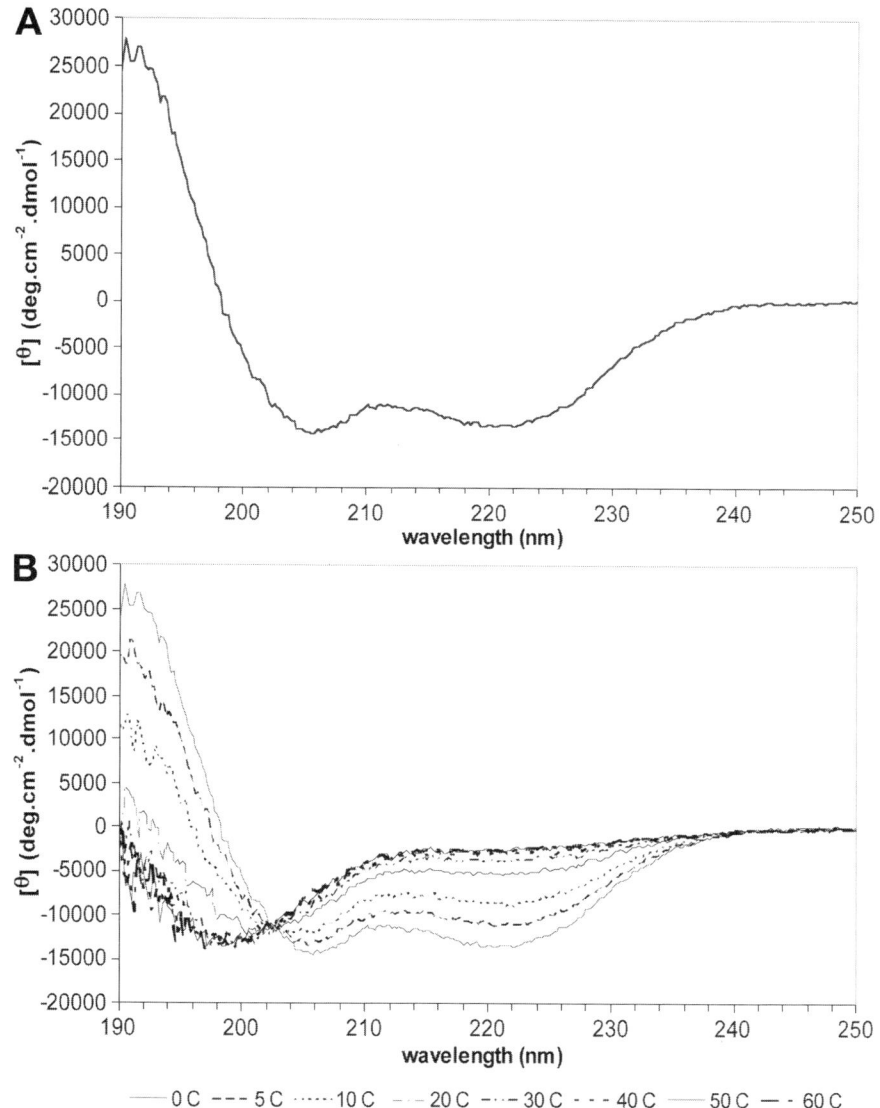

Fig. 2. (**A**) Typical far UV CD spectrum for an α-helical peptide showing minima around 208 and 222 nm and a maximum around 192 nm. Measured using a Jasco J-810 spectrophotometer in a 1-mm path length cell at 0.1°C. The data resolution was 0.2 nm, bandwidth 1nm, scan speed 20 nm/min with four accumulations (averages). The peptide sequence is acetyl-AKAAAAKAAAApSAAAAKAAGY-amide (pS denotes phosphoserine). (**B**) Same peptide as in (**A**) showing isodichroic point at 202 nm when temperature is increased from 0°C to 60°C.

Aggregation does not occur if the mean residual ellipticities of the peptides are independent of concentration.

3.3. pH Titrations

The helicities of peptides containing charged residues (e.g., Asp, Glu, Cys, His) vary with pH. Titrations as a function of pH thus can be used for analyzing possible interactions involving these residues. The pK_a values of the ionizable side chain groups are evaluated between approximately pH 2–10 by curve fitting to the Henderson-Hasselbach equation. The upper range of accessible pH is normally given by Lys deprotonation, which leads to peptide aggregation.

1. Record CD spectra of the buffer at 222 nm. Typical buffer used in pH titration contains 10 mM NaCl, 1 mM sodium phosphate, 1 mM sodium borate, and 1 mM sodium citrate.
2. Record CD spectra of the peptide (typically 20–30 μM or where no aggregation occurs) at 222 nm at initial pH.
3. Record ultraviolet (UV) CD spectra after pH adjustments that are made by the addition of either dilute NaOH or HCl. Allow the solution to equilibrate, typically 5–10 min, and check the pH. Take note of the volume of acid or alkali added.
4. Subtract the average buffer reading from the average peptides readings.
5. Convert the net reading from unit of ellipticity (mdeg) to molar ellipticity $[\theta]_{222}$ or appropriate units. Compensate for the volume change to correct the concentrations.
6. Plot $[\theta]_{222}$ vs. pH
7. Fit the data to a Henderson-Hasselbach equation depending on the number of the apparent titrable groups in the curve. For one apparent pK_a, $[\theta]_{222}$ as a function of pH is given by:

$$[\theta]_{222} = [\theta]_{222,\text{high\_pH}} * \left(1 - \frac{1}{1+10^{pH-pKa}}\right) + [\theta]_{222,\text{low\_pH}} * \left(\frac{1}{1+10^{pH-pKa}}\right)$$

where $[\theta]_{222,\text{ high\_pH}}$ and $[\theta]_{222,\text{ low\_pH}}$ are the molar ellipticities measured at 222 nm at the titration end points at high and low pH.

The equation for two different pK_as is given by:

$$[\theta]_{222} = [\theta]_{222,\text{mid\_pH}} * \left(\frac{1}{1+10^{pH-pKa1}}\right) + [\theta]_{222,\text{low\_pH}} * \left(\frac{1}{1+10^{pH-pKa1}}\right)$$

$$+ [\theta]_{222,(\text{mid\_pH--high\_pH})} * \left(1 - \frac{1}{1+10^{pH-pKa2}}\right)$$

where pK_a1 and pK_a2 are the pK_as measured for the acid–base equilibrium at low and high pH, respectively. $[\theta]_{222,\text{ high\_pH}}$, $[\theta]_{222,\text{ mid\_pH}}$, and $[\theta]_{222,\text{ low\_pH}}$ are the molar ellipticities measured at 222 nm at the titration end points at high, mid, and low pH. $[\theta]_{222 \text{ mid\_pH–high\_pH}}$ is the change in molar ellipticity associated with pK_a2.

3.4. Checking for Isodichroic Points

The most common method for converting CD data into helix content information is based upon the two-state model, i.e., all residues are either helix or coil, and there are no other secondary structures present. This can be examined using circular dichroism. If the two-state model applies then there will be an isodichroic point (a wavelength in which the signal intensity is invariant with a perturbation, such as changing temperature) in the transition from helix to coil at approx 202–204 nm wavelength (*see* **Fig. 2B**). To check for this:

1. Set up a solution at a known concentration of peptide at a low temperature (e.g., 0°C).
2. Leave to equilibrate to temperature (at least 5 min).
3. Measure the far UV CD spectrum (*see* **Subheading 2.3.1.**).
4. Increase the temperature and re-equilibrate.
5. Repeat **steps 3** and **4** until a temperature is reached where there is little or no helical signal from the solution (e.g., 60–70°C).
6. Overlay plots of the raw data to check for the isodichroic point. If there is not such a point or it moves with changing temperature, then there may be another secondary structural type present.

3.5. Conversion of CD Data Into Helix Content Information

For this section, it is assumed that the data from the instrument are in millidegrees. If this is not the case, *see* **Note 17**.

1. Having checked for an isodichroic point the helix content is best derived from the data at 222 nm. Data should first be converted from millidegrees to mean residue ellipticity in standard units (deg/cm^2/dmol^{-1}) (*see* **Note 15**).
2. The following equations *(91)* can then be used to estimate the helix content of the peptide.

$$f_H = (\theta_{OBS} - \theta_C)/(\theta_H - \theta_C)$$

$$\theta_C = 2220 - 53T$$

$$\theta_H = (-44{,}000 + 250T)(1 - 3/N_r)$$

In these equations, f_H is the fraction of helix (0 to 1), θ_C is the signal for 100% coil structure, θ_H is that for 100% helix, θ_{OBS} is the corrected observed signal at 222 nm (in mean residue ellipticity units), and T is the experimental temperature (°C). N_r is the number of amino acid residues in the peptide.

3.6. Aromatic Residues

Aromatic side chains on amino acids in structured regions of a peptide can cause problems with both CD measurement and concentration determination (*see* **Note 16**). This can lead to erroneous estimates of helix content, both through concentration inaccuracies and through effects on the CD signal itself.

3.7. Other Methods

There are several other spectroscopic techniques that can be used for examination of helix content of peptides/proteins. These include, but are not limited to, nuclear magnetic resonance spectroscopy (NMR), Fourier transform infrared spectroscopy, and Raman optical activity (ROA).

Notes 17–19 give some brief information and reference material about the application of these techniques to helix determination but detailed protocols and theory are beyond the scope of this chapter.

4. Notes

1. Amino acid propensities at interior and N- and C-terminal positions of the helix are given in **Table 1**.
2. A summary of currently determined helical side-chain interaction energies determined in helical peptides is given in **Table 2**.
3. Helical wheels are available at the following addresses: http://www.site.uottawa.ca/~turcotte/resources/HelixWheel/ and http://bioinf.man.ac.uk/~gibson/HelixDraw/helixdraw.html. Type in your sequence to see the side-chain distribution.
4. AGADIR can be run in a web browser at http://www.embl-heidelberg.de/Services/serrano/agadir/agadir-start.html. Type in the sequence and follow the instructions given.
5. Both SCINT2 and CAP-HELIX are available for download as UNIX/Irix programs at http://www.bi.umist.ac.uk/users/mjfajdg/HC.ht). Instructions for use are also given at this location.
6. Calibration of CD spectropolarimeters is usually carried out using ammonium d-camphor-10-sulphonate (ACS). Most users calibrate only at 291 nm where a 0.6% (w/v) solution should give a reading of 190.4 ± 1%. The ACS should also give another negative peak at 192.5 nm and calibration using both peaks is generally better than with one alone.
7. Nitrogen flushing is essential for several reasons. The xenon lamps used in CD spectrometers have a quartz envelope so if oxygen is present when running, large quantities of ozone can be generated. This has health and safety implications for users and can also be severely detrimental to mirror surfaces in the instrument. Oxygen also absorbs UV light below 195 nm. For these reasons, a flow of at least 3 L/min of clean, dry nitrogen should be used at higher wavelengths. Increased flow should be used if the wavelength is to approach 180 nm. Below this level (not available in most lab instruments) a flow of up to 50 L/min may be necessary. The flow should be started several minutes before the spectra are to be collected. When opening the sample holder compartment, keep the lid open for only a short time to help with flushing.
8. The combination of cuvette and sample concentration should be chosen to give a maximum absorbance of roughly 0.8, this includes *all* species in the solution,

Table 1
Table of Amino Acid Propensities at Interior and N- and C-Terminal Positions of the Helix

	Interior (104)	Ncap (25)	N1 (27,105,106)	N2 (26,105,106)	N3 (28,105,106)	C3 (107)	C2 (107,108)	C1 (107)	Ccap (25)	C' (109)
A	0	0	0	0	0	0	0	0	0	0
C°	0.68	−1.4							0.2	
C^-			1.0	0.9	—					0.3
D°	0.69	−1.6	0.5	0.7					0.2	
D^-			0	−0.2	1.1					0.3
E°	0.40	−0.7	1.0	−0.2	—				−0.4	
E^-			0.1	−0.4	0.6				−0.5	0.1
F	0.54	−0.7	1.4	0.9	1.3					−1.1
G	1.00	−1.2	0.8	0.4	0.8	2.1	0.6	0.4	0.1	
H°	0.61	−0.7	0.7	0.8	2.6		0.8			
H^+			—	—	—				−0.2	−0.9
I	0.41	−0.5	0.5	0.5	0.6	0.2	0.3	0.5		1.5
K^+	0.26	0.1	0.7	0.9	0.9				−0.1	−0.1
L	0.21	−0.7	0.3	0.5	0.6		0.1		−0.1	
M	0.24	−0.3	0.3	0.5	0.5		0.1		−0.3	0.1
N	0.65	−1.7	—	1.2	0.7	0.5	0.5	0.3	0.1	−0.4
P	3.16	−0.4	0.5	—	—					1.2
Q	0.39	2.5	0.4	0.4	0.7	0.2	0.1	0.05	−0.5	−0.1
R^+	0.21	−0.1	0.7	0.8	—				−0.4	−0.2
S	0.50	−1.2	0.4	0.6	0.8	0.6	0.6	0.5	0.8	0.3
T	0.66	−0.7	0.5	0.5	0.9	0.8	0.5	0.8		1.1
V	0.61	−0.1	0.6	0.4	0.4	0.3	0.5	0.6		1.6
W	0.49	−1.3	0.4	0.8	4.0				0.9	0.7
Y	0.53	−0.9	—	—	1.2				−2.2	

G relative to Ala for transition from coil to the position (kcal/mol). References are given in parentheses.

Table 2
Side-Chain Interaction Energies Determined in Helical Peptides

Interaction	G (kcal/mol)	Source
Ile – Lys ($i,i+4$)	–0.22	*(51)*
Val – Lys ($i,i+4$)	–0.25	*(51)*
Ile – Arg ($i,i+4$)	–0.22	*(51)*
Phe – Met ($i,i+4$)	–0.8	*(47)*
Met – Phe ($i,i+4$)	–0.5	*(47)*
Gln – Asn ($i,i+4$)	–0.5	*(44)*
Asn – Gln ($i,i+4$)	–0.1	*(44)*
Phe – Lys ($i,i+4$)	–0.14	*(49)*
Lys – Phe ($i,i+4$)	–0.10	*(49)*
Phe – Arg ($i,i+4$)	–0.18	*(49)*
Phe – Orn ($i,i+4$)	–0.4	*(50)*
Arg – Phe ($i,i+4$)	–0.1	*(49)*
Tyr – Lys ($i,i+4$)	–0.22	*(49)*
Glu – Phe ($i,i+4$)	–0.5	*(110)*
Asp – Lys ($i,i+3$)	–0.12	*(111)*
Asp – Lys ($i,i+4$)	–0.24	*(111)*
Asp – His ($i,i+3$)	≥0.63	*(112)*
Asp – His ($i,i+4$)	≥0.63	*(112)*
Asp – Arg ($i,i+3$)	–0.8	*(113)*
Glu – His ($i,i+3$)	–0.23	*(111)*
Glu – His ($i,i+4$)	–0.10	*(111)*
Glu – Lys ($i,i+3$)	–0.38	*(41)*
Glu – Lys ($i,i+4$)	–0.44	*(41)*
Phe – His ($i,i+4$)	–1.27	*(42)*
Phe – Met ($i,i+4$)	–0.7	*(48)*
His – Asp ($i,i+3$)	–0.53	*(42)*
His – Asp ($i,i+4$)	–2.38	*(114)*
His – Glu ($i,i+3$)	–0.45	*(111)*
His – Glu ($i,i+4$)	–0.54	*(111)*
Lys – Asp ($i,i+3$)	–0.4	*(111)*
Lys – Asp ($i,i+4$)	–0.58	*(111)*
Lys – Glu ($i,i+3$)	–0.38	*(111)*
Lys – Glu ($i,i+4$)	–0.46	*(111)*
Lys – Lys ($i,i+4$)	+0.17	*(39)*
Leu – Tyr ($i,i+3$)	–0.44	*(115)*
Leu – Tyr ($i,i+4$)	–0.65	*(115)*
Met – Phe ($i,i+4$)	–0.37	*(48)*
Gln – Asp ($i,i+4$)	–0.97	*(43)*
Gln – Glu ($i,i+4$)	–0.31	*(41)*
Trp – Arg ($i,i+4$)	–0.4	*(110)*
Trp – His ($i,i+4$)	–0.8	*(116)*
Tyr – Leu ($i,i+3$)	–0.02	*(115)*
Tyr – Leu ($i,i+4$)	–0.44	*(115)*

Table 2 *(Continued)*
Side-Chain Interaction Energies Determined in Helical Peptides

Interaction	G (kcal/mol)	Source
Tyr – Val (i,i+3)	–0.13	*(115)*
Tyr – Val (i,i+4)	–0.31	*(115)*
Arg (i,i+4) Glu (i,i+4) Arg	–1.5	*(117)*
Arg (i,i+3) Glu (i,i+3) Arg	–1.0	*(117)*
Arg (i,i+3) Glu (i,i+4) Arg	–0.3	*(117)*
Arg (i,i+4) Glu (i,i+3) Arg	–0.1	*(117)*
Phosphoserine – Arg (i,i+4)	–0.45	*(70)*

i.e., solvent, buffer salts and the protein/peptide in question. A good starting point in our hands is a peptide concentration of roughly 20 μ*M* with a path length of 1 mm. The cuvette should be of quartz and of CD quality.

9. Wavelength range. Secondary structural information is obtained in the far UV wavelength range, for a typical laboratory instrument this is between 180 and 260 nm. Noise from excessive absorbance can become a problem at the lower wavelength range. This can be checked for in most instruments by monitoring the detector voltage. As the sample absorbs light, less arrives at the detector and the instrument attempts to compensate by increasing the voltage to the detector. If the absorbance is too high, then very little light will reach the detector leading to greatly increased noise in the signal or no signal at all, just noise. In this case reduce the concentration of the sample *or* reduce the path length of the cuvette.

10. Response time. The signal-to-noise ratio of the system is in proportion to the integration time so the easiest way to improve the quality of spectra is to increase the response time. When measuring model helical peptides response times in the region of 1 second are common.

11. Bandwidth or slit width. Setting the width of the slits can also be used to decrease noise level in spectra. The width should be as large as possible, *but* comparable to the natural bandwidth of the bands to be scanned. Scanning parameters for α-helices would typically have bandwidths in the 1-nm region.

12. Scanning speed selection is important, as speeds that are too high will tend to distort the obtained spectra. The maximum scanning speed is obtained from the bandwidth: response time ratio, so with a 1-nm slit and 1 s response time the maximum speed would be 1 nm/s or 60 nm/min. Slower scan speeds will give better quality spectra but obviously take longer times.

13. Data density/pitch. This is the measurement interval or number of data points per nanometer wavelength. This has no effect on noise in the spectrum, but it is advisable to choose a large number of data points if there will be post accumulation processing such as curve fitting or filtering to reduce noise.

14. Averaging or accumulation. A further way to reduce the signal-to-noise ratio in spectra is to accumulate several spectra and average them. Noise will reduce as the square root of the number of accumulations, e.g., four accumulations will

reduce noise by a factor of two, nine by a factor of three. The main effect is on short-term random noise. Long-term noise such as temperature drifts will not be compensated for in this way.

15. Concentration. Any errors in concentration will propagate throughout the calculations and so this must be known with the best accuracy possible. Colorimetric assays such as Bradford, Lowry, and BCA are not accurate enough for this purpose. Spectroscopic (UV absorbance) methods are much better if a reasonable chromophore is present because they are rapid and nondestructive of sample. However, the absorbance maxima and extinction coefficient of chromophores can be environmentally sensitive and so, for example, a Trp residue in helical or β-sheet conformation may give a different reading to one in random coil conformation. Extinction coefficients usually quoted are for residues in free solution or fully solvent exposed; therefore, for the best accuracy, concentration should be corrected for this, usually involving serial dilution in 6M guanidinium chloride and comparison with the same dilutions in water. An accurate measurement of helix content depends on an accurate spectroscopic measurement and, equally importantly, peptide concentration. This is usually achieved by including a Tyr side chain at one end of the peptide. The extinction coefficient of Tyr at 275 nm is 1450 M^{-1}/cm^{-1} *(92)*. If Trp is present, measurements at 281 nm can be used where the extinction coefficient of Trp is 5690 M^{-1}/cm^{-1} and Tyr 1250 M^{-1}/cm^{-1} *(93)*. Phe absorbance is negligible at this wavelength. These UV absorbance values are ideally made in 6M GuHCl, pH 7.0, 25°C, though we have found very little variation with solvent so measurements in water are identical within error. The main source of error is in pipetting small volumes; this is typically around 2%. Pipetting larger volumes with well-maintained fixed volume pipets can help minimize this error that may well be the largest when measuring a percentage helix content by CD. If aromatic residues are not present then amino acid analysis is a very good method for determination of concentration but is destructive of sample.

16. Peptide aggregation. Peptide aggregation is generally caused by (1) hydrophobic residues in the sequence, or (2) the number of residues added for solubility is not sufficient. Three Lys or Gln residues are normally included to ensure good solubility for a 20-mer peptide, providing they are not involved in any side chain interactions (3) residues added for solubility interact with other residues or they are on the same face of the helix.

17. Data interconversion. CD data is usually expressed as an ellipticity (θ) in millidegrees or delta absorbance (A). These are easily interconvertable as A = θ/ 3298. Ellipticity is also converted to molar ellipticity ([θ]) using the equation [θ] = θ/(10.C.l), where θ is ellipticity in millidegrees, C is molar concentration (moles/dm^3), and l is cell path length (cm). With peptides and proteins the mean residue molar ellipticity is often used. This replaces the C in the previous equation with C_r, where $C_r = (1000.n.c_g)/M_r$. Here M_r is the molecular weight of the species, n is the number of peptide bonds and c_g is the macromolecular concentration (g/mL).

18. Aromatic residues/problems with CD measurements. When an aromatic residue is in a structured region of a protein or peptide, it can change the circular dichro-

ism readings as well as altering the extinction coefficient of the residue itself. The reason for this is a coupling of electronic transitions in the aromatic groups *(94)* to the backbone. There are several computational methods of dealing with this perturbation in order to "correct" the CD signal. The matrix method of Bayley, Nielsen, and Schellman *(95)* is a common method where the expected signal is computed and differences due to the aromatic groups can then be estimated. Though computational methods are improving continually, there is no absolute method to overcome this effect.

19. NMR spectroscopy. Secondary structure determination by NMR techniques does not require a full three-dimensional structural analysis as with X-ray crystallography. Knowledge of the amide and α proton chemical shifts are, in principle, all that is necessary, although if this information is available, it is likely that nearly complete assignments of side-chain protons are also available. Although obtaining the sequential resonance assignments is a laborious task, the NMR method is perhaps the most powerful method of secondary structure determination without a crystal structure. Unlike CD and other kinds of spectroscopy, NMR secondary structure determination does not give average secondary structure content, but assigns secondary structures to different parts of the protein/peptide chain. Several parameters from NMR are needed for unambiguous structure assignments. A mixture of coupling constants, nuclear Overhauser effect distances, amide proton exchange rates and chemical shifts may be necessary. It is also possible to use ^{13}C NMR resonances to obtain secondary structure information *(96–99)*. Extreme signal overlap within Alanine based peptides often reduces the usefulness of NMR, however.

20. Fourier transform infrared spectroscopy. This method is increasing in popularity for secondary structure studies. Peptides and proteins have nine characteristic absorbance bands produced by the amide functionality. These are named amide A, B, I, II ... VII. Of these, amide I and II are the major bands of interest. The amide I band (between 1600 and 1700 cm^{-1}) is associated with the C = O stretching vibration and is directly related to the backbone conformation. Amide II results from the N-H bending vibration and from the C-N stretching vibration. This is a more complex, conformationally sensitive band and is found in the 1510 and 1580 cm^{-1} region. Several amino acid side chains also absorb in the same region as the amide I and II bands (Asp, Asn, Glu, Gln, Lys, Arg, Tyr, Phe, and His). The contributions of these side chains must be accounted for before using the amide I and II bands for secondary structure determination. Fortunately this has been well investigated by Venyaminov and Kalnin *(100)*. It is well accepted that, in aqueous environments, absorbance in the range from 1650 to 1658 cm^{-1} is generally associated with the presence of α-helix. Precise interpretation of bands in this region is difficult because there is significant overlap of the helical structures with random structures. One way to resolve this issue is to exchange the hydrogen from the peptide N-H with deuterium. If the protein contains a significant amount of random structure, the H-D exchange will result in a large shift in the position of the random structure (will now absorb at around 1646 cm^{-1}) and only a minor change in the position of the helical band.

21. ROA. This bears the same relation to conventional Raman spectroscopy that UV CD has to conventional UV spectroscopy. The main area of interest for helical structures is the amide III region. Positive ROA bands in the region of 1340–1345 cm^{-1} are characteristic of α-helix in a hydrated state. A clear positive ROA band in the region of 1295–1310 cm^{-1} is assigned to α-helix in a more hydrophobic environment *(101,102)*. A full review of secondary structure assignments for ROA bands is given in **ref. *103***.

References

1. Barlow, D. J. and Thornton, J. M. (1988) Helix geometry in proteins. *J. Mol. Biol.* **201,** 601–619.
2. Scholtz, J. M. and Baldwin, R. L. (1992) The mechanism of α-helix formation by peptides. *Ann. Rev. Biophys. Biomol. Struct.* **21,** 95–118.
3. Baldwin, R. L. (1995) α-helix formation by peptides of defined sequence. *Biophys. Chem.* **55,** 127–135.
4. Chakrabartty, A. and Baldwin, R. L. (1995) Stability of alpha-helices. *Adv. Protein Chem.* **46,** 141–176.
5. Kallenbach, N. R., Lyu, P., and Zhou, H. (1996) CD spectroscopy and the helix-coil transition in peptides and polypeptides, in *Circular Dichroism and the Conformational Analysis of Biomolecules* (Fasman, G. D., ed.), Plenum Press: New York, pp. 202–257.
6. Rohl, C. A. and Baldwin, R. L. (1998) Deciphering rules of helix stability in peptides. *Meth. Enzymol.* **295,** 1–26.
7. Andrews, M. J. I. and Tabor, A. B. (1999) Forming stable helical peptides using natural and artificial amino acids. *Tetrahedron* **55,** 11711–11743.
8. Serrano, L. (2000) The relationship between sequence and structure in elementary folding units. *Adv. Prot. Chem.* **53,** 49–85.
9. Richardson, J. S. and Richardson, D. C. (1988) Amino acid preferences for specific locations at the ends of α helices. *Science* **240,** 1648–1652.
10. Presta, L. G. and Rose, G. D. (1988) Helix signals in proteins. *Science* **240,** 1632–1641.
11. Serrano, L. and Fersht, A. R. (1989) Capping and alpha-helix stability. *Nature* **342,** 296–299.
12. Bell, J. A., Becktel, W. J., Sauer, U., Baase, W. A., and Matthews, B. W. (1992) Dissection of helix capping in T4 lysozyme by structural and thermodynamic analysis of six amino acid substitutions at Thr 59. *Biochemistry* **31,** 3590–3596.
13. Chakrabartty, A., Doig, A., and Baldwin, R. (1993) Helix capping propensities in peptides parallel those in proteins. *Proc. Natl. Acad. Sci. USA* **90,** 11332–11336.
14. Forood, B., Feliciano, E. J., and Nambiar, K. P. (1993) Stabilization of alpha-helical structures in short peptides via end capping. *Proc. Natl. Acad. Sci. USA* **90,** 838–842.
15. Dasgupta, S. and Bell, J. A. (1993) Design of helix ends. Amino acid preferences, hydrogen bonding and electrostatic interactions. *Int. J. Pept. Prot. Res.* **41,** 499–511.
16. Harper, E. T. and Rose, G. D. (1993) Helix stop signals in proteins and peptides: the capping box. *Biochemistry* **32,** 7605–7609.

17. Seale, J. W., Srinivasan, R., and Rose, G. D. (1994) Sequence determinants of the capping box, a stabilizing motif at the N-termini of α-helices. *Prot. Sci.* **3,** 1741–1745.
18. Munoz, V., Blanco, F. J. and Serrano, L. (1995) The hydrophobic staple motif and a role for loop-residues in α-helix stability and protein folding. *Struct. Biol.* **2,** 380–385.
19. Schellman, C. (1980) The alphaL conformation at the ends of helices. *Protein Folding* 53–61.
20. Aurora, R., Srinivasan, R., and Rose, G. D. (1994) Rules for α-helix termination by glycine. *Science* **264,** 1126–1130.
21. Aurora, R. and Rose, G. D. (1998) Helix capping. *Protein Sci.* **7,** 21–38.
22. Wada, A. (1976) The α-helix as an electric macro-dipole. *Adv. Biophys.* **9,** 1–63.
23. Hol, W. G., van Duijnen, P. T., and Berendsen, H. J. (1978) The alpha-helix dipole and the properties of proteins. *Nature* **273,** 443–446.
24. Aqvist, J., Luecke, H., Quiocho, F. A., and Warshel, A. (1991) Dipoles located at helix termini of proteins stabilize charges. *Proc. Nat. Acad. Sci. USA* **88,** 2026–2030.
25. Doig, A. J. and Baldwin, R. L. (1995) N- and C-capping preferences for all 20 amino acids in α-helical peptides. *Protein Sci.* **4,** 1325–1336.
26. Cochran, D. A. E. and Doig, A. J. (2001) Effects of the N2 residue on the stability of the α-helix for all 20 amino acids. *Protein Sci.* **10,** 1305–1311.
27. Cochran, D. A. E., Penel, S., and Doig, A. J. (2001) Effect of the N1 residue on the stability of the α-helix for all 20 amino acids. *Protein Sci.* **10,** 463–470.
28. Iqbalsyah, T. M. and Doig, A. J. (2004) Effect of the N3 residue on the stability of the alpha-helix. *Protein Sci.* **13,** 32–39.
29. Doig, A.J., MacArthur, M. W., Stapley, B. J., and Thornton, J. M. (1997) Structures of N-termini of helices in proteins. *Protein Sci.* **6,** 147–155.
30. Penel, S., Hughes, E., and Doig, A. J. (1999) Side-chain structures in the first turn of the α-helix. *J. Mol. Biol.* **287,** 127–143.
31. Marqusee, S. and Baldwin, R. L. (1987) Helix stabilization by Glu–...Lys+ salt bridges in short peptides of de novo design. *Proc. Natl. Acad. Sci. USA* **84,** 8898–8902.
32. Park, S. H., Shalongo, W., and Stellwagen, E. (1993) Residue helix parameters obtained from dichroic analysis of peptides of defined sequence. *Biochemistry* **32,** 7048–7053.
33. Marqusee, S., Robbins, V. H., and Baldwin, R. L. (1989) Unusually stable helix formation in short alanine-based peptides. *Proc. Natl. Acad. Sci. USA* **86,** 5286–5290.
34. Lyu, P. C., Marky, L. A., and Kallenbach, N. R. (1989) The role of ion-pairs in α-helix stability: two new designed helical peptides. *J. Am. Chem. Soc.* **111,** 2733–2734.
35. Lyu, P. C., Liff, M. I., Marky, L. A., and Kallenbach, N. R. (1990) Side-chain contributions to the stability of α-helical structure in peptides. *Science* **250,** 669–673.
36. Gans, P. J., Lyu, P. C., Manning, M. C., Woody, R. W., and Kallenbach, N. R. (1991) The helix-coil transition in heterogeneous peptides with specific side chain interactions: Theory and comparison with circular dichroism. *Biopolymers* **31,** 1605–1614.

37. Horovitz, A., Serrano, L., Avron, B., Bycroft, M., and Fersht, A. R. (1990) Strength and co-operativity of contributions of surface salt bridges to protein stability. *J. Mol. Biol.* **216,** 1031–1044.

38. Merutka, G. and Stellwagen, E. (1991) Effect of amino acid ion pairs on peptide helicity. *Biochemistry* **30,** 1591–1594.

39. Stellwagen, E., Park, S. H., Shalongo, W., and Jain, A. (1992) The contribution of residue ion pairs to the helical stability of a model peptide. *Biopolymers* **32,** 1193–1200.

40. Huyghues-Despointes, B. M., Scholtz, J. M., and Baldwin, R. L. (1993) Helical peptides with three pairs of Asp-Arg and Glu-Arg residues in different orientations and spacings. *Protein Sci.* **2,** 80–85.

41. Scholtz, J.M., Qian, H., Robbins, V. H., and Baldwin, R. L. (1993) The energetics of ion-pair and hydrogen-bonding interactions in a helical peptide. *Biochemistry* **32,** 9668–9676.

42. Huyghues-Despointes, B. M. and Baldwin, R. L. (1997) Ion-pair and charged hydrogen-bond interactions between histidine and aspartate in a peptide helix. *Biochemistry* **36,** 1965–1970.

43. Huyghues-Despointes, B. M., Klingler, T. M., and Baldwin, R. L. (1995) Measuring the strength of side-chain hydrogen bonds in peptide helices: the Gln.Asp (i, i + 4) interaction. *Biochemistry* **34,** 13267–13271.

44. Stapley, B. J. and Doig, A. J. (1997) Hydrogen bonding interactions between glutamine and asparagine in α-helical peptides. *J. Mol. Biol.* **272,** 465–473.

45. Padmanabhan, S. and Baldwin, R. L. (1994) Helix-stabilizing interaction between tyrosine and leucine or valine when the spacing is i, i + 4. *J. Mol. Biol.* **241,** 706–713.

46. Padmanabhan, S. and Baldwin, R. L. (1994) Tests for helix-stabilizing interactions between various nonpolar side chains in alanine-based peptides. *Protein Sci.* **3,** 1992–1997.

47. Stapley, B. J., Rohl, A., and Doig, A. J. (1995) Addition of side chain interactions to modified Lifson-Roig helix-coil theory: application to energetics of Phenylalanine-Methionine interactions. *Protein Sci.* **4,** 2383–2391.

48. Viguera, A. R. and Serrano, L. (1995) Side-chain interactions between sulfur-containing amino acids and phenylalanine in α-helices. *Biochemistry* **34,** 8771–8779.

49. Andrew, C. D., Bhattacharjee, S., Kokkoni, N., Hirst, J. D., Jones, G. R., and Doig, A. J. (2002) Stabilizing interactions between aromatic and basic side chains in α-helical peptides and proteins. Tyrosine effects on helix circular dichroism. *J. Am. Chem. Soc.* **124,** 12706–12714.

50. Tsou, L. K., Tatko, C. D., and Waters, M. L. (2002) Simple cation-π interaction between a phenyl ring and a protonated amine stabilizes an α-helix in water. *J. Am. Chem. Soc.* **124,** 14917–14921.

51. Andrew, C. D., Penel, S., Jones, G. R., and Doig, A. J. (2001) Stabilizing nonpolar/polar side-chain interactions in the α-helix. *Proteins* **45,** 449–455.

52. Muñoz, V. and Serrano, L. (1997) Development of the multiple sequence approximation within the AGADIR model of α-helix formation: comparison with Zimm-Bragg and Lifson-Roig formalisms. *Biopolymers* **41,** 495–509.

53. Lacroix, E., Viguera, A. R., and Serrano, L. (1998) Elucidating the folding problem of α-helices: local motifs, long-range electrostatics, ionic-strength dependence and prediction of NMR parameters. *J. Mol. Biol.* **284,** 173–191.
54. Muñoz, V. and Serrano, L. (1994) Elucidating the folding problem of helical peptides using empirical parameters. *Nat. Struct. Biol.* **1,** 399–409.
55. Muñoz, V. and L. Serrano, L. (1995) Elucidating the folding problem of helical peptides using empirical parameters. II. Helix macrodipole effects and rational modification of the helical content of natural peptides. *J. Mol. Biol.* **245,** 275–296.
56. Chakrabarti, P. (1991) Does helix dipole have any role in binding metal ions in protein structures? *Arch. Biochem. Biophys.* **290,** 387–390.
57. Chakrabarti, P. (1994) An assessment of the effect of the helix dipole in protein structures. *Protein Eng.* **7,** 471–474.
58. Edmonds, D. T. (1985) The alpha-helix dipole in membranes: a new gating mechanism for ion channels. *Eur. Biophys. J.* **13,** 31–35.
59. Fairman, R., Shoemaker, K. R., York, E. J., Stewart, J. M., and Baldwin, R. L. (1989) Further studies of the helix dipole model: effects of a free alpha-NH3+ or alpha-COO- group on helix stability. *Proteins* **5,** 1–7.
60. Gilson, M. K. and Honig, B. (1989) Destabilization of an alpha-helix-bundle protein by helix dipoles. *Proc. Natl. Acad. Sci. USA* **86,** 1524–1528.
61. Hol, W. G. J. (1985) Effects of the [alpha]-helix dipole upon the functioning and structure of proteins and peptides. *Adv. Biophys.* **19,** 133–165.
62. Hol, W. G. J., van Duijnen, P. T., and Berendsen, H. J. C. (1978) The α-helix dipole and the properties of proteins. Nature, **273,** 443–446.
63. Lockhart, D. J. and Kim P. S. (1993) Electrostatic screening of charge and dipole interactions with the helix backbone. *Science* **260,** 198–202.
64. Miranda, J. L. (2003) Position-dependent interactions between cysteine residues and the helix dipole. *Protein Sci.* **12,** 73–81.
65. Nicholson, H., Anderson, D. E., Dao-pin, S., and Matthews, B. W. (1991) Analysis of the interaction between charged side chains and the alpha-helix dipole using designed thermostable mutants of phage T4 lysozyme. *Biochemistry* **30,** 9816–9828.
66. Sali, D., Bycroft, M., and Fersht, A. R. (1988) Stabilization of protein structure by interaction of alpha-helix dipole with a charged side chain. *Nature* **335,** 740–743.
67. Sheridan, R. P. and Allen, L. C. (1980) The electrostatic potential of the alpha helix (electrostatic potential/[alpha]-helix/secondary structure/helix dipole). *Biophys. Chem.* **11,** 133–136.
68. Szilak, L., Moitra, J., and Vinson, C. (1997) Design of a leucine zipper coiled coil stabilized 1.4kcal/mol by phosphorylation of a serine in the e position. *Protein Sci.* **6,** 1273–1283.
69. Szilak, L., Moitra, J., Krylov, D., and Vinson, C. (1997) Phosphorylation destabilizes alpha-helices. *Nat. Struct. Biol.* **4,** 112–114.
70. Liehr, S. and Chenault, H. K. (1999) A comparison of the α-helix forming propensities and hydrogen bonding properties of serine phosphate and α-amino-γ-phoshphonobutyric acid. *Bioorg. Med. Chem. Lett.* **9,** 2759–2762.
71. Andrew, C. D., Warwicker, J., Jones, G. R., and Doig, A. J. (2002) Effect of phosphorylation on α-helix stability as a function of position. *Biochemistry* **41,** 1897–1905.

72. Smart, J. L. and McCammon, J. A. (1999) Phosphorylation stabilizes the N-termini of α-helices. *Biopolymers* **49,** 225–233.
73. Jackson, D. Y., King, K. S., Chmielewski, J., Singh, S., and Schultz, P.G. (1991) General approach to the synthesis of short α-helical peptides. *J. Am. Chem. Soc.* **113,** 9391–9392.
74. Berezhkovskiy, L. M., Pham, S., Reich, E. P., and Deshpande, S. (1999) Synthesis and kinetics of cyclization of MHC class II derived cyclic peptide vaccine for diabetes. *J. Pept. Res.* **54,** 112–119.
75. Jernigan, R., Raghunathan, G., and Bahar, I. (1994) Characterisation of interactions and metal-ion binding-sites in proteins. *Curr. Opin. Struct. Biol.* **4,** 256–263.
76. Alberts, I. L., Nadassy K., and Wodak S. J. (1998) Analysis of zinc binding sites in protein crystal structures. *Protein Sci.* **7,** 1700–1716.
77. Ghadiri, M. R. and Choi, C. (1990) Secondary structure nucleation in peptides—transition-metal ion stabilized alpha-helices. *J. Am. Chem. Soc.* **112,** 1630–1632.
78. Kise, K. J. and Bowler, B. E. (2002) Induction of helical structure in a heptapeptide with a metal cross-link: modification of the Lifson-Roig Helix-coil theory to account for covalent cross-links. *Biochemistry* **41,** 15826–15837.
79. Ruan, F., Chen, Y., and Hopkins, P. B. (1990) Metal ion enhanced helicity in synthetic peptides containing unnatural, metal-ligating residues. *J. Am. Chem. Soc.* **112,** 9403–9404.
80. Cline, D. J., Thorpe, C., and Scheider, J. P. (2003) Effects of As(III) Binding on r-helical structure. *J. Am. Chem. Soc.* **125,** 2923–2929.
81. Beyer, R.L., Hoang, H. N., Appleton, T. G., and Fairlie, D. P. (2004) Metal clips induce folding of a short unstructured peptide into an α-helix via turn conformations in water. Kinetic versus thermodynamic products. *J. Am. Chem. Soc.* **126,** 15096–15105.
82. Kemp, D. S., Boyd, J. G., and Muendel C. C. (1991) The helical s-constant for Alanine in water derived from template-nucleated helices. *Nature* **352,** 451–454.
83. Kemp, D. S., Allen, T. J., and Oslick, S. L. (1995) The energetics of helix formation by short templated peptides in aqueous solution. 1. Characterization of the reporting helical template Ac-HE1(1). *J. Am. Chem. Soc.* **117,** 6641–6657.
84. Groebke, K., Renold, P., Tsang, K. Y., Allen, T. J., McClure, K. F., and Kemp, D. S. (1996) Template-nucleated alanine-lysine helices are stabilized by position-dependent interactions between the lysine side chain and the helix barrel. *Proc. Natl. Acad. Sci. USA* **93,** 4025–4029.
85. Kemp, D. S., Allen, T. J., Oslick, S. L., and Boyd, J. G., (1996) The structure and energetics of helix formation by short templated peptides in aqueous solution. 2. Characterization of the helical structure of Ac-Hel(1)-Ala(6)-OH. *J. Am. Chem. Soc.* **118,** 4240–4248.
86. Kemp, D. S., Oslick, S. L., and Allen, T. J. (1996) The structure and energetics of helix formation by short templated peptides in aqueous solution. 3. Calculation of the helical propagation constant s from the template stability constants t/c for Ac-Hel1-Alan-OH, n = 1–6. *J. Am. Chem. Soc.* **118,** 4249–4255.
87. Austin, R. E., Maplestone, R. A., Sefler, A. M., Liu, K., et al. (1997) Template for stabilization of a peptide alpha-helix: Synthesis and evaluation of conformational effects by circular dichroism and NMR. *J. Am. Chem. Soc.* **119,** 6461–6472.

88. Arrhenius, T. and Sattherthwait, A. C. (1989) *Peptides: Chemistry, Structure and Biology: Proceedings of the 11th American Peptide Symposium*, July, La Jolla, CA.
89. Muller, K., Obrecht, D., Knierzinger, A., et al. (1993) *Perspectives in Medicinal Chemistry*. pp. 513–531.
90. Gani, D., Lewis, A., Rutherford, T., et al. (1998) Design, synthesis, structure and properties of an alpha-helix cap template derived from N-[(2S)-2-chloropropionyl]-(2S)-Pro-(2R)-Ala-(2S,4S)-4-thioPro-OMe which initiates alpha-helical structures. *Tetrahedron* **54**, 15793–15819.
91. Luo, P. and Baldwin, L. (1997) Mechanism of helix induction by trifluoroethanol: A framework for extrapolating the helix-forming properties of peptides from trifluoroethanol/water mixtures back to water. *Biochemistry* **36**, 8413–8421.
92. Brandts, J. R. and Kaplan, K. J. (1973) Derivative spectroscopy applied to tyrosyl chromophores. Studies on ribonuclease, lima bean inhibitor, and pancreatic trypsin inhibitor. *Biochemistry* **10**, 470–476.
93. Edelhoch, H. (1967) Spectroscopic determination of tryptophan and tyrosine in proteins. *Biochemistry* **6**, 1948–1954.
94. Woody, R. W., Dunker, A. K., and Fasman, G. D. (1996) Theory of circular dichroism of proteins, in *Circular Dichroism and the Conformational Analysis of Biomolecules* (Fasman, G. D., ed.), Plenum Press, New York, pp. 109–157.
95. Bayley, P. M., Nielsen, E. B., and Schellman, J. A. (1969) The rotatory properties of molecules containing two peptide groups. *J. Phys. Chem.* **73**, 228–243.
96. Spera, S. and Bax, A. (1991) Empirical correlation between protein backbone conformation and $C\alpha$ and $C\beta$ ^{13}C Nuclear magnetic resonance chemical shifts. *J. Am. Chem. Soc.* **113**, 5490–5492.
97. Shalongo, W., Dugad, L., and Stellwagen, E. (1994) Analysis of thermal transitions of a model helical peptide using ^{13}C NMR. *J. Am. Chem. Soc.* **116**, 2500–2507.
98. Shalongo, W., Dugad, L., and Stellwagen, E. (1994) Distribution of helicity within the model peptide Acetyl(AAQAA)$_3$amide. *J. Am. Chem. Soc.* **116**, 8288–8293.
99. Park, S.-H., Shalongo, W., and Stellwagen, E. (1998) Analysis of N-terminal Capping using carbonyl-carbon chemical shift measurements. *Proteins Struct. Funct. Genet.* **33**, 167–176.
100. Venyaminov, S.-Y. and Kalnin, N. N. (1990) Quantitative IR Spectrophotometry of peptide compounds in water (H_2O) solutions. 2. Amide absorption-bands of polypeptides and fibrous proteins in α-coil, β-coil, and random coil conformations. *Biopolymers* **30**, 1243–1257.
101. Blanch, E. W., Kasarda, D. D., Hecht, L., Nielsen, K., and Barron, L. D. (2003) New insight into the solution structures of wheat gluten proteins from Raman optical activity. *Biochemistry* **42**, 5665–5673.
102. Blanch, E. W., Morozova-Roche, L. A., Cochran, D. A. E., Doig, A. J., Hecht, L., and Barron, L. D. (2000) Is polyproline II helix the killer conformation? A Raman optical activity study of the amyloidogenic prefibrillar intermediate of human lysozyme. *J. Mol. Biol.* **301**, 553–563.
103. Barron, L. D., Hecht, L., Blanch, E. W., and Bell, A. F. (2000) Solution structure and dynamics of biomolecules from Raman optical activity. *Prog. Biophys. Mol. Biol.* **73**, 1–49.

104. Pace, C. N. and Scholtz, J. M. (1998) A helix propensity scale based on experimental studies of peptides and proteins. *Biophys. J.* **75,** 422–427.
105. Petukhov, M., Muñoz, V., Yumoto, N., Yoshikawa, S., and Serrano, L. (1998) Position dependence of non-polar amino acid intrinsic helical propensities. *J. Mol. Biol.* **278,** 279–289.
106. Petukhov, M., Uegaki, K., Yumoto, N., Yoshikawa, S., and Serrano, L. (1999) Position dependence of amino acid intrinsic helical propensities II: Non-charged polar residues: Ser, Thr, Asn, and Gln. *Protein Sci.* **8,** 2144–2150.
107. Petukhov, M., et al. (2002) Amino acid intrinsic α-helical propensities III: Positional dependence at several positions of C-terminus. *Protein Sci.* **11,** 766–777.
108. Ermolenko, D. N., Richardson, J. M., and Makhatadze, G. I. (2003) Noncharged amino acid residues at the solvent-exposed positions in the middle and at the C terminus of the α-helix have the same helical propensity. *Protein Sci.* **12,** 1169–1176.
109. Thomas, S. T., Loladze, V. V., and Makhatadze, G. I. (2001) Hydration of the peptide backbone largely defines the thermodynamic propensity scale of residues at the C' position of the C-capping box of α-helices. *Proc. Nat. Acad. Sci. USA* **98,** 10670–10675.
110. Shi, Z., Olson, C. A., Bell, A. J., Jr., and Kallenbach, N. R. (2002) Non-classical helix-stabilizing interactions: C-H...O H-bonding between Phe and Glu side chains in α-helical peptides. *Biophys. Chem.* **101–102,** 267–279.
111. Smith, J. S. and Scholtz, J. M. (1998) Energetics of polar side-chain interactions in helical peptides: Salt effects on ion pairs and hydrogen bonds. *Biochemistry* **37,** 33–40.
112. Luo, R., David, L., Hung, H., Devaney, J., and Gilson, M. K., Strength of solvent-exposed salt bridges. (1999) *J. Phys. Chem. B* **103,** 727–736.
113. Marqusee, S. and Sauer, R. T. (1994) Contributions of a hydrogen bond/salt bridge network to the stability of secondary and tertiary structure in λ repressor. *Protein Sci.* **3,** 2217–2225.
114. Luo, P. and Baldwin, R. L. (1999) Interaction between water and polar groups of the helix backbone: An important determinant of helix propensities. *Proc. Nat. Acad. Sci. USA* **96,** 4930–4935.
115. Shalongo, W. and Stellwagen, E. (1995) Incorporation of pairwise interactions into the Lifson-Roig model for helix prediction. *Protein Sci.* **4,** 1161–1166.
116. Fernández-Recio, J., Vazquez, A., Civera, C., Sevilla, P., and Sancho, J. (1997) The tryptophan/histidine interaction in α-helices. *J. Mol. Biol.* **267,** 184–197.
117. Shi, Z., Olson, C. A., Bell, A. J. Jr., and Kallenbach, N. R. (2001) Stabilization of α-helix structure by polar side-chain interactions: complex salt bridges, cation π interactions and C-H...O-H bonds. *Biopolymers* **60,** 366–380.

2

De novo Design of Monomeric
β-Hairpin and β-Sheet Peptides

David Pantoja-Uceda, Clara M. Santiveri, and M. Angeles Jiménez

Summary

Since the first report in 1993 (JACS 115, 5887-5888) of a peptide able to form a monomeric β-hairpin structure in aqueous solution, the design of peptides forming either β-hairpins (two-stranded antiparallel β-sheets) or three-stranded antiparallel β-sheets has become a field of intense interest. These studies have yielded great insights into the principles governing the stability and folding of β-hairpins and antiparallel β-sheets. This chapter reviews briefly those principles and describes a protocol for the *de novo* design of β-sheet-forming peptides based on them. Criteria to select appropriate turn and strand residues and to avoid aggregation are provided. Because nuclear magnetic resonance is the most appropriate technique to check the success of new designs, the nuclear magnetic resonance parameters characteristic of β-hairpins and three-stranded antiparallel β-sheets are given.

Key Words: Antiparallel β-sheet; β-hairpin; NMR; peptide structure; β-sheet propensities; side chain–side chain interactions; solubility; β-turn prediction; β-turn propensities.

1. Introduction

Protein structures consist of a limited set of secondary structure elements, namely, helices, β-strands, and β-turns, which are organized in different numbers and orientations to produce an extraordinary diversity of protein tertiary structures. Therefore, a reasonable approach to understand protein folding and stability is the study of the conformational behavior of protein fragments and designed peptides. A large amount of information on α-helix folding and stability has been gathered since the early 1980s (*see* Chapter 1). In contrast, early efforts on studying β-sheet forming peptides in aqueous solution did not succeed, likely as a consequence of the strong tendency of sequences with high β-sheet propensity to self-associate. The first peptide able to adopt a monomeric

From: *Methods in Molecular Biology, vol. 340: Protein Design: Methods and Applications*
Edited by: R. Guerois and M. López de la Paz © Humana Press Inc., Totowa, NJ

β-hairpin in aqueous solution was reported in 1993 *(1)*. A β-hairpin is the simplest antiparallel β-sheet motif, and antiparallel arrangements of β-strands are
more easily studied in model peptides than parallel ones, since the latter requires
a lengthy connector. In nature an α-helix acts as linker between most adjacent
parallel β-strands. Peptides forming parallel β-sheets that incorporate unnatural templates *(2,3)*, and the use of peptidomimetics at the β-turn or at the strands
to induce β-hairpin structures *(4,5)* are beyond the scope of this chapter.

Since the report of the first β-hairpin forming peptide, the forces involved in
the stability and folding of two- and three-stranded antiparallel β-sheets have
been extensively investigated by several research groups (for reviews *see* **refs.
6–15**). Based on their conclusions, it is now possible to establish a general
protocol for the design of new β-sheet–forming peptides that we will describe
here (**Subheading 2.**). Previous to that description, the structural characteristics of β-hairpins and three-stranded antiparallel β-sheets that are relevant for
the design will be illustrated (**Subheading 1.1.**). Next, the principles used in
the design of the reported β-sheet peptides will be explained (**Subheading 1.2.**),
and the main conclusions derived from the extensive studies on β-hairpin and
β-sheet stability using peptide models will be summarized (**Subheading 1.3.**).

1.1. Characteristics of β-Hairpin and Three-Stranded Antiparallel β-Sheet Structures

A β-hairpin consists of two antiparallel hydrogen-bonded (H-bonded) β-
strands linked by a loop region (**Figs. 1** and **2**). Characteristic average values
for the ϕ and ψ angles of β-strand residues in antiparallel β-sheets are −139°
and +135°, respectively *(16)*. β-Hairpin motifs differ in the length and shape of
the loop and are classified according to the number of residues in the turn and
the number of interstrand hydrogen bonds between the residues flanking the
turn (n − 1 and c + 1 in **Fig. 1**). This β-hairpin classification uses a X:Y nomenclature *(17)*, with X being the number of residues in the turn region and either
Y = X if the CO and NH groups of the two residues that precede and follow the
turn form two hydrogen bonds (for example, in 2:2 and 4:4 β-hairpins; **Figs.
1A** and **1D**, respectively) or Y = X + 2 if these residues form only one hydrogen bond (as in 3:5 β-hairpins; **Fig. 1C**). In protein β-hairpins with short loops
(2:2, 3:5, and 4:4), the loop conformation corresponds to β-turns with geom-
etries adequate for the characteristic right-handed twist of antiparallel β-sheets.
Thus, for 2:2 β-hairpins, the most frequent β-turn is type I', followed by type
II', with type I more rarely found. This statistical occurrence is explained by
the fact that type I' and II' β-turns have a right-handed twist suitable for the β-
strand pairing, whereas type I and II β-turns are left-handed twisted, with the
degree of twist being larger in the type I and I' β-turns than in the type II and II'
turns. Type II β-turns are very uncommon in protein β-hairpins. A 3:5 β-hair-

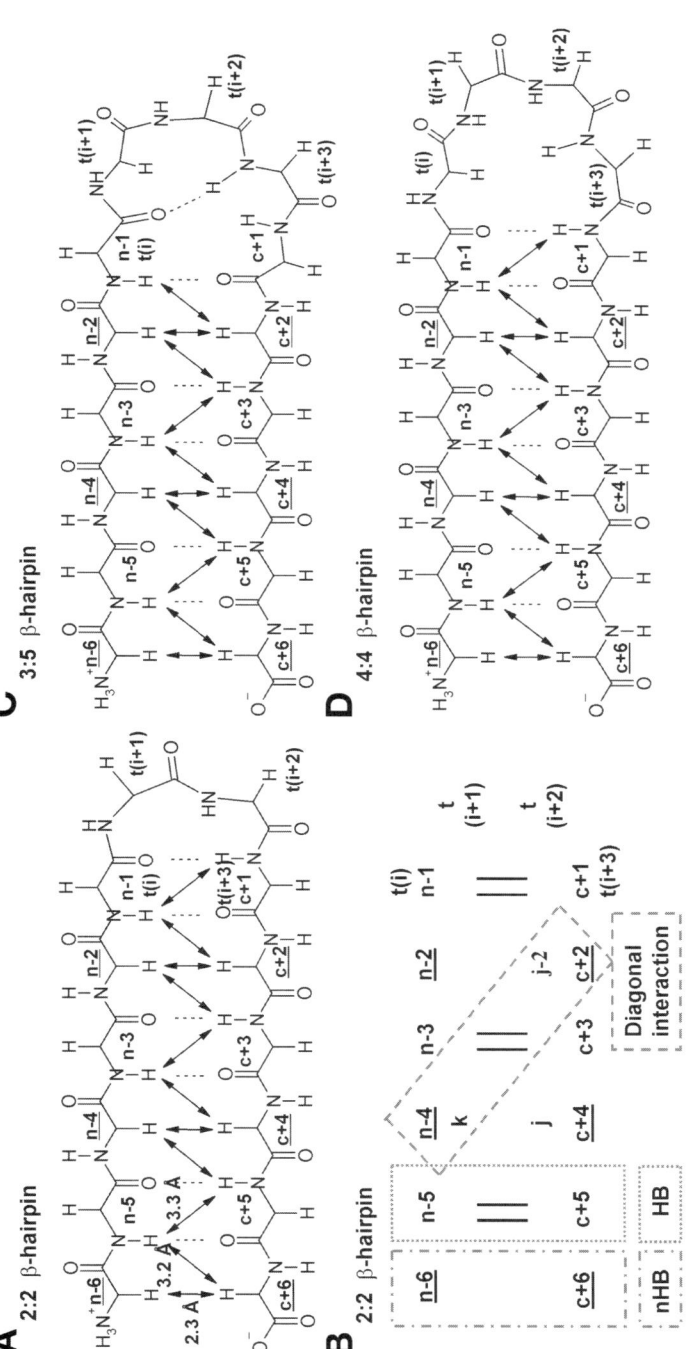

Fig. 1. Schematic representation of the peptide backbone conformation of 2:2 (**A,B**), 3:5 (**C**), and 4:4 (**D**) β-hairpins. Turn residues are labeled as t. Residues in the N-terminal and C-terminal strand are named as n and c, respectively. Side chains of underlined strand residues are pointing toward the same β-sheet face, and those not underlined toward the other. Dotted lines link the NH proton and the acceptor CO oxygen of the β-sheet hydrogen bonds. Black arrows indicate the observable long-range nuclear Overhauser enhancements (NOEs) involving H_α and NH backbone protons (**Subheading 2.7.**). The corresponding average distances in protein antiparallel β-sheets are shown in A. In B, two vertical lines connect H-bonded residues and rectangles indicate a pair of facing residues in a H-bonded site (HB) and in a non–H-bonded site (nHB), and a diagonal interaction between the side chains of residues located in non–H-bonded sites of adjacent strands (they are labeled k and j-2).

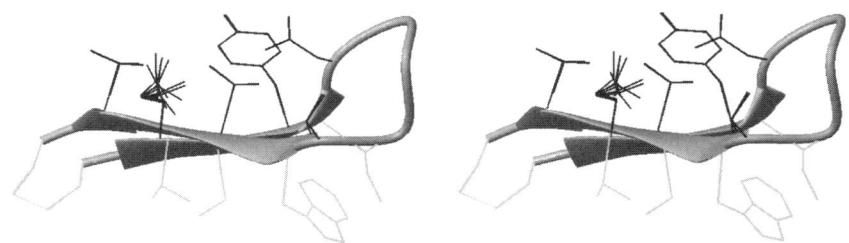

Fig. 2. β-Hairpin structure calculated for a designed 15-residue peptide (Santiveri and Jiménez, unpublished). Backbone atoms are displayed as a ribbon, side chains pointing upward in black and downward in gray.

pin normally exhibits a G1 bulge type I β-turn, a type I β-turn with a Gly residue at the i + 3 position forming a sort of bulge in the hairpin, whereas most of the 4:4 β-hairpins have a canonical type I β-turn. The ϕ and ψ dihedral angles characteristic of the i + 1 and i + 2 residues in types I, I', and II' β-turn are given in **Table 1**.

Two kinds of β-strand positions can be distinguished for facing residues according to whether they form hydrogen bonds or not, i.e., H-bonded sites and non–H-bonded sites (**Fig. 1**). In the β-hairpin, the side chains of consecutive residues in a strand point toward opposite sides of the β-sheet plane, whereas the side chains of facing residues corresponding to adjacent strands are on the same side of the β-sheet (**Figs. 1** and **2**). Because the average distances between the side chains of facing residues are 2.4 Å in non–H-bonded sites and 2.8 Å in H-bonded sites *(18)*, the contribution of a given side chain–side chain interaction to β-hairpin stability depends on the site (**Subheading 1.3.2.2.**). As a consequence of the right-handed twist of β-sheets, the side chains of residues in two consecutive non–H-bonded sites (labeled as k and j-2 in **Fig. 1**) are also quite close (3.0 Å; *[18]*). The interaction between these side chains, referred to as a diagonal interaction (**Fig. 1B**), also contributes to β-hairpin stability (**Subheading 1.3.2.3.**).

After β-hairpins, the next simplest kind of β-sheet motifs are three-stranded antiparallel β-sheets with topology β1-β2-β3. They can be regarded as being formed by two β-hairpins with a common β-strand (β2) that is the C-terminal strand of the hairpin 1 and the N-terminal strand of the hairpin 2 (**Fig. 3**).

1.2. Design of β-Sheet Peptides

Early β-sheet peptides were designed to understand protein β-sheet folding and stability *(6–15)*. However, in the last several years some β-hairpin peptides have been designed with a biological functionality, such as, DNA-, ATP-, metal-, or flavin-binding properties in mind *(11,19–21)*. According to design

Table 1
Residues With High Statistical Probabilities to Be in Each Position of Type I, Type I′, and Type II′ β-Turns Are Ordered From the Most Favorable to the Least (22)

Residue	Type I β-turn	Type I′ β-turn	Type II′ β-turn
i	D>N>>H≈C≈S>P	Y>H>>I≈V	Y>V≈S≈H≈F
i+1	P>>E≈S	N>H>>D>G	p>G
φ, ψ angles	−60, −30	+60, +30	+60, −120
i+2	D>N>>T>S≈W	G	N>S>D>H
φ, ψ angles	−90, 0	90, 0	−80, 0
i+3	G>>C≈T≈D≈R≈N	K>>N≈R≈E≈Q	T>G>N≈R≈F≈K
Type of β-hairpin	3:5 G1 I 4:4 I	2:2 I′	2:2 II
Turn sequences in designed β-hairpins	NPDG AKDG[a] AKAG NSDG[a] EPDG APDG[a] DATK APDG[a] NSDG[a] PATG AKDG[a]	VNGK VDGK INGK IDGK YNGK ENGK VNGO	VpGL EGNK VpGO NGKT SpGK EpNK VpGK SpAK IpGK EpGK

Average values of the φ and ψ dihedral angles are given for the i+1 and i+2 residues of each type of β-turn. Ornithine is indicated by "O" and DPro by a lower case "p."

[a] The reported peptides with these turn sequences adopt a mixture of two β-hairpins, one a 3:5 with a G1 bulge type I β-turn and the other a 4:4 (*54,100*).

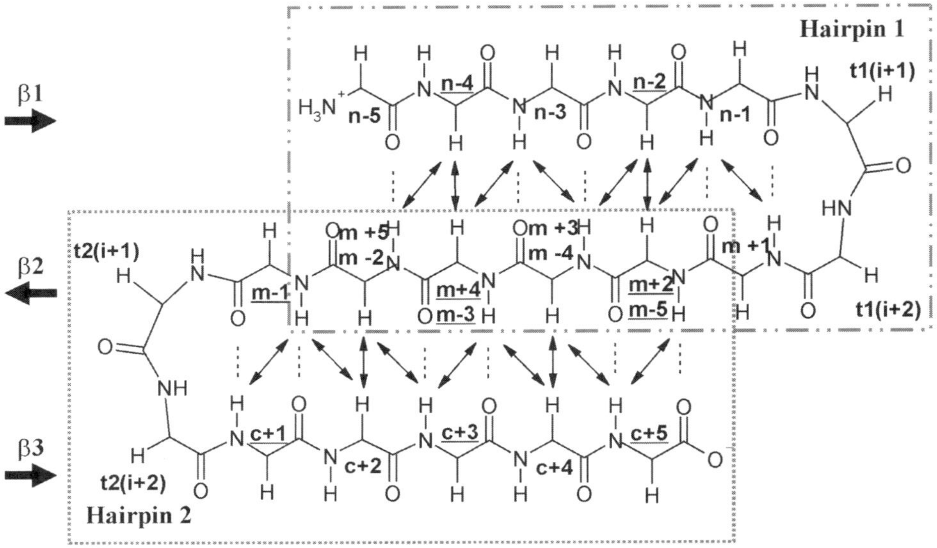

Fig. 3. Schematic representation of the peptide backbone conformation of a three-stranded antiparallel β-sheet motif. The β1-β2-β3 topology is indicated by black arrows (N to C direction) on the left site of the scheme. The two large rectangles surround the residues belonging to each of the 2:2 β-hairpins that compose this β-sheet motif. Turn residues of β-hairpin 1 and of β-hairpin 2 are named t1 and t2, respectively. Residues in the N-terminal strand are named as n, in the middle strand as m, and in the C-terminal strand as c. Side chains of underlined strand residues are pointing toward the same β-sheet face, and those not underlined toward the other. Dotted lines link the NH proton and the acceptor CO oxygen of the β-sheet hydrogen bonds. Double black arrows indicate the observable long-range NOEs involving backbone H_α and NH protons (**Subheading 2.7.**).

strategy, β-sheet forming peptides (*see* **Notes 1** and **2**) can be divided into two groups: (1) those derived from the sequence of native protein β-hairpins or β-sheets and (2) completely *de novo* sequences.

1.2.1. Peptides Derived From Protein β-Sheets

1.2.1.1. PEPTIDES DESIGNED BY MODIFICATION OF THE TURN SEQUENCE IN PROTEIN β-HAIRPINS

The earliest successful strategy in the design of β-hairpin peptides consisted in substituting the native turn sequence for those residues with the highest intrinsic probability to occupy the corresponding β-turn positions *(22)*. Such strategy was used for a nonapeptide *(1)* and a 16-residue peptide *(23)* derived from residues 15–23 of Tendamistat and from the N-terminal hairpin of ubiquitin, respectively. Both peptides adopt 3:5 β-hairpins with a G1 bulge type I β-turn. β-Hairpin formation had not been detected for the peptides encompassing any of the corresponding native protein sequences. Later, the same strategy was applied for the same ubiquitin hairpin by using a different turn sequence, one containing a DPro residue at position i + 1 of the turn, so that the resulting β-hairpin is a 2:2 with a type II' β-turn, instead of a 3:5 β-hairpin with a G1 bulge type I β-turn *(24)*. Furthermore, a single residue substitution of the i + 1 β-turn residue (Thr by Asp) of that ubiquitin hairpin leads to a 3:5 β-hairpin *(25)*.

1.2.1.2. PEPTIDES DESIGNED BY MODIFICATION
OF STRAND RESIDUES IN PROTEIN β-HAIRPINS

Two β-hairpin peptides have been derived from the C-terminal hairpin of the B1 domain of protein G: a 16-residue peptide that adopts a 4:4 β-hairpin designed by substituting three strand residues by Trp to produce a cluster of four Trp residues in non–H-bonded sites of the β-strands *(26)*, and a 10-residue 4:4 β-hairpin peptide designed using the consensus sequence found by statistical analysis of structurally aligned homologues for residues 41–56 of that domain *(27)*.

1.2.1.3. PEPTIDES FORMED BY LINKING TWO NONSEQUENTIAL
ANTIPARALLEL β-STRANDS BY MEANS OF A SHORT LOOP

This strategy has only been used in the case of a designed DNA binding peptide that adopts a 2:2 β-hairpin *(28)*. The DNA-binding motif of the met repressor protein dimer is formed by two antiparallel β-strands at the dimer interface. These two β-strands were linked by the Asn-Gly sequence which is the most favorable one for a type I' β-turn, appropriate for 2:2 β-hairpins (**Subheading 1.1.**; **Table 1**). In one of the two strands, an Ile residue was substituted by a Tyr to facilitate the nuclear magnetic resonance (NMR) assignment (**Subheading 2.7.2.**).

1.2.2. Completely de novo Sequences
1.2.2.1. β-HAIRPINS

The first reported *de novo*–designed β-hairpins were a decapeptide that adopts a 3:5 β-hairpin *(29)* and three dodecapeptides that form 2:2 β-hairpins *(30,31)*. The 3:5 β-hairpin was designed exclusively by selecting strand residues with high intrinsic β-sheet propensities (**Subheading 2.3.**) and a turn sequence with high individual statistical probabilities for each position in a type I β-turn (NPDG as in the 9-residue β-hairpin derived from Tendamistat *(1)*; **Subheading 1.2.1.1.**). The three 2:2 β-hairpins contain either the Asn-Gly turn sequence *(30,31)*, which is the most optimal sequence for a type I' β-turn or the DPro-Gly *(30)* that leads to a type II' β-turn, with both types of β-turn being appropriate for 2:2 β-hairpins (**Subheading 1.1.**). Strand residues in the 12-residue peptide designed by Ramírez-Alvarado et al. *(30)* are those statistically favorable in the corresponding β-strand positions according to the examination of the structure protein database included in the WHAT IF program *(32)*. To prevent aggregation and to increase solubility, this peptide contains two positively charged Arg residues at the peptide ends separated from the residues involved in the β-hairpin structure by one Gly residue (**Subheading 2.4.**).

A recently *de novo*–designed β-hairpin system is a series of disulfide-cyclized 10-residue peptides *(33,34)*, being the Cys residues placed in a non–H-bonded site. By examination of disulfide-bonded Cys residues that connect adjacent antiparallel strands in proteins, side chains of either Leu or aromatic residues were those providing better packing with the disulfide bond; therefore, this type of residues were selected. The turn sequences are Gly-Asn, DPro-Asn, and DPro-Gly for 2:2 β-hairpins with type II' β-turns and Asn-Gly for 2:2 β-hairpins with type I' β-turns. Very stable monomeric β-hairpins were derived from these peptides by removing the Cys residues, conserving the turn sequences and incorporating a cluster of four Trp residues *(26)*.

Finally, an ATP-binding β-hairpin peptide *(19)* has been designed by incorporating a Trp-Trp diagonal interaction (**Fig. 1B**) into a 12-residue β-hairpin peptide previously designed by Gellman's group *(31)*. This Trp-Trp diagonal interaction should allow nucleobase intercalation because in one β-hairpin peptide designed by Cochran et al. *(26)* two Trp diagonal residues seem to form a cleft. Lys residues were incorporated on the same face as the Trp residues to afford electrostatic interactions with nucleotide phosphates. The peptide has a net charge of +4 to increase solubility.

1.2.2.2. THREE-STRANDED β-SHEETS

After the success in designing β-hairpin peptides, several research groups addressed the design of three-stranded antiparallel β-sheets (*see* **Note 3**), the next step up in motif complexity. Almost simultaneously, four different peptides that adopt monomeric three-stranded β-sheets in aqueous solution were reported *(35–38)*. The two β-hairpins that compose the β-sheet motif in the four peptides are type 2:2 (**Fig. 3**). Taking into account the crucial role of the turn in β-hairpin folding and stability (**Subheading 1.3.1.**), the incorporation of sequences optimal to form either type I' β-turns or type II' β-turns was a key factor for successful designs. Although the design strategies differ in the procedure followed to select the strand residues, all of them took into account β-sheet propensities *(39–44)* and statistical preferences for interstrand residue pairs *(18,45–47)*. The four β-sheet models also differ in the number of residues per strand. Criteria to prevent aggregation and to aid solubility were also important and consequently all of them incorporate from two to five positively charged residues with their side chains pointing toward different sides of the β-sheet plane (*see* **Note 4**). In this way, positive charge is distributed over both faces of the β-sheet and self-association is minimized. Apart from the principles described here, the sequence of the 24-residue β-sheet *(38)* was chosen by statistically analyzing the Protein Data Bank (PDB) *(48)* as was previously done for a β-hairpin peptide *(30; see* **Subheading 1.2.2.1.**). The sequence of a 20-residue β-sheet *(36)* was selected by evaluating the van der Waals energies of several sequences using a template backbone structure derived from two proteins with antiparallel β-sheets (a dehydrogenase fragment and a WW domain). Later, the β-sheet stability was improved in two of these peptides, in one of them *(36)* by using a protein engineering rotamer library algorithm to evaluate the effect of amino acid substitutions on the parent β-sheet peptide *(49)*, and in the other *(35)* by including a DPro residue at the β-turn *(50); see* **Subheading 1.3.1.**

Another β-sheet design consists in extending the sequence of a β-hairpin peptide by adding a type I' β-turn, Asn-Gly sequence and a third strand to its C-terminus. This third strand contains Phe and Trp residues in non–H-bonded sites which face Tyr and Val residues in the second strand giving rise to a stabilizing hydrophobic cluster *(51)*.

1.3. Main Conclusions About Contributions to β-Sheet Folding and Stability

Analysis of the conformational behavior of peptides able to adopt β-sheet motifs in solution has provided information about their formation and stability *(6–15,52,53)*. These conclusions are summarized here with emphasis on their applicability as design rules rather than on the physical-chemical basis of β-hairpin and β-sheet stability (*see* **Note 5**).

1.3.1. Essential Role of the Turn Sequence

The peptides derived from Tendamistat *(1)* and ubiquitin *(23)* by incorporating the NPDG turn sequence (favorable for a type I β-turn; **Table 1**) adopt β-hairpins whose strand registers surprisingly differ from the one in the corresponding protein. These results were the first evidence indicating the importance of the turn region in β-hairpin formation. Later, the essential role played by the turn in directing β-hairpin formation *(24,54,55)* and also in its final stability has been demonstrated in several β-hairpin peptide systems (*see* reviews, **refs. 6–15**). A strong correlation between β-hairpin population and the statistical occurrence of the residues at the position i+1 of type I' β-turns in protein 2:2 β-hairpins has been found *(56; see* **Fig. 1A**). As a general rule, a good β-turn sequence is a necessary, but not sufficient condition for a sequence to adopt a β-hairpin structure *(55)*. The crucial role of the β-turn sequence in dictating the β-strand register has also been confirmed in three-stranded antiparallel β-sheet systems *(57)* using the β-sheet model designed by Gellman's group *(37)*. The incorporation of a DPro as i + 1 turn residue greatly stabilizes β-hairpins as well as multistranded β-sheets while substitution of DPro by LPro prevents the β-sheet formation *(31,37,49,50,57–59)*.

The observation that turns and strand–strand interactions appear to make independent and additive contributions to β-hairpin stability *(34)* provides a solid basis for the design protocol described below in which β-turn and β-strand residues are selected independently (**Subheading 2.**).

1.3.2. Contributions to Stability of β-Strand Residues

1.3.2.1. β-Sheet Propensities

Strand residues with high intrinsic β-sheet propensities help to stabilize the β-hairpins and three-stranded β-sheets, whereas those with low intrinsic β-sheet propensities destabilize them. For example, the incorporation of a Gly residue either in an edge strand or in the central strand led to a large destabilization of a three-stranded antiparallel β-sheet peptide *(60)*. That residues with high intrinsic β-sheet propensities (**Subheading 2.3.**) are generally hydrophobic accounts for the high tendency of β-sheet peptides to aggregate.

1.3.2.2. Cross-Strand Side Chain–Side Chain Interactions of Facing Residues

Side chain–side chain interactions of cross-strand facing residues contribute to β-hairpin and β-sheet stability (**Fig. 1**). Some of these stabilizing interactions, either hydrophobic or electrostatic, have been identified and quantified *(6–8,10–12,26,33,34,61–71)* (**Subheading 2.3.** and **Table 2**).

As occurs in α-helices, ionic interactions were demonstrated to be stabilizing in several β-hairpin peptides *(10,11)*. The contributions of Glu-Lys salt bridges placed in non–H-bonded sites are pH dependent decreasing at low pH, where the Glu residue is neutral *(63)*. Two Glu-Lys salt bridges produce a much larger overall contribution to β-hairpin stability than the sum of the two individual Glu-Lys interactions, indicating that co-operativity plays an important role in determining the energetics *(63)*. A Glu-Lys ion pair located at the β-hairpin ends turned out to be more stabilizing than a Lys-Glu pair at the same position *(62)*. The interaction between the N-terminal positively charged amino group and the negatively charged carboxylate group at the C-terminus also contributes to β-hairpin stability *(72)*. Furthermore, a 2:2 β-hairpin peptide is stabilized by a salt bridge interaction between the N-terminal positively charged Lys residue and the C-terminal carboxylate *(10)*.

A series of cross-strand interactions involving mainly hydrophobic and aromatic residues in either a non–H-bonded site *(33,34)* or in a H-bonded site *(61)* has been investigated using a 10-residue disulfide-cyclized β-hairpin system and a method to evaluate β-hairpin population based on the rate of disulfide bond formation (**Subheading 2.7.4.**). Contributions to β-hairpin stability of favorable cross-strand facing interactions (Ile-Trp at a non–H-bonded site and

Table 2
Favorable Cross-Strand Side Chain–Side Chain Interactions
Between Facing and Diagonal β-Strand Residues (Fig. 1)

	Facing residues Non–H-bonded sites	H-bonded site
Statistical data	C-C>>**E-K**>D-H>N-N>**W-W**>C-W≈ D-G≈D-R>K-N≈N-S>H-P≈Q-R *(47)*	C-C>E-K≈E-R>H-H≈Q-R≈ D-N≈F-F≈C-H≈S-S≈D-K≈ K-Q≈**N-T** *(47)*
Experimental[a]	**W-W**>>W-F>W-Y>W-L>W-M>W-I> W-V>>Y-L>M-L>F-L>L-L>I-L≈V-L *(33,34)* **C-C** *(33,34)*; Y-W *(64)*; I-W *(65)* F-F *(66)*; Y-F *(69)*; Y-L *(70)* **K-E** *(62,63)*	V-V>H-V≈V-H *(61)* S-T, T-T *(64)* **N-T** *(54)* S-T *(65)*
	Diagonal interactions	
Statistical data	W-Y, K-E, E-R, R-E *(18)*	
Experimental[a]	Y-K *(70)* W-K, W-R, F-K, F-R *(67)*	

Pair-wise interactions found to be favorable statistically and experimentally are shown in bold.
[a]As deduced from experimental studies on β-hairpin peptides (**Subheading 1.2.**).

Ser-Thr in a H-bonded site *[65]*) as well as a hydrophobic cluster (Trp/Val-Tyr/Phe in non–H-bonded sites *[7]*) have been shown to be dependent on their proximity to the turn region. They make larger contributions when they are closer to the turn.

Aromatic interactions have also been investigated by using the 12-residue peptide designed by Gellman's group *(31,74)*. The Phe-Phe cross-strand interaction in a non–H-bonded site was found to have edge-face geometry *(66–68)*. This implies that the interaction is not driven by the hydrophobic effect, which would favour the maximum burial of surface area. Edge-face interactions are proposed to be driven by electronic or van der Waals forces or a combination of both between the partially positive hydrogen on one aromatic ring and the π-cloud of the other ring. In the same peptide model, the Phe-Phe pair interaction in a non–H-bonded site has been shown to be more stabilizing than the Glu-Lys interaction *(75)*.

1.3.2.3. DIAGONAL SIDE CHAIN–SIDE CHAIN INTERACTIONS

Diagonal side chain–side chain interactions also contribute to β-hairpin stability (**Fig. 1B**). Evidence for the contribution to stability of a favorable Tyr-Lys diagonal interaction in a 12-residue β-hairpin peptide was found *(70)*. Afterwards, the diagonal cation-π interactions between Phe or Trp with either Lys or Arg were investigated *(67)*. The Lys residue was found to interact through the polarized C_ε carbon, whereas the Arg guanidinium moiety is stacked against the aromatic ring of Phe or Trp.

1.3.2.4. HYDROPHOBIC CLUSTERS

Hydrophobic clusters stabilise β-hairpin structures. These stabilizing effects have been demonstrated by incorporating the Trp/Val-Tyr/Phe hydrophobic cluster taken from the C-terminal β-hairpin of protein G B1 domain into non–H-bonded sites of two designed 12-residue β-hairpin peptides *(71)*. A remarkable high stability has also been reported for designed β-hairpins containing a cluster of four Trp residues in non–H-bonded sites *(26)*.

1.3.3. Co-Operativity

The question of whether the folding of β-hairpin and β-sheet peptides is co-operative or not still remains open. Evidence for co-operativity has been reported for some β-hairpin and β-sheet peptides *(37,38,51,76)*, but not found for others *(35)*. Two types of co-operativity can be distinguished in antiparallel β-sheet peptides, the longitudinal or parallel to the strand axis, and the perpendicular to the strand direction. The increase of β-hairpin stability observed upon strands lengthening indicates the existence of longitudinal co-operativity *(77)*. Regarding to perpendicular co-operativity, it has been shown that adding a fourth strand to a three-stranded β-sheet peptide leads to further β-sheet stability *(59)*.

1.3.4. β-Sheet Twist

The right-handed twist characteristic of β-sheets (**Subheading 1.1.**) seems to be related to β-sheet stability, being generally the most twisted β-hairpins the most stable ones. This is the case for the 3:5 β-hairpins, which are more twisted and also more stable than the 4:4 β-hairpins *(78)*. The existence of a correlation between the degree of twist and the buried hydrophobic surface has been found in a three-stranded β-sheet *(50)*. Thus, β-sheet twist appears to contribute to β-sheet stability by increasing hydrophobic surface burial.

1.3.5. Hydrogen Bonds

As occurs in proteins, contribution of hydrogen bonds to β-hairpin stability is not clear. The infrared spectrum of a 16-residue β-hairpin peptide showed no features in the amide I region to suggest a significant contribution from interstrand hydrogen bonds *(79)*, whereas in another 16-residue β-hairpin peptide, an amide I band at approx 1617 cm^{-1} was identified and attributed to hydrogen-bonding across the strand *(80)*. That the rare protein 2:2 β-hairpins with the unsuitable type I β-turn usually exhibit longer strands than those with the appropriate type I' β-turns might be explained by the additional hydrogen bonds compensating for the unfavorable conformation of the type I β-turn *(45)*.

1.3.6. Disulfide Bonds

In general, disulfide bonds stabilize the β-hairpin structures because the unfolded state becomes more rigid and a decrease in the loss of entropy on folding is observed. Some natural antimicrobial peptides are β-hairpins stabilized by disulfide bonds *(81)*. Disulfide cyclization of designed peptides, as well as backbone cyclization, have been used as references for the folded state *(74)*. In addition, one of the peptide systems used to investigate β-hairpin stability is a series of disulfide-cyclized decapeptides *(33,34)*.

2. Methods

The protocol proposed here for the *de novo* design of monomeric β-sheet peptides consists of the following steps.

1. Selection of goal β-sheet and peptide length (**Subheading 2.1.**).
2. Selection of β-turn sequence (**Subheading 2.2.**).
3. Selection of β-strand residues. Solubility criteria (**Subheadings 2.3. and 2.4.**).
4. *In silico* validation of the sequence resulting from **steps 1–3** (**Subheading 2.6.**).
5. Peptide preparation.
6. Experimental validation of the designed β-sheet (**Subheading 2.7.**).

The two structurally different regions that constitute a β-hairpin, the β-turn (**step 2**), and the two antiparallel β-strands (**step 3**) are considered independently. In most reported β-sheet forming peptides, **step 4** was not performed. Methods to carry out peptide preparation (**step 5**) either by chemical synthesis or by biotechnology are beyond the scope of the current review. From the design perspective, the procedure used to obtain the designed peptide is important only regarding to the possibilities for the protection of peptide ends or the incorporation of non-natural amino acids. Concerning **step 6**, the method used for confirming that the designed peptide adopts the goal structure has also to be considered. Thus, NMR assignment is facilitated by nonrepetitive sequences.

2.1. Selection of Goal β-Sheet Structure and Peptide Length

The design of a protein or peptide structure consists in finding a sequence able to form a selected goal structure. The choice of this structure is the starting point in any design project. The principles described on the following sections are applicable for β-hairpins with short loops and can be extended to three-stranded antiparallel β-sheets. If the aim of the project is to re-design a protein domain or fragment of known β-sheet structure (for example, to improve certain structural characteristics of a natural biologically active sequence), the rules given in **Subheadings 2.2.–2.4.** can be applied by maintaining the residues important for the biological activity of the molecule, and substituting some other residues to improve its stability. Sequence alignment procedures are very useful in these redesign cases (**Subheading 2.5.**).

The length of a β-hairpin peptide is n + c + t, where n and c are the number of residues at N- and C-terminal strands, respectively, and t the number of residues in the turn. This last value depends on the type of β-hairpin, being two in a 2:2 β-hairpin, three in a 3:5 β-hairpin, and four in a 4:4 β-hairpin (**Subheading 1.1.**; **Fig. 1**). If n and c are different, residues at the end of the longest strand are not paired. Strand length in the reported β-hairpin peptides ranges from two to nine, most of them having three to seven strand residues. In terms of stability, the strand length increases stability up to seven residues long, but no stability increment is observed with further strand lengthening (*77*; *see* **Subheading 1.3.3.**).

To design a three-stranded antiparallel β-sheet, the procedure followed for β-hairpins (**Subheadings 2.2.–2.6.**) is applied twice, once for the N-terminal hairpin and then again for the C-terminal hairpin (hairpins 1 and 2 in **Fig. 3**). These two β-hairpins can be of the same type and thus have the same number of turn residues, or they can be of a different type. It is necessary to bear in mind that the sequences of the C-terminal strand of hairpin 1 and the N-terminal strand of hairpin 2 are the same and that every residue at this middle strand that is located in an H-bonded site in hairpin 1 is in a non–H-bonded site in hairpin 2 and vice versa (**Fig. 3**). Designed three-stranded β-sheets have three to seven residues per strand.

2.2. Selection of β-Turn Residues

Because the turn region plays an essential role in determining β-hairpin conformation and its stability (**Subheading 1.3.**), the selection of an adequate β-turn sequence is crucial to ensure that the designed peptide will adopt the target β-hairpin. **Table 1** has been built using β-turn positional potentials statistically derived from protein structures *(22)*. **Table 1** is very useful in selecting the β-turn residues by taking into account which is the most appropriate turn for the desired β-hairpin, namely type I for 3:5 and 4:4 β-hairpins, and type I' or II' for 2:2 β-hairpins. The final designed β-hairpin not only must have an optimal β-turn sequence but also it should not contain any sequence likely to form an alternative β-turn.

2.3. Selection of β-Strand Residues

The following principles must be considered to select β-strand residues:

2.3.1. Intrinsic β-Sheet Propensities

As a general rule, residues in the strands should have high intrinsic β-sheet propensities. Because intrinsic β-sheet propensities seem to be context dependent *(41)*, the reported scales of β-sheet propensities show differences. Nevertheless, they are useful as a guide because of their concordance with respect to which residues are good β-sheet formers. The differences are in the rank order of the residues. Thus, according to statistical data *(39)*, the residues with high intrinsic β-sheet propensities are V>I>T>Y>W>F>L, whereas C>M>Q>S>R are more or less neutral to adopt β-sheet conformations. Among experimental scales *(40–44)*, the β-sheet favorable residues are Y>T>I>F>W>V and the neutral ones are S>M>C>L>R, for example *(43)*.

2.3.2. Residues Adjacent to the Turn

Apart from having high intrinsic β-sheet propensities, the best residues to be in positions preceding the β-turn in 2:2 and 3:5 β-hairpins (n – 1 in **Figs. 1A** and **1C**) should also have high intrinsic probability to be at the position i of the corresponding type of β-turn (**Table 1**), and the best one to follow the β-turn in 2:2 β-hairpins (c + 1 in **Fig. 1C**) should have high intrinsic probability to be at the position i + 3 of type I' or type II' β-turns (**Table 1**).

2.3.3. Pair-Wise Cross-Strand Interactions

Facing residues should correspond to pairs with favorable side chain–side chain interactions. **Table 2** that lists the most favorable pair-wise interactions according to statistical data *(47)* is useful as a guide to select them, though the statistical data on cross-strand pair-wise interactions does not always coincide with the experimental results regarding to β-sheet stability and statistical analysis by different authors show some discrepancies. Other statistical data on pair-wise interactions, which are not included in **Table 2**, might also be used *(18,45,46)*. The cross-strand side chain–side chain interactions found to be stabilizing in model β-hairpin peptides are also given in **Table 2**.

2.3.4. Diagonal Interactions

In addition to pair-wise cross-strand interactions between facing residues, favorable diagonal interactions and hydrophobic clusters will contribute to β-hairpin stability (**Subheading 1.3.2.3.**). **Table 2** lists preferred diagonal interactions according to statistical data *(18)* as well as those found experimentally and servicable as aids for selecting favorable diagonal interactions.

2.3.5. Peptide Ends

Peptide ends can be nonprotected or protected by acetylation of the N-terminal and by amidation of the C-terminal. If they are not protected and the two terminal amino acids are facing residues, the interaction between the positively charged amino group and the negatively charged carboxylate makes a favorable contribution to β-hairpin stability *(72)*. To increase solubility, it may be advisable to protect one of the peptide ends (**Subheading 2.4.**). The method to be used for obtaining the peptide can also determine the possibilities for peptide ends; for example, protection by acetylation and amidation is easy and convenient if the peptide is going to be prepared by chemical synthesis. Some residues at the N-terminus will be added in the case of cloning and expression of the peptide.

2.4. Solubility Criteria (see Note 6)

Because aggregation and solubility problems in peptides and proteins seem to be higher close to the isoelectric point, a peptide with either a net positive charge or a net negative charge will be likely more soluble. Incorporation of an electrostatic interaction resulted in peptide aggregation in one case *(78)*. Thus, it can be advisable to protect the N-termini in Asp/Glu-containing peptides and the C-termini in those containing positively charged residues (Lys, Arg, Orn). The distribution of the charged and polar residues also plays an important role in solubility. Because amphipathic sheets with a hydrophilic face and a hydrophobic one are more prone to aggregate than the nonamphipathic ones, the charged and polar side chains should be pointing toward both faces of the β-sheet. Other strategy to avoid aggregation used in designed β-hairpin peptides consists in the incorporation of charged residues at the peptide ends separated from the hairpin by spacing linkers consisting of Gly residues *(30)*.

2.5. Alignment to Sequences That Adopt the Target β-Sheet Motif

Naturally occurring peptides or protein domains having the desired β-sheet motif or designed peptides previously reported to adopt the target structure can be used as the starting point for the design. In this case the alignment of all the known peptides or protein domains that have the same structure or function that the target one can be used to get a consensus sequence. A 10-residue β-hairpin peptide *(27)* has been designed using this strategy (**Subheading 1.2.2.**).

2.6. In silico Validation of Designed β-Sheet Sequences (see Note 7)

The probability of the sequence designed according to the principles in **Subheadings 2.1.–2.5.** to adopt the goal β-sheet structure can be examined by applying a β-hairpin prediction program developed in our laboratory (*see* **Note 8**). The program contains two principal subroutines: one predicts the β-turn residues and the other deals with the β-strand residues. In a first step, a normalized version of the β-turn positional potentials published by Hutchinson and Thornton *(22)* is employed to predict the residues most likely to form a β-turn and the type of turn they will adopt. In a second independent step, the prediction of most favorable residues in β-sheet conformation is determined by a linear combination of terms derived from intrinsic β-sheet propensities *(43)*, cross-strand pair interactions *(47)*, and number of hydrogen bonds formed. Finally, the prediction of the β-hairpin type adopted by the amino acid sequence is a combination of both results, which are ranked numerically together with rules based on the most favorable type of protein β-hairpins.

2.7. Checking the Success of the Design

Monomeric state of the β-hairpin peptides is usually confirmed by analytical ultracentrifugation and by dilution experiments monitored by circular dichroism (CD) or NMR.

2.7.1. CD Spectroscopy

Characteristic far-ultraviolet CD spectra for β-sheets exhibit one minimum at approx 216 nm and one maximum at approx 195 nm *(82)*. Thus, CD spectroscopy provides a quick way to confirm whether or not a designed peptide adopts a β-sheet structure. However, no information about strand register can be gathered from a CD spectrum.

2.7.2. NMR Spectroscopy

NMR is the best technique to demonstrate that a particular peptide adopts its target β-sheet structure. The observation of NOEs between backbone protons of residues in adjacent strands provides unambiguous evidence about the strand register, and thus the type of β-hairpin. Type of β-turn can also be determined by NMR.

2.7.2.1. NOEs

In antiparallel β-sheets, the backbone protons that are close enough to give rise to NOEs are (1) the H_α protons of residues facing each other in a non–H-bonded site (2.3 Å; **Fig. 1A**), (2) the NH amide protons of facing residues in H-bonded sites (3.3 Å; **Fig. 1A**), and (3) the H_α protons in a non–H-bonded site and NH protons in an H-bonded site of residues in adjacent strands, when these two sites are consecutive (3.2 Å; **Fig. 1A**). Because some H_α signals may be obscured by that of water, to observe H_α–H_α NOEs is convenient to dissolve the peptide in D_2O. Further information on the β-hairpin adopted can be gathered from the NOEs between protons of side chains located at the same β-sheet face.

In three-stranded β-sheets, the set of NOEs involving backbone protons is also compatible with the independent formation of β-hairpin 1 and β-hairpin 2 (**Fig. 3**). Only the observation of at least one long-range NOE involving side chain protons of residues at the N-terminal strand and at the C-terminal one demonstrates the formation of the three-stranded β-sheet motif *(50; see* **Fig. 3**).

2.7.2.2. Conformational Shifts

The patterns of $^1H_\alpha$, $^{13}C_\alpha$, and $^{13}C_\beta$ conformational shifts ($\Delta\delta = \delta^{observed} - \delta^{random\ coil}$, ppm; *see* **Note 9**) are also useful for identifying β-hairpins and three-stranded β-sheets. β-Strands can be delineated by the stretches of at least two consecutive residues having positive $\Delta\delta_{H\alpha}$ and $\Delta\delta_{C\beta}$ values and negative $\Delta\delta_{C\alpha}$ values. Two of such stretches are observed in β-hairpins and three in three-stranded β-sheets. These stretches are separated by two to four residues with $\Delta\delta_{H\alpha}$ negative or very small in absolute value, with at least one of them having a positive $\Delta\delta_{C\alpha}$ value and a negative $\Delta\delta_{C\beta}$ value *(83)*. A more thorough analysis of the $\Delta\delta_{C\alpha}$ and $\Delta\delta_{C\beta}$ profiles at the turn region allows us to identify the particular type of β-hairpin and β-turn (for details *see* **ref. 83**).

2.7.3. Quantification of β-Hairpin and β-Sheet Populations

In contrast to helical peptides in which CD spectroscopy provides a method for quantifying helix population, there is not a well-established method to quantify β-sheet populations. Assuming a two-state behavior for β-hairpin or β-sheet formation, populations can be evaluated from different NMR parameters, such as H_α–H_α NOEs and $^1H_\alpha$, $^{13}C_\alpha$, and $^{13}C_\beta$ chemical shifts. Apart from the validity of the two-state assumption, the absence of accurate reference values for the completely folded and random coil states limits the precision and accuracy of the quantification of β-sheet populations (for details on this question *see* **refs. 6–8,10,74,83**).

2.7.4. Method Based on Disulfide Bond Formation

A recently reported nonspectroscopic method to measure β-hairpin stability is based on thiol-disulfide equilibrium in cystine-cyclized peptides *(33,34)*. β-Hairpin stabilities in disulfide-cyclized decapeptides were compared on the basis of the changes in the thiol-disulfide equilibrium constant upon residue mutation.

3. Notes

1. Two protein fragments that adopt their native β-hairpin structures have been reported *(84,85)*.
2. Non-water-soluble peptides that adopt β-hairpin structures in chloroform, benzene, and alcoholic solvents have also been designed *(86)*. As in many water-soluble β-hairpins, their β-turn sequences are either DPro-Gly or Asn-Gly. Some of these short and very hydrophobic β-hairpin peptides have been crystallized *(87)*.
3. A designed three-stranded antiparallel β-sheet with a β2-β1-β3 topology instead of the β1-β2-β3 of the meander β-sheets reviewed here has been reported *(88)*. Several antiparallel β-sheets with more than three strands have also been designed: two four-stranded antiparallel β-sheet peptides, a 26-residue peptide that adopts the β-sheet in either pure methanol or in water-methanol solutions *(89,90)*, and a 50-residue molecule composed of two BPTI-derived β-hairpin modules that are connected by a cross-link between two Lys residues in the inner strands *(91)*; a 34-residue peptide that forms a five-stranded β-sheet and contains a metal binding site *(92)*; and an eight-stranded antiparallel β-sheet formed by connecting two four-stranded β-sheets with a disulfide bond *(93)*. Dimeric and tetrameric β-

sheets *(94)* and a trimer composed of three β-hairpin modules *(95)* have also been designed.

4. N-methyl amino acids were incorporated in a designed three-stranded β-sheet to prevent aggregation *(96)*. A non-water-soluble three-stranded antiparallel β-sheet containing DPro-Gly sequences in the two turns has also been designed *(89)*.

5. β-Hairpins are stabilized by alcohol cosolvents, trifluoroethanol, and methanol. For deeper discussion on the physicochemical origin of the contributions to β-hairpin and β-sheet stability, see reviews *(6–15)* and references therein and here.

6. A recently published program that predicts the tendency of a given sequence to aggregate may be used to screen out sequences with the strongest tendency to self-associate *(97)*.

7. Methods developed for predicting β-turns in a protein from its amino acid sequence could also be used to validate the designed β-sheet sequences. However, so far the use of such methods has not been reported in the case of any designed β-hairpin or β-sheet peptide. A web server that predicts β-turns from sequence by implementing several reported statistical algorithms is available (http://imtech.res.in/raghava/betapred/home.html; *[98]*). A method reported for the recognition of β-hairpins in proteins that combines secondary structure predictions and threading methods by using a database search and a neural network approach could also be used *(99)*. Another possibility is to construct a model structure and evaluate its stability by rotamer library algorithms, as done in the case of a 20-residue three-stranded β-sheet (*49*; *see* **Subheading 1.2.2.2.**).

8. BEHAIRPRED is accessible from http://bionmr.cipf.es/software/behairpred.htm (Pantoja-Uceda and Jiménez, unpublished). It accepts the non-natural residues Orn and DPro as input. In addition, it is able to identify type II β-turns, even though they are quite uncommon in β-hairpins.

9. $^{1}H_{\alpha}$, $^{13}C_{\alpha}$, and $^{13}C_{\beta}$ conformational shifts ($\Delta\delta_{H\alpha}$, $\Delta\delta_{C\alpha}$, and $\Delta\delta_{C\beta}$, respectively) are defined as the deviation of the experimentally measured chemical shift ($\delta^{observed}$, ppm) from reference δ values for the random coil state ($\delta^{random\ coil}$, ppm; *[83]*).

Acknowledgments

We thank Prof. M. Rico and Dr. D. V. Laurents for critical readings of the manuscript and financial support from the CSIC Intramural Project 200580F0162.

References

1. Blanco, F. J., Jiménez, M. A., Herranz, J., Rico, M., Santoro, J., and Nieto, J. L. (1993) NMR evidence of a short linear peptide that folds into a β-hairpin in aqueous solution. *J. Am. Chem. Soc.* **115**, 5887–5888.

2. Nowick, J. S., Cary, J. M., and Tsai, J. H. (2001) A triply templated artificial β-sheet. *J. Am. Chem. Soc.* **123**, 5176–5180.

3. Fisk, J. D. and Gellman, S. H. (2001) A parallel β-sheet model system that folds in water. *J. Am. Chem. Soc.* **123**, 343–344.

4. Nowick, J. S. (1999) Chemical models of protein β-sheets. *Acc. Chem. Res.* **32**, 287–296.

5. Robinson, J. A. (1999) The design, synthesis and conformation of some new β-hairpin mimetics: novel reagents for drug and vaccine discovery. *Synlett* **4,** 429–441.
6. Blanco, F., Ramírez-Alvarado, M., and Serrano, L. (1998) Formation and stability of β-hairpin structures in polypeptides. *Curr. Opin. Struct. Biol.* **8,** 107–111.
7. Gellman, S. H. (1998) Minimal model systems for β-sheet secondary structure in proteins. *Curr. Opin. Chem. Biol.* **2,** 717–725.
8. Lacroix, E., Kortemme, T., López de la Paz, M., and Serrano, L. (1999) The design of linear peptides that fold as monomeric β-sheet structures. *Curr. Opin. Struct. Biol.* **9,** 487–493.
9. Ramírez-Alvarado, M., Kortemme, T., Blanco, F. J., and Serrano, L. (1999) β-Hairpin and β-sheet formation in designed linear peptides. *Bioorg. Med. Chem.* **7,** 93–103.
10. Searle, M. S. (2001) Peptide models of protein β-sheets: design, folding and insights into stabilizing weak interactions. *J. Chem. Soc. Perkin Trans.* **2,** 1011–1020.
11. Searle, M. S. (2004) Insights into stabilizing weak interactions in designed peptide β-hairpins. *Biopolymers* **76,** 185–195.
12. Searle, M. S. and Ciani, B. (2004) Design of β-sheet systems for understanding the thermodynamics and kinetics of protein folding. *Curr. Opin. Struct. Biol.* **14,** 458–464.
13. Serrano, L. (2000) The relationship between sequence and structure in elementary folding units. *Adv. Protein. Chem.* **53,** 49–85.
14. Smith, C. K. and Regan, L. (1997) Construction and design of β-sheets. *Acc. Chem. Res.* **30,** 153–161.
15. Venkatraman, J., Shankaramma, S. C., and Balaram, P. (2001) Design of folded peptides. *Chem. Rev.* **101,** 3131–3152.
16. Richardson, J. S. (1981) The anatomy and taxonomy of protein structure. *Adv. Protein Chem.* **34,** 167–339.
17. Sibanda, B. L., Blundell, T. L., and Thornton, J. M. (1989) Conformation of β-hairpins in protein structures. A systematic classification with applications to modelling by homology, electron density fitting and protein engineering. *J. Mol. Biol.* **206,** 759–777.
18. Cootes, A. P., Curmi, P. M., Cunningham, R., Donnelly, C., and Torda, A. E. (1998) The dependence of amino acid pair correlations on structural environment. *Proteins* **32,** 175–189.
19. Butterfield, S. M. and Waters, M. L. (2003) A designed β-hairpin peptide for molecular recognition of ATP in water. *J. Am. Chem. Soc.* **125,** 9580–9581.
20. Butterfield, S. M., Goodman, C. M., Rotello, V. M., and Waters, M. L. (2004) A peptide flavoprotein mimic: flavin recognition and redox potential modulation in water by a designed β-hairpin. *Angew. Chem. Int. Ed. Engl.* **43,** 724–727.
21. Butterfield, S. M., Cooper, W. J., and Waters, M. L. (2005) Minimalist protein design: A β-hairpin peptide that binds ssDNA. *J. Am. Chem. Soc.* **127,** 24–25.
22. Hutchinson, E. G. and Thornton, J. M. (1994) A revised set of potentials for β-turn formation in proteins. *Protein Sci.* **3,** 2207–2216.
23. Searle, M. S., Williams, D. H., and Packman, L. C. (1995) A short linear peptide derived from the N-terminal sequence of ubiquitin folds into a water-stable non-native β-hairpin. *Nat. Struct. Biol.* **2,** 999–1006.

24. Haque, T. S. and Gellman, S. H. (1997) Insights into β-hairpin stability in aqueous solution from peptides with enforced type I' and type II' β-turns. *J. Am. Chem. Soc.* **119,** 2303–2304.

25. Zerella, R., Chen, P. Y., Evans, P. A., Raine, A., and Williams, D. H. (2000) Structural characterization of a mutant peptide derived from ubiquitin: implications for protein folding. *Protein Sci.* **9,** 2142–2150.

26. Cochran, A. G., Skelton, N. J., and Starovasnik, M. A. (2001) Tryptophan zippers: stable, monomeric β-hairpins. *Proc. Natl. Acad. Sci. USA* **98,** 5578–5583.

27. Honda, S., Yamasaki, K., Sawada, Y., and Morii, H. (2004) 10 residue folded peptide designed by segment statistics. *Structure (Camb.)* **12,** 1507–1518.

28. Maynard, A. J. and Searle, M. S. (1997) NMR structural analysis of a β-hairpin peptide designed for DNA binding. *Chem. Commun.* 1297–1298.

29. de Alba, E., Jiménez, M. A., Rico, M., and Nieto, J. L. (1996) Conformational investigation of designed short linear peptides able to fold into β-hairpin structures in aqueous solution. *Fold Des.* **1,** 133–144.

30. Ramírez-Alvarado, M., Blanco, F. J., and Serrano, L. (1996) *De novo* design and structural analysis of a model β-hairpin peptide system. *Nat. Struct. Biol.* **3,** 604–612.

31. Stanger, H. E. and Gellman, S. H. (1998) Rules for antiparallel β-sheet design: D-Pro-Gly is superior to L-Asn-Gly for β-hairpin nucleation. *J. Am. Chem. Soc.* **120,** 4236–4237.

32. Vriend, G. (1990) WHAT IF: a molecular modeling and drug design program. *J. Mol. Graph.* **8,** 52–56.

33. Russell, S. and Cochran, A. G. (2000) Designing stable β-hairpins: energetics contributions from cross-strand residues. *J. Am. Chem. Soc.* **122,** 12600–12601.

34. Cochran, A. G., Tong, R. T., Starovasnik, M. A., Park, E. J., McDowell, R. S., Theaker, J. E., et al. (2001) A minimal peptide scaffold for β-turn display: optimizing a strand position in disulfide-cyclized β-hairpins. *J. Am. Chem. Soc.* **123,** 625–632.

35. de Alba, E., Santoro, J., Rico, M., and Jiménez, M. A. (1999) De novo design of a monomeric three-stranded antiparallel β-sheet. *Protein Sci.* **8,** 854–865.

36. Kortemme, T., Ramírez-Alvarado, M., and Serrano, L. (1998) Design of a 20-amino acid, three-stranded β-sheet protein. *Science* **281,** 253–256.

37. Schenck, H. L. and Gellman, S. H. (1998) Use of a designed triple-stranded antiparallel β-sheet to probe β-sheet cooperativity in aqueous solution. *J. Am Chem. Soc.* **120,** 4869–4870.

38. Sharman, G. J. and Searle, M. S. (1998) Cooperative interaction between the three strands of a designed antiparallel β-sheet. *J. Am. Chem. Soc.* **120,** 5291–5300.

39. Fasman, G. D. (1989) The development of the prediction of protein structure, in *Prediction of Protein Structure and the Principles of Protein Conformation* (Fasman, G. D., ed.), Plenum Press, New York, pp. 193–316.

40. Kim, C. A. and Berg, J. M. (1993) Thermodynamic β-sheet propensities measured using a zinc-finger host peptide. *Nature* **362,** 267–270.

41. Minor, D. L., Jr. and Kim, P. S. (1994) Context is a major determinant of β-sheet propensity. *Nature* **371,** 264–267.

42. Minor, D. L., Jr. and Kim, P. S. (1994) Measurement of the β-sheet-forming propensities of amino acids. *Nature* **367,** 660–663.
43. Smith, C. K., Withka, J. M., and Regan, L. (1994) A thermodynamic scale for the β-sheet forming tendencies of the amino acids. *Biochemistry* **33,** 5510–5517.
44. Street, A. G. and Mayo, S. L. (1999) Intrinsic β-sheet propensities result from van der Waals interactions between side chains and the local backbone. *Proc. Natl. Acad. Sci. USA* **96,** 9074–9076.
45. Gunasekaran, K., Ramakrishnan, C., and Balaram, P. (1997) β-Hairpins in proteins revisited: lessons for *de novo* design. *Protein Eng.* **10,** 1131–1141.
46. Hutchinson, E. G., Sessions, R. B., Thornton, J. M., and Woolfson, D. N. (1998) Determinants of strand register in antiparallel β-sheets of proteins. *Protein Sci.* **7,** 2287–2300.
47. Wouters, M. A. and Curmi, P. M. (1995) An analysis of side chain interactions and pair correlations within antiparallel β-sheets: the differences between backbone hydrogen-bonded and non-hydrogen-bonded residue pairs. *Proteins* **22,** 119–131.
48. Berman, H. M., Westbrook, J., Feng, Z., Gilliland, G., Bhat, T. N., Weissig, H., et al. (2000) The Protein Data Bank. *Nucleic Acids Res.* **28,** 235–242.
49. López de la Paz, M., Lacroix, E., Ramírez-Alvarado, M., and Serrano, L. (2001) Computer-aided design of β-sheet peptides. *J. Mol. Biol.* **312,** 229–246.
50. Santiveri, C. M., Santoro, J., Rico, M., and Jiménez, M. A. (2004) Factors involved in the stability of isolated β-sheets: turn sequence, β-sheet twisting, and hydrophobic surface burial. *Protein Sci.* **13,** 1134–1147.
51. Griffiths-Jones, S. R. and Searle, M. S. (2000) Structure, folding, and energetics of cooperative interactions between β-strands of a de novo designed three-stranded antiparallel β-sheet peptide. *J. Am. Chem. Soc.* **122,** 8350–8356.
52. Dhanasekaran, M., Prakash, O., Gong, Y. X., and Baures, P. W. (2004) Expected and unexpected results from combined β-hairpin design elements. *Org. Biomol. Chem.* **2,** 2071–2082.
53. Fesinmeyer, R. M., Hudson, F. M., and Andersen, N. H. (2004) Enhanced hairpin stability through loop design: the case of the protein G B1 domain hairpin. *J. Am. Chem. Soc.* **126,** 7238–7243.
54. de Alba, E., Jiménez, M. A., and Rico, M. (1997) Turn residue sequence determines β-hairpin conformation in designed peptides. *J. Am. Chem. Soc.* **119,** 175–183.
55. Santiveri, C. M., Santoro, J., Rico, M., and Jiménez, M. A. (2002) Thermodynamic analysis of β-hairpin-forming peptides from the thermal dependence of ^{1}H NMR chemical shifts. *J. Am. Chem. Soc.* **124,** 14903–14909.
56. Ramírez-Alvarado, M., Blanco, F. J., Niemann, H., and Serrano, L. (1997) Role of β-turn residues in β-hairpin formation and stability in designed peptides. *J. Mol. Biol.* **273,** 898–912.
57. Chen, P. Y., Lin, C. K., Lee, C. T., Jan, H., and Chan, S. I. (2001) Effects of turn residues in directing the formation of the β-sheet and in the stability of the β-sheet. *Protein Sci.* **10,** 1794–1800.
58. Raghothama, S. R., Awasthi, S. K., and Balaram, P. (1998) β-Hairpin nucleation by Pro-Gly β-turns. Comparison of DPro-Gly and LPro-Gly sequences in an apolar octapeptide. *J. Chem. Soc. Perkin Trans.* **2,** 137–143.

59. Syud, F. A., Stanger, H. E., Mortell, H. S., Espinosa, J. F., Fisk, J. D., Fry, C. G., and Gellman, S. H. (2003) Influence of strand number on antiparallel β-sheet stability in designed three- and four-stranded β-sheets. *J. Mol. Biol.* **326,** 553–568.
60. Santiveri, C. M., Rico, M., Jiménez, M. A., Pastor, M. T., and Pérez-Payá, E. (2003) Insights into the determinants of β-sheet stability: ^{1}H and ^{13}C NMR conformational investigation of three-stranded antiparallel β-sheet-forming peptides. *J. Pept. Res.* **61,** 177–188.
61. Russell, S. J., Blandl, T., Skelton, N. J., and Cochran, A. G. (2003) Stability of cyclic β-hairpins: asymmetric contributions from side chains of a hydrogen-bonded cross-strand residue pair. *J. Am. Chem. Soc.* **125,** 388–395.
62. Ramírez-Alvarado, M., Blanco, F. J., and Serrano, L. (2001) Elongation of the BH8 β-hairpin peptide: Electrostatic interactions in β-hairpin formation and stability. *Protein Sci.* **10,** 1381–1392.
63. Ciani, B., Jourdan, M., and Searle, M. S. (2003) Stabilization of β-hairpin peptides by salt bridges: role of preorganization in the energetic contribution of weak interactions. *J. Am. Chem. Soc.* **125,** 9038–9047.
64. de Alba, E., Rico, M., and Jiménez, M. A. (1997) Cross-strand side-chain interactions versus turn conformation in β-hairpins. *Protein Sci.* **6,** 2548–2560.
65. Santiveri, C. M., Rico, M., and Jiménez, M. A. (2000) Position effect of cross-strand side-chain interactions on β-hairpin formation. *Protein Sci.* **9,** 2151–2160.
66. Tatko, C. D. and Waters, M. L. (2002) Selective aromatic interactions in β-hairpin peptides. *J. Am. Chem. Soc.* **124,** 9372–9373.
67. Tatko, C. D. and Waters, M. L. (2003) The geometry and efficacy of cation-π interactions in a diagonal position of a designed β-hairpin. *Protein Sci.* 12, 2443–2452.
68. Tatko, C. D. and Waters, M. L. (2004) Comparison of C-H...π and hydrophobic interactions in a β-hairpin peptide: impact on stability and specificity. *J. Am. Chem. Soc.* **126,** 2028–2034.
69. Kobayashi, N., Honda, S., Yoshii, H., and Munekata, E. (2000) Role of side-chains in the cooperative β-hairpin folding of the short C-terminal fragment derived from streptococcal protein G. *Biochemistry* **39,** 6564–6571.
70. Syud, F. A., Stanger, H. E., and Gellman, S. H. (2001) Interstrand side chain–side chain interactions in a designed β-hairpin: significance of both lateral and diagonal pairings. *J. Am. Chem. Soc.* **123,** 8667–8677.
71. Espinosa, J. F., Syud, F. A., and Gellman, S. H. (2002) Analysis of the factors that stabilize a designed two-stranded antiparallel β-sheet. *Protein Sci.* **11,** 1492–1505.
72. de Alba, E., Blanco, F. J., Jiménez, M. A., Rico, M., and Nieto, J. L. (1995) Interactions responsible for the pH dependence of the β-hairpin conformational population formed by a designed linear peptide. *Eur. J. Biochem.* **233,** 283–292.
73. Espinosa, J. F., Munoz, V., and Gellman, S. H. (2001) Interplay between hydrophobic cluster and loop propensity in β-hairpin formation. *J. Mol. Biol.* **306,** 397–402.
74. Syud, F. A., Espinosa, J. F., and Gellman, S. H. (1999) NMR-based quantification of β-sheet populations in aqueous solution through use of reference peptides for the folded and unfolded states. *J. Am. Chem. Soc.* **121,** 11578–11579.

75. Kiehna, S. E. and Waters, M. L. (2003) Sequence dependence of β-hairpin structure: comparison of a salt bridge and an aromatic interaction. *Protein Sci.* **12,** 2657–2667.

76. Honda, S., Kobayashi, N., and Munekata, E. (2000) Thermodynamics of a β-hairpin structure: evidence for cooperative formation of folding nucleus. *J. Mol. Biol.* **295,** 269–278.

77. Stanger, H. E., Syud, F. A., Espinosa, J. F., Giriat, I., Muir, T., and Gellman, S. H. (2001) Length-dependent stability and strand length limits in antiparallel β-sheet secondary structure. *Proc. Natl. Acad. Sci. USA* **98,** 12015–12020.

78. de Alba, E., Rico, M., and Jiménez, M. A. (1999) The turn sequence directs β-strand alignment in designed β-hairpins. *Protein Sci.* **8,** 2234–2244.

79. Colley, C. S., Griffiths-Jones, S. R., George, M. W., and Searle, M. S. (2000) Do interstrand hydrogen bonds contribute to β-hairpin stability in solution? IR analysis of peptide folding in water. *Chem. Commun.* 593–594.

80. Arrondo, J. L., Blanco, F. J., Serrano, L., and Goni, F. M. (1996) Infrared evidence of a β-hairpin peptide structure in solution. *FEBS Lett.* **384,** 35–37.

81. Andreu, D. and Rivas, L. (1998) Animal antimicrobial peptides: an overview. *Biopolymers* **47,** 415–433.

82. Johnson, W. C. J. (1988) Secondary structure of proteins through circular dichroism. *Annu. Rev. Biophys. Chem.* **17,** 145–166.

83. Santiveri, C. M., Rico, M., and Jiménez, M. A. (2001) $^{13}C_\alpha$ and $^{13}C_\beta$ chemical shifts as a tool to delineate β-hairpin structures in peptides. *J. Biomol. NMR* **19,** 331–345.

84. Blanco, F. J., Rivas, G., and Serrano, L. (1994) A short linear peptide that folds into a native stable β-hairpin in aqueous solution. *Nat. Struct. Biol.* **1,** 584–590.

85. Searle, M. S., Zerella, R., Williams, D. H., and Packman, L. C. (1996) Native-like β-hairpin structure in an isolated fragment from ferredoxin: NMR and CD studies of solvent effects on the N-terminal 20 residues. *Protein Eng.* **9,** 559–565.

86. Awasthi, S. K., Raghothama, S., and Balaram, P. (1995) A designed β-hairpin peptide. *Biochem. Biophys. Res. Commun.* **216,** 375–381.

87. Karle, I. L., Awasthi, S. K., and Balaram, P. (1996) A designed β-hairpin peptide in crystals. *Proc. Natl. Acad. Sci. USA* **93,** 8189–8193.

88. Ottesen, J. J. and Imperiali, B. (2001) Design of a discretely folded mini-protein motif with predominantly β-structure. *Nat. Struct. Biol.* **8,** 535–539.

89. Das, C., Raghothama, S., and Balaram, P. (1998) A designed three stranded β-sheet peptide as a multiple β-hairpin model. *J. Am. Chem. Soc.* **120,** 5812–5813.

90. Das, C., Nayak, V., Raghothama, S., and Balaram, P. (2000) Synthetic protein design: construction of a four-stranded β-sheet structure and evaluation of its integrity in methanol-water systems. *J. Pept. Res.* **56,** 307–317.

91. Carulla, N., Woodward, C., and Barany, G. (2002) BetaCore, a designed water soluble four-stranded antiparallel β-sheet protein. *Protein Sci.* **11,** 1539–1551.

92. Venkatraman, J., Naganagowda, G. A., Sudha, R., and Balaram, P. (2001) *De novo* design of a five-stranded β-sheet anchoring a metal-ion binding site. *Chem. Commun.* 2660–2661.

93. Venkatraman, J., Nagana Gowda, G. A., and Balaram, P. (2002) Design and construction of an open multistranded β-sheet polypeptide stabilized by a disulfide bridge. *J. Am. Chem. Soc.* **124,** 4987–4994.
94. Mayo, K. H. and Ilyina, E. (1998) A folding pathway for betapep-4 peptide 33mer: from unfolded monomers and β-sheet sandwich dimers to well-structured tetramers. *Protein Sci.* **7,** 358–368.
95. Meier, S., Guthe, S., Kiefhaber, T., and Grzesiek, S. (2004) Foldon, the natural trimerization domain of T4 Fibritin, dissociates into a monomeric A-state form containing a stable β-hairpin: atomic details of trimer dissociation and local β-hairpin stability from residual dipolar couplings. *J. Mol. Biol.* **344,** 1051–1069.
96. Doig, A. J. (1997) A three-stranded β-sheet peptide in aqueous solution containing N-methyl amino acids to prevent aggregation. *Chem. Commun.* 2153–2154.
97. Fernández-Escamilla, A. M., Rousseau, F., Schymkowitz, J., and Serrano, L. (2004) Prediction of sequence-dependent and mutational effects on the aggregation of peptides and proteins. *Nat. Biotechnol.* **22,** 1302–1306.
98. Kaur, H. and Raghava, G. P. (2002) BetaTPred: prediction of beta-TURNS in a protein using statistical algorithms. *Bioinformatics* **18,** 498–499.
99. de la Cruz, X., Hutchinson, E. G., Shepherd, A., and Thornton, J. M. (2002) Toward predicting protein topology: an approach to identifying β-hairpins. *Proc. Natl. Acad. Sci. USA* **99,** 11157–11162.
100. Carulla, N., Woodward, C., and Barany, G. (2000) Synthesis and characterization of a β-hairpin peptide that represents a 'core module' of bovine pancreatic trypsin inhibitor (BPTI). *Biochemistry* **39,** 7927–7937.

3

De novo Proteins From Binary-Patterned Combinatorial Libraries

Luke H. Bradley, Peter P. Thumfort, and Michael H. Hecht

Summary

Combinatorial libraries of well-folded *de novo* proteins can provide a rich source of reagents for the isolation of novel molecules for biotechnology and medicine. To produce libraries containing an abundance of well-folded sequences, we have developed a method that incorporates both rational design and combinatorial diversity. Our method specifies the "binary patterning" of polar and nonpolar amino acids, but allows combinatorial diversity of amino acid side chains at each polar and nonpolar site in the sequence. Protein design by binary patterning is based on the premise that the appropriate arrangement of polar and nonpolar residues can direct a polypeptide chain to fold into amphipathic elements of secondary structures, which anneal together to form a desired tertiary structure. A designed binary pattern exploits the periodicities inherent in protein secondary structure, while allowing the identity of the side chain at each polar and nonpolar position to be varied combinatorially. This chapter provides an overview of the considerations necessary to design binary patterned libraries of novel proteins.

Key Words: Protein design; binary patterning; combinatorial library; *de novo* proteins; library design.

1. Introduction

The amino acid sequences in a combinatorial library can be drawn from an enormous number of possibilities. For example, for a chain of 100 residues composed of the 20 standard amino acids, there are 20^{100} possible sequences. Because sequence space is so enormous, neither nature nor laboratory studies can explore all possibilities.

Although the quantity of sequences in a randomly generated library may be enormous, the quality of those sequences is likely to be low. Indeed, libraries of randomly generated sequences yield proteins with desired properties only very rarely (*1–5*). Powerful methods for screening and selection can enable the

From: *Methods in Molecular Biology, vol. 340: Protein Design: Methods and Applications*
Edited by: R. Guerois and M. López de la Paz © Humana Press Inc., Totowa, NJ

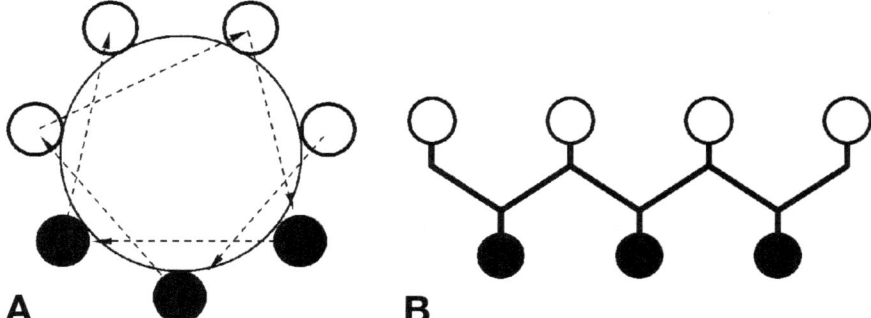

Fig. 1. The designed binary pattern of polar (open circles) and nonpolar (closed circles) amino acids for α-helices and β-strands exploits the inherent periodicities of the respective secondary structures. (**A**) α-helices have a repeating periodicity of 3.6 residues per turn. By placing a nonpolar amino acid at every third or fourth position, an amphipathic helix can be encoded in which one face is polar and the opposite face is nonpolar. Note that this figure shows the positioning of seven amino acids. For longer α-helices, the binary pattern can be adjusted to allow the periodicity of 3.6 residues per turn to maintain the amphipathic nature through the entire length of the helix. (**B**) β-strands have an alternating periodicity of polar and nonpolar amino acids. This pattern would cause one face of the strand to be polar and the opposite face to be nonpolar.

isolation of rare "winners" from vast libraries of inactive candidates; however, the success of these methods depends on the quality of the library being screened or selected. To enhance the likelihood of success, combinatorial libraries must be focused into regions of sequence space that are most likely to yield well-folded proteins. Thus, the numerical power of the combinatorial approach must be tempered by features of rational design.

To focus a library into the most productive regions of sequence space, we are guided by the observation that natural proteins typically fold into structures that (1) contain abundant secondary structures and (2) expose polar side chains to solvent while burying nonpolar side chains in the interior. Our strategy for protein design draws on these two features to rationally design focused libraries of *de novo* sequences in ways that favor folded structures.

The strategy—called the binary code strategy—is based on the premise that the appropriate patterning of polar (P) and nonpolar (N) residues can direct a polypeptide chain to form amphiphilic secondary structure *(6–9)*. A designed binary pattern exploits the periodicities inherent in protein secondary structure: α-helices have a periodicity of 3.6 residues per turn, whereas β-strands have an alternating periodicity (**Fig. 1**). Thus, a binary patterned sequence designed to form amphipathic α-helices would place nonpolar residues at

Middle Position

		T	C	A	G
		TTT Phe	TCT Ser	TAT Tyr	TGT Cys
	T	TTC Phe	TCC Ser	TAC Tyr	TGC Cys
		TTA Leu	TCA Ser	TAA Stop	TGA Stop
		TTG Leu	TCG Ser	TAG Stop	TGG Trp
		CTT Leu	CCT Pro	CAT His	CGT Arg
	C	CTC Leu	CCC Pro	CAC His	CGC Arg
		CTA Leu	CCA Pro	CAA Gln	CGA Arg
		CTG Leu	CCG Pro	CAG Gln	CGG Arg
First Position		ATT Ile	ACT Thr	AAT Asn	AGT Ser
	A	ATC Ile	ACC Thr	AAC Asn	AGC Ser
		ATA Ile	ACA Thr	AAA Lys	AGA Arg
		ATG Met	ACG Thr	AAG Lys	AGG Arg
		GTT Val	GCT Ala	GAT Asp	GGT Gly
	G	GTC Val	GCC Ala	GAC Asp	GGC Gly
		GTA Val	GCA Ala	GAA Glu	GGA Gly
		GTG Val	GCG Ala	GAG Glu	GGG Gly

Fig. 2. The organization of the genetic code allows sequence degeneracy for polar and nonpolar amino acids to be incorporated into a combinatorial library of synthetic genes by defining the middle position of the codon. For a nonpolar amino acid position, the degenerate codon NTN would encode Phe, Leu, Ile, Met, or Val. For positions requiring polar amino acids, the degenerate codon VAN would encode His, Gln, Asn, Lys, Asp, or Glu. (N represents an equimolar mixture of A, C, T, and C; V represents an equimolar mixture A, C, or G.).

every third or fourth position. In contrast, the binary pattern for an amphipathic β-strand would alternate between polar and nonpolar residues (*see* **Note 1**). In the binary code strategy, the precise three-dimensional packing of the side chains is not specified *a priori*. Consequently, within a library of binary patterned sequences, the identity of the side chain at each polar and nonpolar position can be varied, thereby facilitating enormous combinatorial diversity.

A combinatorial library of binary patterned proteins is expressed from a combinatorial library of synthetic genes. Each gene encodes a different amino acid sequence, but all sequences within a given library have the same patterning of polar and nonpolar residues. This sequence degeneracy is made possible by the organization of the genetic code (**Fig. 2**). The degenerate codon NTN encodes nonpolar amino acids, whereas the degenerate codon VAN encodes polar amino acids. (V = A, G, or C; N = A, G, C, or T; *see* **Subheading 3.2.** on codon usage.) With these degenerate codons, positions requiring a nonpolar amino acid are filled by phenylalanine, leucine, isoleucine, methionine, or valine, whereas positions requiring a polar amino acid are filled by glutamate, aspartate, lysine, asparagine, glutamine, or histidine (**Fig. 2**; *see* **Note 2**).

This chapter outlines the methodology for using binary patterning to design libraries of *de novo* proteins. Using examples from our laboratory, we describe the design and construction of both α-helical and β-sheet proteins.

2. Materials

Oligonucleotides were purchased from commercial vendors (e.g., IDT; www.idtdna.com). All oligonucleotides should be PAGE purified (*see* **Note 3**). For overlap extensions, we use a thermostable polymerase that leaves blunt ends (such as Deep Vent [New England Biolabs] or Pfu polymerase). All other enzymes and reagents are commercially available.

3. Methods

3.1. Design of a Structural Template

Binary patterning can be applied to any amphipathic α-helical or β-stranded segment of a protein. Although our laboratory has focused on *de novo* proteins, the binary code strategy can also be applied to local areas of existing proteins such as the active site, part of the core, or an interface *(10)*. For the design of *de novo* proteins, the success of the strategy depends primarily on how well the template is designed. Several factors important for template design are described.

3.1.1. Binary Patterned Regions

3.1.1.1. α-Helical Designs

Binary patterning exploits the periodicities inherent in secondary structures. α-Helices have a repeating periodicity of 3.6 residues per turn (**Fig. 1A**). To design an amphipathic segment of α-helical secondary structure, a binary pattern of P-N-P-P-N-N-P is used. Our initial α-helical design focused on the four-helix bundle motif (**Fig. 3**). In this structure, the hydrophobic face of each helix is oriented toward the central core of the bundle, whereas the hydrophilic faces of the helices are exposed to aqueous solvent. The P-N-P-P-N-N-P pattern favors formation of an amphiphilic α-helical structure that can bury all nonpolar amino acids upon formation of the desired tertiary structure. From our designed four-helix bundle libraries, more than 60 proteins have been purified and characterized. All have shown typical α-helical circular dichroism spectra. Additionally, the collection yielded several proteins with native-like properties, including nuclear magnetic resonance (NMR) chemical shift dispersion, co-operative chemical and thermal denaturations, and slow hydrogen/deuterium exchange rates *(11–15)*. Recently, the structure of protein S-824 from a second-generation binary patterned library was determined by NMR spectroscopy, and shown to be a four-helix bundle as specified by the binary code design *(16)*.

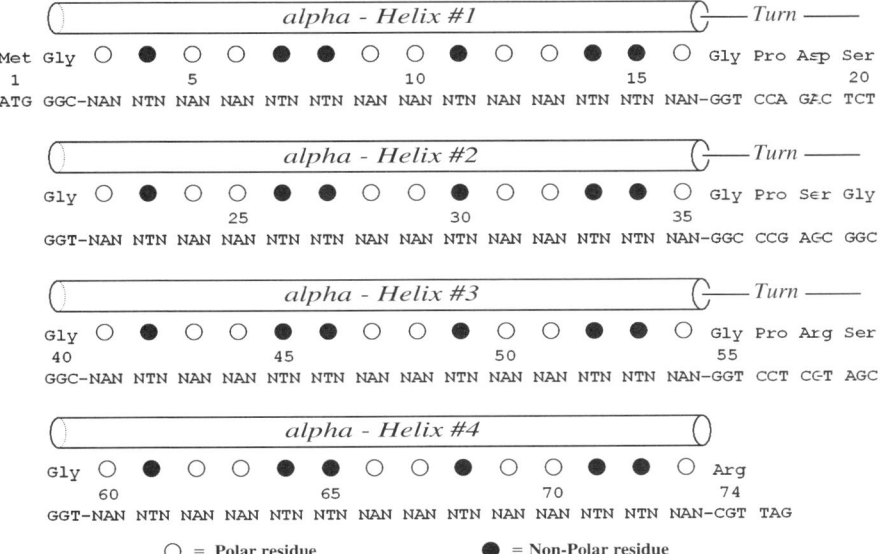

Fig. 3. The design template for the initial four-helix bundle library. Nonpolar positions (closed circles) were encoded by the degenerate NTN codon, and polar positions (open circles) were encoded by the degenerate VAN codon. The defined residue positions were located at the N- and C-terminal regions as well as the interhelical turns of the designed protein.

3.1.1.2. β-Sheet Design

Amphiphilic β-strands have an alternating periodicity of …P-N-P-N… (**Fig. 1B**). Based on this periodicity, a combinatorial library of synthetic genes can be created to encode β-sheet structures in which polar residues comprise one face and nonpolar residues comprise the opposing face of the resulting β-sheet. The sequences in our first β-sheet library were designed to have six β–strands with each strand having the binary pattern P-N-P-N-P-N-P (7). Proteins from this library were expressed from synthetic genes cloned into *Escherichia coli*. All proteins from this collection that have been analyzed thus far indeed form β-sheet secondary structure, displaying circular dichroism spectra with the characteristic minimum at approx 217 nm. In aqueous solution, the β-sheet proteins from this initial library self-assembled into amyloid-like fibrils, with nonpolar side chains forming a hydrophobic core and polar side chains exposed to solvent *(7)*.

When these same β-sheet sequences are placed in a heterogeneous environment with a polar/nonpolar interface, they form a different structure. For example, at an air/water interface, they self-assemble into flat β-sheet monolayers with the nonpolar residues pointing up toward air and polar side chains pointing down toward water (*17*). Alternatively, at an interface between water and the nonpolar surface of graphite, binary patterned β-sheet sequences undergo template-directed assembly on the graphite surface to yield highly ordered structures (*18*).

The formation of fibrils in aqueous solution, and monolayers at polar/non-polar interfaces is consistent with the inherent tendency of β-strands to assemble into oligomeric structures (*19,20*). Designed β-strands with sequences that adhere rigorously to the alternating polar/nonpolar binary pattern are especially prone to aggregate because of their need to bury their "sticky" hydrophobic face (*8,21*). To favor monomeric β-sheet proteins, the alternating binary pattern must be modified: The pattern P-N-P-N-P-N-P on the edge strand of a β-sheet can be changed to P-N-P-**K**-P-N-P (where K denotes lysine). The four methylene groups of the lysine side chain can substitute for the replaced non-polar residue, whereas the charged amine at the end of the lysine side chain will seek solvent and thereby prevent aggregation. This strategy has been used successfully to convert binary patterned *de novo* proteins from amyloid-like structures to monomeric β-sheet proteins (*21*).

3.1.2. Fixed Regions

In practice, it is often necessary to keep part of the protein sequence fixed (i.e., not combinatorially diverse), especially when the target sequence is long. When assembling a library of synthetic genes, these constant regions serve as sites for single-stranded synthetic oligonucleotides to anneal together and prime the enzymatic synthesis of complementary strands (**Fig. 4**). (Assembly of full-length genes from single-stranded oligonucleotides is discussed in **Subheading 3.3.**).

Single-stranded oligonucleotides are typically used to encode the binary patterning of individual segments of secondary structure. Nondegenerate fixed regions on the 5' and 3' termini of these oligonucleotides are typically used to encode fixed turn regions between units of secondary structure (**Figs. 3** and **4**) (*6,7*). The amino acid sequences chosen to occupy these turn regions are based on statistical and rational design criteria outlined as follows.

1. Sequences in the turn regions are chosen based on positional preferences. For example, in the initial four-helix bundle library, glycine residues were placed at N-cap and C-cap positions at the termini of the helices (**Fig. 3**) (*6*). Glycine residues are frequently found at these positions in natural proteins (*22*). At the position after the C-cap, proline residues were used because they are strong helix breakers. In some situations, however, proline may be undesirable because *cis/trans* isomerism could lead to multiple (rather than unique) conformations. For the β-sheet library (*7*), design of the turn regions was based on the positional turn potentials of various amino acids in the known structures of natural proteins (*23*).

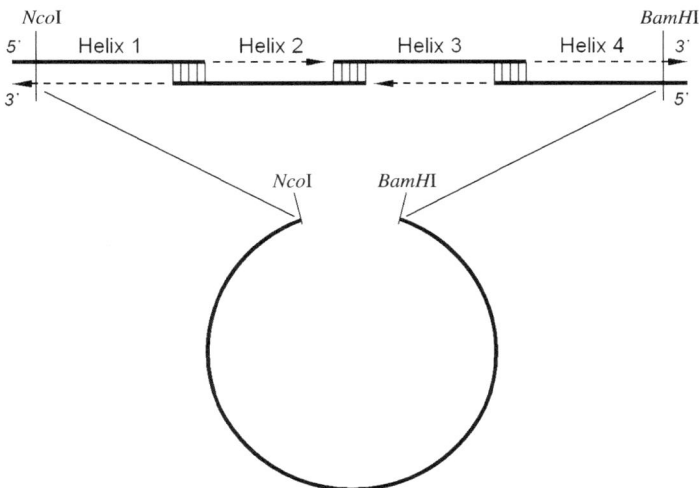

Fig. 4. Assembly of full-length genes from four single-stranded oligonucleotides. Constant regions at the 5' and 3' ends of the single-stranded oligonucleotides serve as sites for annealing and for priming enzymatic synthesis (using DNA polymerase) of complementary strands. The full-length gene is then ready for insertion into an expression vector.

2. Sequences of the turns can be designed to incorporate restriction sites, which can facilitate the assembly of full-length genes *(6)* *(see* **Subheading 3.3.**).
3. The lengths of constant regions must be sufficient to promote sequence-specific annealing. Pairs of oligonucleotides with overlaps of 12 to 15 nucleotides are typically used for annealing. To further enhance annealing, one or two nucleotides in the codons immediately preceding and following the turn regions may also be held constant. For example, the synthetic oligonucleotide (5'…NAN-NTN-NTN-NAN-GGT-CCT-CGT-AGC-3') has a constant gene segment (under-lined) in which 12 nucleotides encode a four residue turn. The previous codon (NAN) encodes a polar amino acid residue. By defining the third position, for example with a G, the codon now becomes NAG, thereby yielding two additional constant bases (bold) for sequence-specific annealing (5' …NAN-NTN-NTN-NAG-GGT-CCT-CGT-AGC-3'). By defining only the second and third (but not the first) positions of the codon, amino acid diversity is maintained.

Besides the turn regions, the N- and C- termini of the *de novo* sequences can also be held constant. Fixed sequences in these regions are typically necessary for cloning into expression vectors. Some design criteria for the termini are:

1. An initiator methionine is placed at the N-terminus of the *de novo* sequence. This is required for expression in vivo.
2. An aromatic chromophore (tyrosine or tryptophan) can be incorporated at a constant site in the sequence to aid in protein purification and concentration deter-

mination *(7,15)*. This aromatic residue could be placed either in a constant turn or at one of the chain termini. In some of our libraries, we inserted a tyrosine immediately after the initiator methionine. This provides a chromophore, while also preventing cleavage of the initiator methionine in vivo *(24–27)*.

3. The C-terminal residue of the designed proteins should be charged and polar. The C-terminal sequence of a protein can affect its rate of intracellular proteolysis, and the presence of a charged residue at the C-terminus can extend half-life in vivo *(28–30)*. In addition, an exhaustive statistical analysis of protein amino acid sequences found that polar and charged amino acids are overrepresented at the C-termini of proteins *(31)*. For the four-helix bundle libraries, an arginine residue was designed to occupy this terminal position *(6,15)*. Moreover, positively charged side chains at the C-termini of α-helices (and negatively charges residues at the N-termini) can enhance stability by interacting with the helix dipole *(32)*.

3.1.3. Considerations for the Design of Tertiary Structure

A successful binary patterned template must be long enough to encode well-folded structures, but at the same time short enough to be accessible to strategies for assembling large libraries of error-free genes. Many proteins from our first generation (74-residue) four-helix bundle library formed dynamic structures resembling molten globules *(6,11–14)*. To investigate the potential of the binary code strategy to encode collections of native-like tertiary structures, a second-generation library of binary-patterned α-helical proteins was prepared *(15)*. This new library was based on protein #86, a preexisting sequence from the original 74-residue library. The major change to protein #86 was the addition of six combinatorially diverse residues to each of the four helices. These residues continued to follow the binary patterning. Overall, the second-generation proteins were 102 amino acids in length, which is similar to a number of natural four-helix bundles.

Characterization of five sequences chosen arbitrarily from this second-generation library showed that all were substantially more stable than the parental protein #86 *(15)*. In addition, most of them yielded NMR spectra that were well dispersed and exhibited well-resolved nuclear Overhauser effect cross peaks, indicative of well-folded tertiary structures *(15)*. Recently, the solution structure of a protein (S-824) from this second-generation library was solved *(16)*. The structure was indeed a four-helix bundle in accordance with its binary patterned design. Moreover, the protein was not a molten globule: the interior side chains were well ordered—even by the standards of natural proteins *(16)*.

3.2. Codon Usage
3.2.1. The VAN (Polar) Codon

1. The first base of the VAN codon is occupied by an equimolar mixture of G, C, and A. By excluding T, two stop codons (TAG and TAA) and two tyrosine codons are eliminated (*see* **Note 4**).

2. The mixture of nucleotides at the third base of the VAN codon can be altered to favor some amino acids over others. An equimolar mixture of G, C, A, and T would yield an equal likelihood of histidine, glutamine, asparagine, lysine, aspartate, and glutamate. However, some of the residues of the VAN codon have higher intrinsic propensities than others to form α-helices *(33–36)*. By omitting T from the third position (i.e., VAV), glutamine, lysine, and glutamate are favored over histidine, asparagine, and aspartate. This increases the percentage of residues with high propensities to form α-helices *(33–36)*.

3.2.2. The NTN (Nonpolar) Codon

Equimolar mixtures of all four bases at the first and third positions of the NTN codon would encode six times as many leucines as methionines. In addition, an equimolar mixture would encode protein sequences in which one quarter of the hydrophobic residues would be valine. Because valine has a relatively low α-helical propensity *(33–36)*, this may be undesirable for some designs. By altering the molar ratio of the mixture at the first and third N positions, the relative abundance of hydrophobic residues can be altered. For example, in the initial four-helix bundle library, the first base of the NTN codon contained A:T:C:G in a molar ratio of 3:3:3:1 and the third base mixture contained an equimolar mixture of G and C *(6)*.

3.2.3. Codon Usage of the Host Expression System

The DNA sequences in both the constant and the combinatorial regions of the library should be biased to favor those codons used most frequently in the host expression system. For example, including only C and G (rather than all four bases) in the third position of a degenerate codon favors those codons preferred by *E. coli (37)*. Codons that are used only rarely in the host expression system, such as CGA, AGA, and AGG (arginine); CTA (leucine); CCC (proline); and ATA (isoleucine) in *E. coli* should be avoided wherever possible, because genes containing rare codons may express poorly *(38)*. Other (non–*E. coli*) expression systems have different codon preferences, and these must be considered in the design.

3.2.4. Restriction Digest Analysis in Silico

To ensure that those restriction sites used for ligating the library into the host vector occur only at the ends of the genes and not within the (degenerate) gene sequences, restriction digests are performed *in silico* before finalizing the design of the gene library. We have developed a program that reads the binary patterned sequence—including the degenerate wildcard bases (*see* **Note 5**)—and randomly generates a large number (typically 10^4) of gene sequences from the library master template. This pool of gene sequences is subsequently analyzed against a restriction enzyme database, and the data for all restriction sites are sorted by cutting frequency. This enables a quick comparison among different design choices and provides a list of sites that will be absent from the library and can therefore be used for cloning into the appropriate vectors.

For example, the NdeI restriction site, CATATG, includes the ATG initiator codon and is commonly used as a cloning site at the 5' end of genes. To assess whether this site will appear in a combinatorial library, one must consider the degenerate codons used to construct the library. For example, the combinatorial sequence VAN.NTN, which encodes a polar residue followed by a nonpolar residue, would contain the NdeI site in 1 of 192 cases. In contrast, the combinatorial sequence VAV.NTS, which also encodes a polar residue followed by a nonpolar residue, will never encode an NdeI site. (N denotes an equimolar mixture of all four bases; V, an equimolar mixture of A, G, and C; and S, an equimolar mixture of G and C.)

3.3 Assembly of Full-Length Genes

Full-length genes are typically assembled from smaller single-stranded oligonucleotides (**Fig. 4**). This is done for two reasons: (1) to minimize the inherent errors (mostly deletions and frameshifts) associated with the synthesis of long degenerate oligonucleotides and (2) to increase the diversity of full-length sequences via combinatorial assembly (*see* **Note 6**).

When synthesizing the semirandom oligonucleotides, some are made as coding (sense) strands and others as noncoding (antisense) strands (*see* **Fig. 4**). Typically, each oligonucleotide is designed to encode an individual segment of secondary structure. Assembly of full-length genes from such segments allows individual α-helices or β-strands to be designed and manipulated as independent modules, thereby enhancing the versatility of the binary code strategy.

In the design of our initial library of four-helix bundles, four synthetic oligonucleotides were used to construct the full-length gene. Each oligonucleotide was designed to encode a single helix and turn. As described above, the turn regions were defined precisely (i.e., not degenerate), thereby allowing them to serve as priming sites for DNA polymerase to synthesize complementary strands (**Fig. 4**) *(6)*.

Various methods can be used to assemble the full-length genes. In some cases, we have made two libraries of half genes, and then ligated them together to produce a library of genes encoding full-length proteins *(6)*. To ensure correct head to tail ligation, nonpalindromic restriction sites, which produce directional "sticky ends" for ligation, can be designed into the constant regions *(6)*. Other methods for assembling full-length genes include various polymerase chain reaction strategies (e.g., overlap extension), and these have been used in constructing several of our libraries *(7,15,39)*.

3.4. Optimization of Gene Assembly

3.4.1. Avoidance of Incorrect Annealing

The correct assembly of full-length genes (**Subheading 3.3.**) can be hindered by the presence of alternate annealing sites in the synthetic oligonucleotides. Such sites would result from internal repeats or inverted repeats and thus give rise to problems such as hairpins or mispriming. To minimize these undesired sequences, the gene template sequence (containing both fixed and degenerate bases), should be analyzed computationally, and optimized prior to synthesis.

Although algorithms (e.g., dotplots) are available to search DNA sequences for alternate annealing sites or possible secondary structure *(40–43)*, the standard methods are not suitable for analyzing the degenerate sequences required to assemble combinatorial libraries. Therefore we devised a new algorithm, called *Designer DotPlot* to analyze degenerate DNA sequences. In contrast to previous methods, which search DNA sequences containing only the four standard bases A, T, C, and G, *Designer DotPlot* also handles sequences containing the wildcards N, B, D, H, V, M, S, W, Y, H, and K (*see* **Note 5**). *Designer DotPlot* calculates and displays probability-weighted matches between all compatible degenerate bases. In addition to standard base-pairings (A-T, G-C), an optional feature of *Designer DotPlot* also allows G-T base pairs. For example, the degenerate bases R (A/G) and N (A/C/G/T) have a match probability of 0.25. Allowing G-T base pairing increases it to 0.375. *Designer DotPlot* uses a 15×15 lookup table containing the probabilities for all 225 combinations of the 4 standard DNA bases and all 11 wildcards.

Figure 5A shows the dotplot for a sequence encoding a binary patterned α-helix bracketed by constant sequences encoding the helix caps. The darkness of each dot shows the probability of a match. For example, an A-T match has a probability of 1 and is shaded black, whereas an A-B match has a probability of 0.33 and is shown in light gray. Two such plots are generated: one forward (**Fig. 5A**), analyzing the interstrand pairings between the upper and lower strand; and one reverse (not shown), analyzing the intrastrand pairings within the upper strand itself. Each plot is then filtered (window size, probability) and overlaid. Probabilities above a threshold are transformed into a log-scale and rescaled to gray values with the highest value being plotted for every position.

Stretches of sequence that contain a high probability of complementary bases are plotted as lines in **Fig. 5B**, with the problem areas for misannealing and hairpins outlined. Based on the analysis of such plots, the base sequence (including both fixed and degenerate regions) of each design can be improved manually by simply changing the problem regions to alternate codons and replotting, or automatically via a genetic algorithm. The optimized sequence (**Fig. 5C**)

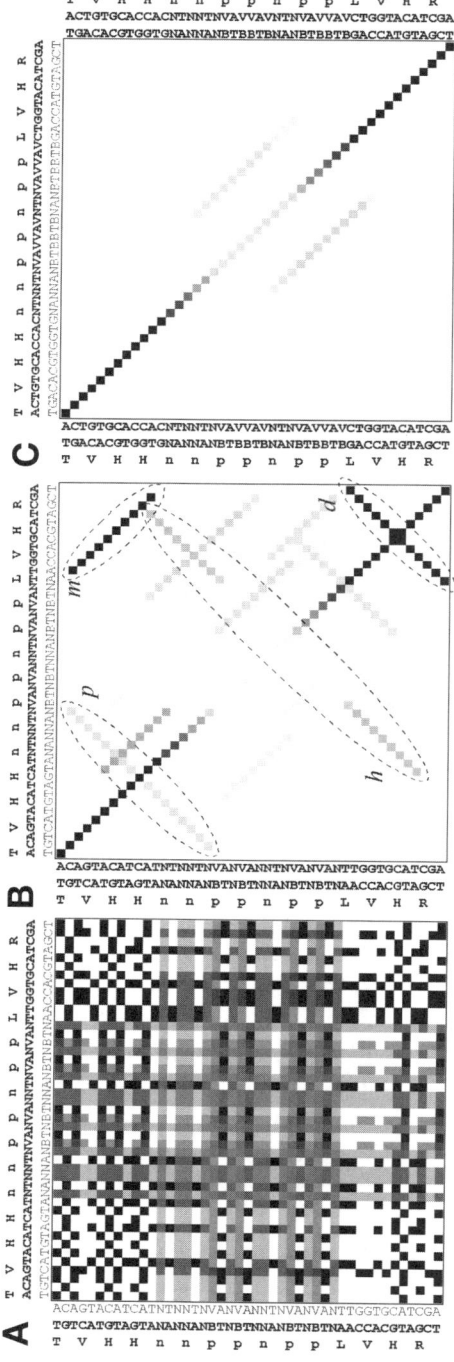

Fig. 5. A sequence analysis and optimization of a binary patterned library using *Designer Dotplot*. For a desired amino acid sequence, the upper (sense) strand and the complementary lower (anti-sense) strand gene templates are labeled on the x and y axes, respectively. (**A**) *Designer dotplot* shows single base matches between the upper and lower strand. The gray scale corresponds to base-pairing probability (darker = greater probability). (**B**) After filtering (window size = 8; cutoff probability = 0.01) and overlaying the *Designer Dotplot* with intrastrand base pairings, problem regions within the sequence are identified as lines. These highlighted assembly problem areas are from palindromic (p), hairpin (h), mispriming (m), and primer-dimer sites (d) present in the DNA library template sequence. (**C**) The highlighted problem areas identified in (B) have been minimized following sequence optimization. By using different codon choices while maintaining same binary pattern, the library template has been optimized for gene assembly.

minimizes possible cross-annealing among oligonucleotides, misannealing or hairpin formation within the same oligonucleotide, and reverse complementarity during assembly of the oligonucleotides (*see* **Note 7**).

3.4.2. Preselection for Open Reading Frames

Constructing large and diverse libraries that are free of frameshifts or stop codons can be extremely challenging. The presence of interrupted sequences complicates screening strategies by increasing the burden on the selection system, while failing to provide additional valid candidate genes for evaluation. It is therefore important to remove incorrect sequences from these libraries before screening for function.

To construct a high-quality combinatorial library of *de novo* genes, we have developed a method to screen libraries of gene segments for open reading frames before assembly of the full-length genes *(39)*. In our system, synthetic DNA from a library is inserted upstream from a selectable intein/thymidylate synthase (TS) fusion (*see* **Note 8**). Gene segments that are in-frame and devoid of stop codons produce a tripartite precursor protein. Subsequent cleavage by the intein releases and activates the TS enzyme. This enables TS deficient *E. coli* host cells to survive in selective medium. Libraries of error-free gene segments are isolated from the surviving cells, and the individual segments are subsequently assembled combinatorially into full-length genes (**Subheading 3.3.**; *also see* **Note 6**). This preselection system has recently enabled the construction of a large library of 102 amino acid sequences in which virtually all sequences are free of frameshifts and deletions *(39)*.

The availability of large, diverse, and error-free libraries of binary patterned sequences encoding well-folded and native-like structures sets the stage for experiments aimed at the isolation of novel proteins with functions that may ultimately find use in biotechnology and medicine.

4. Notes

1. The overall amino acid compositions are similar for the α-helical and β-sheet binary patterned libraries. Therefore, the different properties of the resulting proteins *(6,7)* are *not* from differences in amino acid composition *(8)*. Moreover, the observed differences are *not* from differences in sequence length: irrespective of length, sequences with the P-N-P-P-N-N-P periodicity form α-helical proteins, whereas those with the P-N-P-N-P-N-P periodicity form β-sheet proteins. Thus, it is the binary patterning itself that dictates whether a *de novo* protein forms α or β structure.

2. The binary codons, VAN and NTN, encode six polar amino acids (glutamate, aspartate, lysine, asparagine, glutamine, and histidine), and five nonpolar amino acids (valine, methionine, isoleucine, leucine, and phenylalanine), respectively. In addition to these 11 variable amino acids, a variety of other residues can be incorporated into the constant regions of the sequences. For example, our recent

library of 102 residue four-helix bundles contains 17 of the 20 amino acids *(15)*. Only alanine, proline, and cysteine were omitted. In natural proteins, alanine occurs both in surface and core positions. Thus, its role in the binary code as polar or nonpolar is somewhat ambiguous. Proline is a special case because its restricted *phi* angle makes it useful only in certain well-defined regions of structure. Cysteine should be used only in designs wherein a disulfide bond or metal binding is planned.

3. PAGE purification of synthetic oligonucleotides is essential. This reduces the likelihood of truncated oligonucleotides being incorporated into the library. Although this purification step reduces the quantity of DNA (and potentially the diversity), the quality of the genes, and resulting libraries is enhanced significantly.

4. By excluding T from the first position of the VAN (polar) codon, tyrosine codons are avoided. This is desirable because tyrosine is not a completely polar residue and frequently occurs in the hydrophobic cores of natural proteins. Therefore only the most polar residues (histidine, glutamine, asparagine, lysine, aspartate, and glutamate) are incorporated into the designed surface positions.

5. According to the International Union of Biochemistry, the degenerate base symbols and their nucleotide base compositions are as follows *(44)*: K = G/T, M = A/C, R = A/G, S = C/G, W = A/T, Y = C/T, B = C/G/T, D = A/G/T, H = A/C/T, V = A/C/G, N = A/C/G/T.

6. Our method for constructing full-length *de novo* gene libraries relies on combinatorial assembly using libraries of shorter fragments *(6,7)*. Among the advantages of this approach is an increase in the diversity of the full-length library. For example, four individual segment libraries containing only 10^4 sequences per library can be combined to yield a full-length library with a theoretical diversity of 10^{16} sequences *(39)*.

7. The purpose of the *Designer Dotplot* analysis of library sequences is not merely to eliminate all undesired base-pairings, because this may not be possible, but rather to identify problem regions and design assembly protocols accordingly. For example, library oligonucleotides with significant cross-annealings should be assembled separately and joined by restriction digest and ligation (*see* **Subheading 3.3.**).

8. The preselection for gene fragments that are in frame must be independent of the structure and solubility of the encoded polypeptide fragments. Because binary patterned segments of secondary structure are designed to fold only in the context of the full tertiary structure *(8)*, it is crucial that these segments not be weeded out of a library before assembly of the full-length genes. The intein-thymidylate synthase system (**Subheading 3.4.2.**) selects for in-frame gene segment sequences, regardless of the structure or solubility of the expressed polypeptide. This is made possible by a poly-asparagine linker designed to separate the inserted gene segment from the intein-TS fusion *(39)*. This linker permits the reporter TS enzyme to fold and function independently of the polypeptide encoded by the inserted gene segment.

Acknowledgments

Supported by NIH R01-GM62869 (MHH). LHB was supported by a postdoctoral fellowship from the Princeton University Council on Science and Technology.

References

1. Mandecki, W. (1990) A method for construction of long randomized open reading frames and polypeptides. *Protein Eng.* **3,** 221–226.
2. Davidson, A. R. and Sauer, R. T. (1994) Folded proteins occur frequently in libraries of random amino acid sequences. *Proc. Natl. Acad. Sci. USA* **91,** 2146–2150.
3. Davidson, A. R., Lumb, K. J., and Sauer, R. T. (1995) Cooperatively folded proteins in random sequence libraries. *Nat. Struct. Biol.* **2,** 856–864.
4. Prijambada, I. D., Yomo, T., Tanaka, F., Kawama, T., Yamamoto, K., Hasegawa, A., et al. (1996) Solubility of artificial proteins with random sequences. *FEBS Lett.* **382,** 21–25.
5. Keefe, A. D. and Szostak, J. W. (2001) Functional proteins from a random sequence library. *Nature* **410,** 715–718.
6. Kamtekar, S., Schiffer, J. M., Xiong, H., Babik, J. M., and Hecht, M. H. (1993) Protein design by binary patterning of polar and nonpolar amino acids. *Science* **262,** 1680–1685.
7. West, M. W., Wang, W., Patterson, J., Mancias, J. D., Beasley, J. R., and Hecht, M. H. (1999) *De novo* amyloid proteins from designed combinatorial libraries. *Proc. Natl. Acad. Sci. USA* **96,** 11211–11216.
8. Xiong, H., Buckwalter, B. L., Shieh, H. M., and Hecht, M. H. (1995) Periodicity of polar and nonpolar amino acids is the major determinant of secondary structure in self-assembling oligomeric peptides. *Proc. Natl. Acad. Sci. USA* **92,** 6349–6353.
9. Hecht, M. H., Das, A., Go, A., Bradley, L. H., and Wei, Y. (2004) De novo proteins from designed combinatorial libraries. *Protein Sci.* **13,** 1711–1723.
10. Taylor, S. V., Walter, K. U., Kast, P., and Hilvert, D. (2001) Searching sequence space for protein catalysts. *Proc. Natl. Acad. Sci. USA* **98,** 10596–10601.
11. Roy, S., Ratnaswamy, G., Boice, J. A., Fairman, R., McLendon, G., and Hecht, M.H. (1997) A protein designed by binary patterning of polar and nonpolar amino acids displays native-like properties. *J. Am. Chem. Soc.* **119,** 5302–5306.
12. Roy, S., Helmer, K. J., and Hecht, M. H. (1997) Detecting native-like properties in combinatorial libraries of *de novo* proteins. *Folding Des.* **2,** 89–92.
13. Roy, S. and Hecht, M. H. (2000) Cooperative thermal denaturation of proteins designed by binary patterning of polar and nonpolar amino acids. *Biochemistry* **39,** 4603–4607.
14. Rosenbaum, D. M., Roy, S., and Hecht, M. H. (1999) Screening combinatorial libraries of *de novo* proteins by hydrogen-deuterium exchange and electrospray mass spectrometry. *J. Am. Chem. Soc.* **121,** 9509–9513.
15. Wei, Y., Liu, T., Sazinsky, S. L., Moffet, D. A., Pelczer, I., and Hecht, M. H. (2003) Stably folded and well-ordered structures from a designed combinatorial library of *de novo* proteins. *Protein Sci.* **12,** 92–102.

16. Wei, Y., Kim, S., Fela, D., Baum, J., and Hecht, M. H. (2003) Solution structure of a de novo protein from a designed combinatorial library. *Proc. Natl. Acad. Sci. USA* **100,** 13270–13273.

17. Xu, G., Wang, W., Groves, J. T., and Hecht, M. H. (2001). Self-assembled mono-layers from a designed combinatorial library of *de novo* β-sheet proteins. *Proc. Natl. Acad. Sci. USA* **98,** 3652–3657.

18. Brown, C .L., Aksay, I. A., Saville, D. A., and Hecht, M. H. (2002) Template-directed assembly of a *de novo* designed protein. *J. Am. Chem. Soc.* **124,** 6846–6848.

19. Hecht, M. H. (1994) *De novo* design of β-sheet proteins. *Proc. Natl. Acad. Sci. USA* **91,** 8729–8730.

20. Richardson, J. S. and Richardson, D. C. (2002) Natural beta-sheet proteins use negative design to avoid edge-to-edge aggregation. *Proc. Natl. Acad. Sci. USA* **99,** 2754–2759.

21. Wang, W. and Hecht, M. H. (2002) Rationally designed mutations convert *de novo* amyloid-like fibrils into soluble monomeric β-sheet proteins. *Proc. Natl. Acad. Sci. USA* **99,** 2760–2765.

22. Richardson, J. S. and Richardson, D. C. (1988) Amino acid preferences for spe-cific locations at the ends of alpha helices. *Science* **240,** 1648–1652.

23. Hutchinson, E. G. and Thornton, J. M. (1994) A revised set of potentials for β-turn formation in proteins. *Protein Sci.* **3,** 2207–2216.

24. Hirel, P. H., Schmitter, M. J., Dessen, P., Fayat, G., and Blanquet, S. (1989) Extent of N-terminal methionine excision from *Escherichia coli* proteins is governed by the side-chain length of the penultimate amino acid. *Proc. Natl. Acad. Sci. USA* **86,** 8247–8251.

25. Dalboge, H., Bayne, S., and Pedersen, J. (1990) *In vivo* processing of N-terminal methionine in *E. coli*. *FEBS Lett.* **266,** 1–3.

26. Tsunasawa, S., Stewart, J. W., and Sherman, F. (1985) Amino-terminal process-ing of mutant forms of yeast iso-1-cytochrome c. The specificities of methionine aminopeptidase and acetyltransferase. *J. Biol. Chem.* **260,** 5382–5391.

27. Huang, S., Elliott, R. C., Liu, P. S., Koduri, R. K., Weickmann, J. L., Lee, J. H., et al. (1987) Specificity of cotranslational amino-terminal processing of proteins in yeast. *Biochemistry* **26,** 8242–8246.

28. Bowie, J. U. and Sauer, R. T. (1989) Identification of C-terminal extensions that protect proteins from intracellular proteolysis. *J. Biol. Chem.* **264,** 7596–7602.

29. Parsell, D. A., Silber, K. R., and Sauer, R. T. (1990) Carboxy-terminal determi-nants of intracellular protein degradation. *Genes Dev.* **4,** 277–286.

30. Milla, M. E., Brown, B. M., and Sauer, R. T. (1993) P22 Arc repressor: enhanced expression of unstable mutants by addition of polar C-terminal sequences. *Pro-tein Sci.* **2,** 2198–2205.

31. Berezovsky, I. N., Kilosanidze, G. T., Tumanyan, V. G., and Kisselev, L. L. (1999) Amino acid composition of protein termini are biased in different manners. *Pro-tein Eng.* **12,** 23–30.

32. Shoemaker, K. R., Kim, P. S., York, E. J., Stewart, J. M., and Baldwin, R. L. (1987) Tests of the helix dipole model for stabilization of alpha-helices. *Nature* **326,** 563–567.

33. Chou, P. Y. and Fasman, G. D. (1978) Empirical predictions of protein conformation. *Annu. Rev. Biochem.* **47,** 251–276.
34. Fasman, G. D. (1989) Prediction of Protein Structure and the Principles of Protein Conformation. Plenum, New York.
35. Creighton, T. E. (1993) Proteins: Structures and Molecular Properties (2nd ed.). Freeman, New York.
36. Pace, C. N. and Scholtz, J. M. (1998) A helix propensity scale based on experimental studies of peptides and proteins. *Biophys. J.* **75,** 422–427.
37. DeBoer, H. A. and Kastelein, R. A. (1986) Biased codon usage: an exploration of its role in optimization of translation, in *Maximizing Gene Expression* (Rezinikoff, W. and Gold, L., eds.) Butterworth, Stoneham, MA, pp. 225–285.
38. Kane, J. F. (1995) Effects of rare codon clusters on high-level expression of heterologous proteins in *Escherichia coli. Curr. Opin. Biotechnol.* **6,** 494–500.
39. Bradley, L. H., Kleiner, R. E., Wang, A. F., Hecht, M. H., and Wood, D. W. (2005) An intein-based genetic selection enables construction of a high-quality library of binary patterned *de novo* sequences. *Protein Eng. Des. Sel.* **18,** 201–207.
40. Maizel, J. V., Jr. and Lenk R. P. (1981) Enhanced graphic matrix analysis of nucleic acid and protein sequences. *Proc. Nat. Acad. Sci. USA* **78,** 7665–7669.
41. Tinoco, I., Jr., Uhlenbeck, O. C., and Levine, M. D. (1971) Estimation of secondary structure in ribonucleic acids. *Nature* **230,** 363–367.
42. Zuker, M. (2003) Mfold web server for nucleic acid folding and hybridization prediction. *Nucleic Acids Res.* **31,** 3406–3415.
43. Hofacker, I .L. (2003) Vienna RNA secondary structure server. *Nucleic Acids Res.* **31,** 3429–3431.
44. Cornish-Bowden, A. (1985) Nomenclature for incompletely specified bases in nucleic acid sequences: recommendations 1984. *Nucleic Acids Res.* **13,** 3021–3030.

4

Non-Protein Amino Acids in the Design of Secondary Structure Scaffolds

Radhakrishnan Mahalakshmi and Padmanabhan Balaram

Summary

The use of stereochemically constrained amino acids permits the design of short peptides as models for protein secondary structures. Amino acid residues that are restrained to a limited range of backbone torsion angles (ϕ-ψ) may be used as folding nuclei in the design of helices and β-hairpins. α-Amino-isobutyric acid (Aib) and related $C^{\alpha\alpha}$ dialkylated residues are strong promoters of helix formation, as exemplified by a large body of experimentally determined structures of helical peptides. ^{D}Pro-Xxx sequences strongly favor type II' turn conformations, which serve to nucleate registered β-hairpin formation. Appropriately positioned ^{D}Pro-Xxx segments may be used to nucleate the formation of multistranded antiparallel β-sheet structures. Mixed (α/β) secondary structures can be generated by linking rigid modules of helices and β-hairpins. The approach of using stereochemically constrained residues promotes folding by limiting the local structural space at specific residues. Several aspects of secondary structure design are outlined in this chapter, along with commonly used methods of spectroscopic characterization.

Key Words: Peptide design; helical peptides; peptide hairpins; D-amino acids; peptide scaffolds; NMR of peptides; peptide crystal structure.

1. Introduction

The design and construction of complex folds using first principles relies on a complete understanding of the factors that facilitate formation of protein secondary structural elements; namely, helices, strands and turns. Subsequent assembly of secondary structures, driven by noncovalent interactions or covalent constraints such as disulfides, can lead to designed tertiary folds. This chapter describes approaches to the design of secondary structures such as helices, β-hairpins, and β-sheets using conformationally constrained amino acids to nucleate local folding of synthetic peptides.

From: *Methods in Molecular Biology, vol. 340: Protein Design: Methods and Applications*
Edited by: R. Guerois and M. L. de la Paz © Humana Press Inc., Totowa, NJ

The stereochemically allowed backbone conformations of polypeptide chains were first defined by Ramachandran and coworkers in the early 1960s, when they introduced the concept of a two-dimensional (ϕ-ψ) plot, which permitted mapping the allowed conformations in torsion angle space *(1–3)*. Substitution at the C^α carbon atom dramatically limits the range of allowed local conformations at an amino acid residue. This is clearly exemplified by a comparison of the Ramachandran plots generated for Ac-Gly-NHMe and Ac-L-Ala-NHMe (**Figs. 1A** and **1B**). For all L-residues, the vast majority of allowed conformations lie in the region of the Ramachandran plot with negative ϕ-values. The corresponding map for Ac-D-Ala-NHMe is generated by the inversion of the L-Ala map about the origin. **Figure 1C** shows the superposition of the conformational maps for l- and D-Ala. The zones of overlap of the two conformational maps are extremely limited and are restricted to the right-(α_R) and left-(α_L) handed helical regions. This restriction is a consequence of the local nonbonded interactions involving the C^β atoms and the preceding and succeeding peptide units. As a corollary, the range of allowed conformations of achiral $C^{\alpha\alpha}$ dialkylated amino acid residues (Aib, α,α-dimethylglycine or α-methylalanine and its higher homologues) (**Fig. 2**) are limited to the left-(α_L) and right-(α_R) handed helical regions of conformational space *(4–6)*. Consequently, Aib and related C^α tetra-substituted amino acids are stereochemically constrained to adopt local helical conformations. Incorporation of Aib and related residues into a peptide chain can thus provide a folding nucleus. The expectations of this analysis are borne out by the observed ϕ-ψ values for Aib residues in a large number of experimentally determined crystal structures of synthetic and natural peptides. **Figure 1D** shows a cluster plot for the distribution of Aib conformations obtained from the Cambridge Structural Database *(7)*, which clearly emphasizes the overwhelming preference of this residue for α_R and α_L conformations *(8)*.

Of the 20 amino acids that occur in proteins, L-Pro is naturally constrained by pyrrolidine ring formation to adopt ϕ-values approx $-60° \pm 20°$. **Figure 3A** shows a cluster plot for the distribution of the observed ϕ-values of L-Pro residue in protein crystal structures determined at a resolution $\leq 2\text{Å}$. The two densely populated regions correspond to local helical (α_R, $\psi \sim -30° \pm 20°$) and polyproline II (P_{II}, $\psi \sim +120° \pm 30°$) structures. For D-Pro, the favored conformation would then correspond to $\phi \sim +60°$, $\psi \sim +30°$ and $\phi \sim +60°$, $\psi \sim -120°$ (**Fig. 3B**) *(4)*. In subsequent sections, we detail the use of Aib and ^{D}Pro residues in the construction of designed secondary structures.

2. Methods

2.1. β-Turns

The β-turn is an ubiquitous structural element in proteins and peptides. The β-turn is a structural feature which is determined by the conformations of two

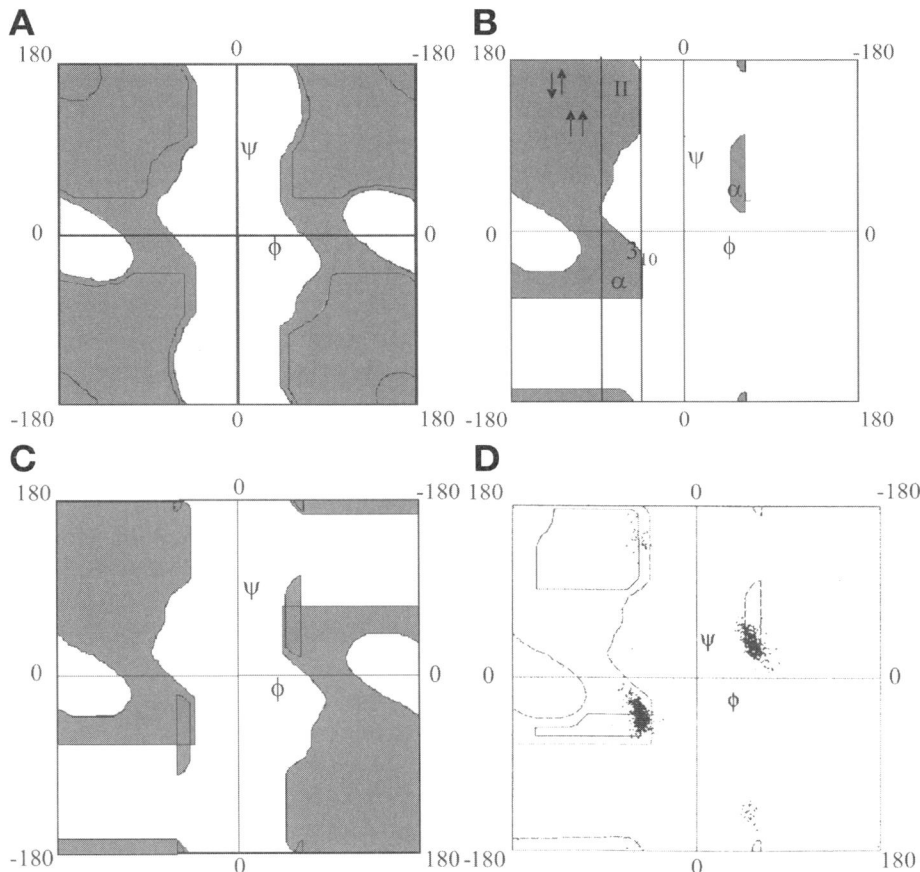

Fig. 1. Ramachandran plots of amino acid residues. **(A)** Sterically allowed regions of the achiral amino acid glycine (Ac-Gly-NHMe) indicating that >60% of the available φ-ψ space can be occupied by this residue. **(B)** Allowed regions for L-Ala (Ac-L-Ala-NHMe) showing that substitution at the Cα carbon constrains the conformationally accessible space to about 30% of the Ramachandran map. **(C)** Superposition of the allowed regions for L- and D-Ala residues. Overlapping regions indicate the theoretically allowed regions for dialkyl substitution at the Cα carbon. **(D)** Plot of the distribution of allowed conformations of 1104 Aib residues from 367 crystal structures, establishing that Aib can accommodate itself into both the 3_{10} and α-helical conformations and can form right- and left-handed helices. Figures adapted from **refs.** *4* and *8.*

successive residues in a peptide chain *(9,10)*. Different turn types are classified on the basis of the φ-ψ values at the corner residues i + 1 and i + 2 (**Fig. 4**) *(11,12)*. A 4→1 hydrogen bond between C = O$_i$ and N–H$_{i+3}$ stabilizes β-turns. Of the turns illustrated in **Fig. 4**, the type I and type III turns are very closely

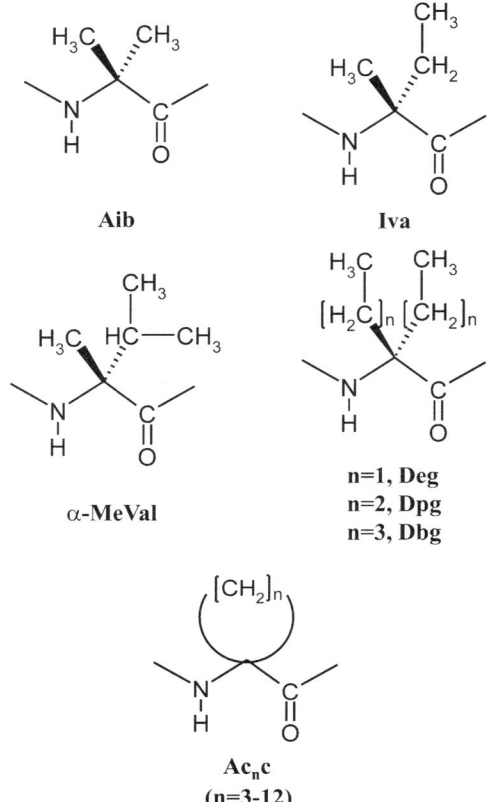

Aib

Iva

α-MeVal

n=1, **Deg**
n=2, **Dpg**
n=3, **Dbg**

Ac$_n$c
(n=3-12)

Fig. 2. Schematic representation of some α,α-dialkylglycines. Two kinds of residues that have been extensively investigated for helix nucleation are the achiral dialkylglycine derivatives having linear side chains, which includes diethylglycine (Deg), dipropylglycine (Dpg), and dibutylglycine (Dbg) and the cyclic derivatives having cycloalkane side chains of n = 3–12 residues. Adapted from **ref. 4**.

Fig. 3. *(opposite page)* φ-ψ values of L-Pro (**A**) and L-Asn (**C**) obtained from 538 independent protein crystal structures derived from the PDB using a resolution cutoff of 2 Å and a sequence homology cutoff of 40%. (**A**) It is evident that L-Pro takes up a restricted φ value because the ψ value is also largely limited to the α-helical and polyproline region of the Ramachandran map. (**B**) The allowed regions for D-Pro can be obtained by inversion of the Ramachandran plot of L-Pro. It is noteworthy that favored conformational angles for residue i+1 in prime β-turns are adopted by ^{D}Pro. Hence ^{D}Pro-Xxx sequences might be anticipated to nucleate β-hairpins in synthetic peptides. (**C**) l-Asn shows a significant cluster of positive φ values in the left-handed α$_L$ region, suggesting that ^{L}Asn-Xxx sequences can also nucleate turns in peptides. Comparative analyses, however, clearly demonstrate the superiority of the ^{D}Pro-Gly sequences over Asn-Gly segments in nucleating tight turns and forming stable β-hairpin structures. Adapted from **ref. 4**.

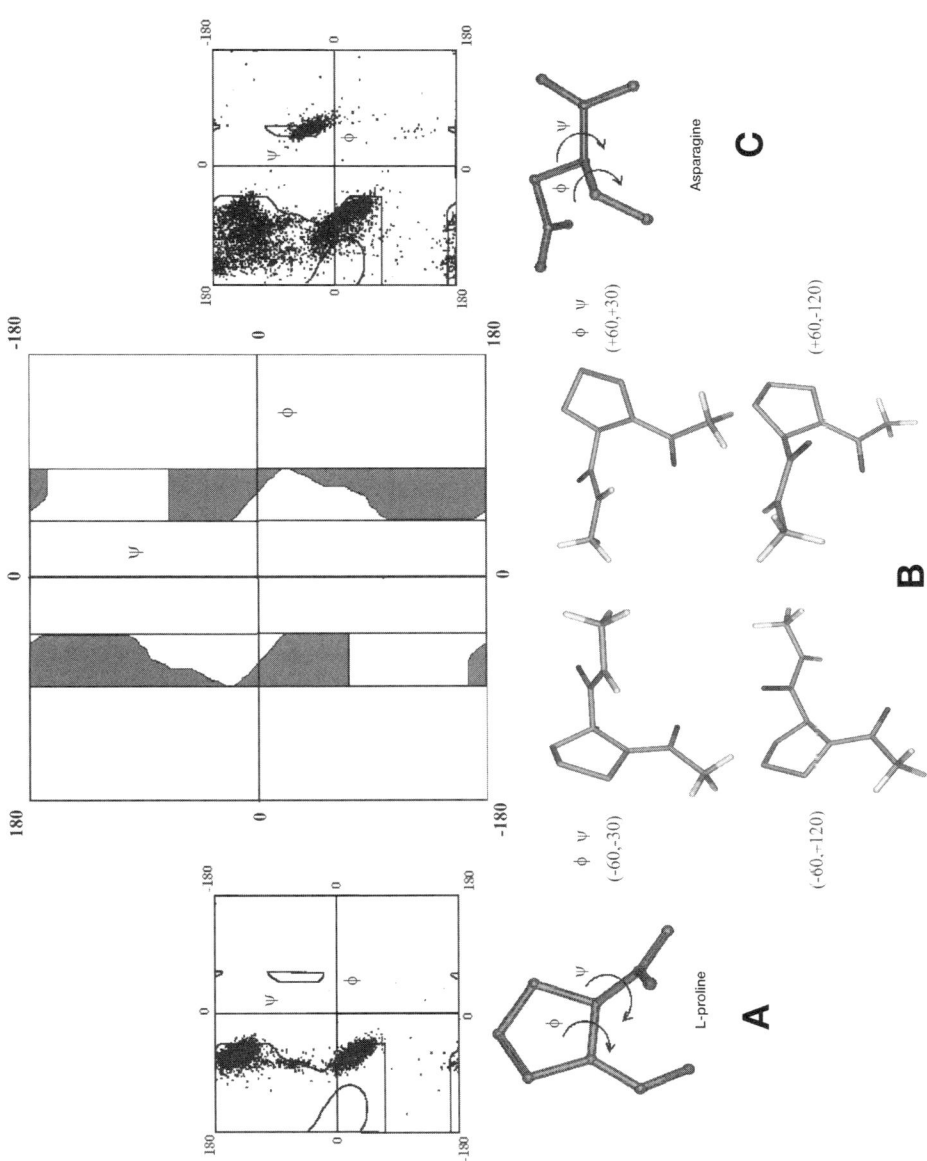

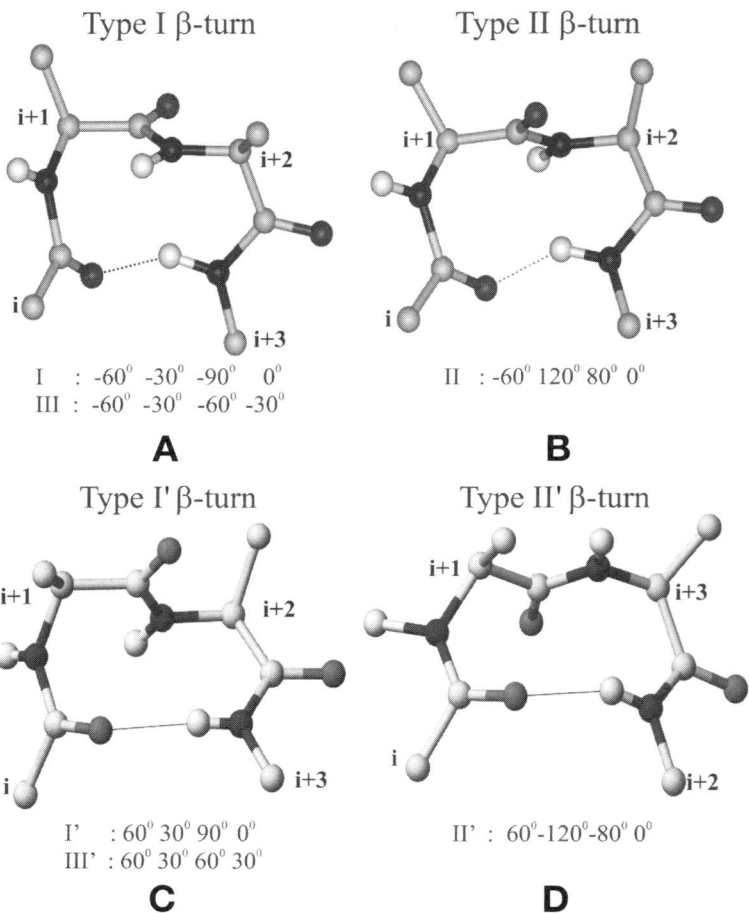

Type I β-turn

i+1
i+2
i
i+3

I   : -60° -30° -90°   0°
III : -60° -30° -60° -30°

A

Type II β-turn

i+1
i+2
i
i+3

II : -60° 120° 80° 0°

B

Type I' β-turn

i+1
i+2
i
i+3

I'   : 60° 30° 90° 0°
III' : 60° 30° 60° 30°

C

Type II' β-turn

i+1
i+3
i
i+2

II' : 60°-120°-80° 0°

D

Fig. 4. **(A,B)** Common turn types formed by two residues in the turn segment with the ideal φ-ψ values indicated below each perspective drawing. **(C,D)** Prime turns are essentially obtained by inversion of the signs of φ and ψ values of their respective counterparts. Protein β-hairpins are generally nucleated by prime turns, types I' and II' being the most common, as observed by Thornton and her colleagues *(12)*.

related and may be considered as largely indistinguishable. The formation of successive (consecutive) type III β-turns results in the generation of a 3_{10} helical structure. The 3_{10} and α-helices lie in the same region of φ-ψ space, and a transition between these two helix types can be achieved by small changes of φ-ψ values and an easily surmountable energy barrier. The type II β-turn is an isolated structural feature that facilitates chain reversal. The prime turns of types I'/III' and type II' are generated by inversion of the signs of the φ-ψ values, listed in **Fig. 4**. Registered β-hairpins in proteins are often nucleated by type I' and II' turns *(13,14)*. In proteins, the positive ϕ value required at the i + 2 position of type II β-turns and in the prime turns are often obtained using Gly residues. Asn, which has been termed as the "least chiral of amino acids" *(15)*, is another residue with a propensity to adopt α_L conformations *(16)* necessary for placement at specific positions in turns (**Fig. 3C**). In designed peptides, turn formation can be achieved by the use of ^{L}Pro/^{D}Pro residues and also by the use of constrained residues such as Aib, which can be accommodated at both positions of type I(III)/I'(III') β-turns and the i + 2 position of type II and II' turns *(5)*. β-turn design and the development of β-turn mimetic templates have been extensively reviewed elsewhere *(4,17)*. In the subsequent section β-turn type conformations are considered only in the context of their occurrence as folding nuclei in helices and hairpins.

2.2. Helices

The nucleation of helical secondary structures in sequences of α-amino acids is facilitated by insertion of residues that are strongly constrained to adopt ϕ-ψ angles in the "helical region" of conformational space. Incorporation of even a single Aib residue into heptapeptide/octapeptide sequences can facilitate helical folding. Centrally positioned Aib residues can promote formation of an incipient 3_{10} helix by recruiting both the preceding and succeeding residues to the formation of a consecutive type III β-turn structure. For example, sequences of the type Boc-Xxx-Aib-Yyy-Zzz-OMe fold into one turn of the 3_{10} helix stabilized by two successive 4→1 hydrogen-bonded type III β-turns *(5,8,18)*. Insertion of Aib residues at the N-terminal sequence or at the center of a sequence has generally been successful in nucleating helical folds. Although many designed peptides have been shown to form helical secondary structures in the crystalline state, nuclear magnetic resonance (NMR) studies in solution suggest that long (7–9 residues) sequences, which contain only a single Aib residue, do not form stable helices in highly solvating media such as DMSO. Helix stability can be improved by increasing the Aib content. A very large number of examples exist in the literature in which sequences with an Aib content of 20 to 30% have formed helical structures in the solid state. Aib content also appears to have an influence on the precise nature of the helix that has formed; 3_{10} helices are preferred when the Aib content is >40%, whereas α-helices are more often found at lower Aib content *(19)*. It must be stressed that both 3_{10} and α-helices occur in the same region of conformational space and can readily interconvert. Careful examination of the available crystal structures in the Cambridge Structural Database reveals that most of the Aib containing helical peptides contain both 4→1 and 5→1 intramolecular hydrogen bonds and may therefore be classified as mixed 3_{10}/α structures. **Figure 5** provides an illustration of the nature

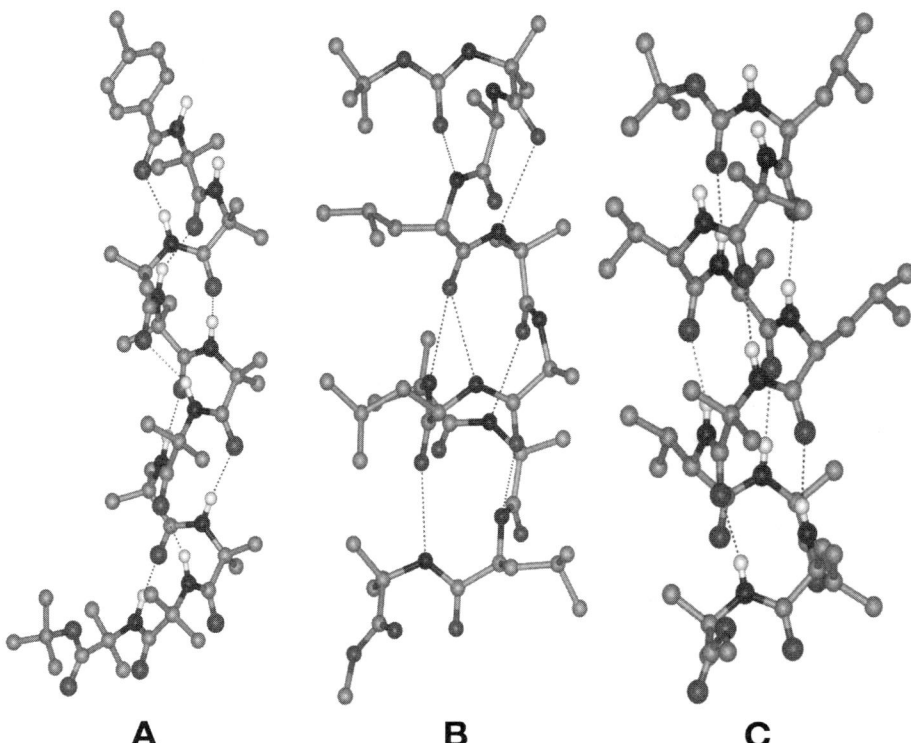

A   **B**   **C**

Fig. 5. Molecular conformation of three synthetic peptide helices in the crystal. **(A)** pBrBz-(Aib)$_{10}$-OtBu forms a 3_{10} helix in the crystal. **(B)** Boc-Aib-(Ala-Leu-Aib)$_3$-OMe adopts a $3_{10}/\alpha$-helical conformation. **(C)** Boc-Leu-Aib-Val-Ala-Leu-Aib-Val-Ala-Leu-Aib-OMe is α-helical. Dotted lines in all three figures represent hydrogen bonds. These examples clearly illustrate that percent of Aib residues determines the helical form that would predominate in peptide structures. Presence of a large number of Aib residues tilts the balance in favor of 3_{10} helices, whereas, in cases of fewer Aib residues in the sequence, $3_{10}/\alpha$-helical and α-helical structures are preferentially adopted. Adapted from **ref. 8**.

of helical structures that can be obtained in Aib-containing sequences. The structures of peptides such as Boc-Val-Ala-Leu-Aib-Val-Ala-Leu-OMe *(20,21)* provide examples of cases in which one centrally positioned Aib is able to nucleate a helical fold. Examples of peptide helices nucleated by the incorporation of the higher homologues of Aib such as α,α-di-n-propylglycine have also been reported *(22)*.

A notable feature of an overwhelming majority of published crystal structures is the tendency of Aib residues placed at the C-terminus to adopt ϕ-ψ values of the opposite sign as compared with the N-terminal helix *(5)*. Place-

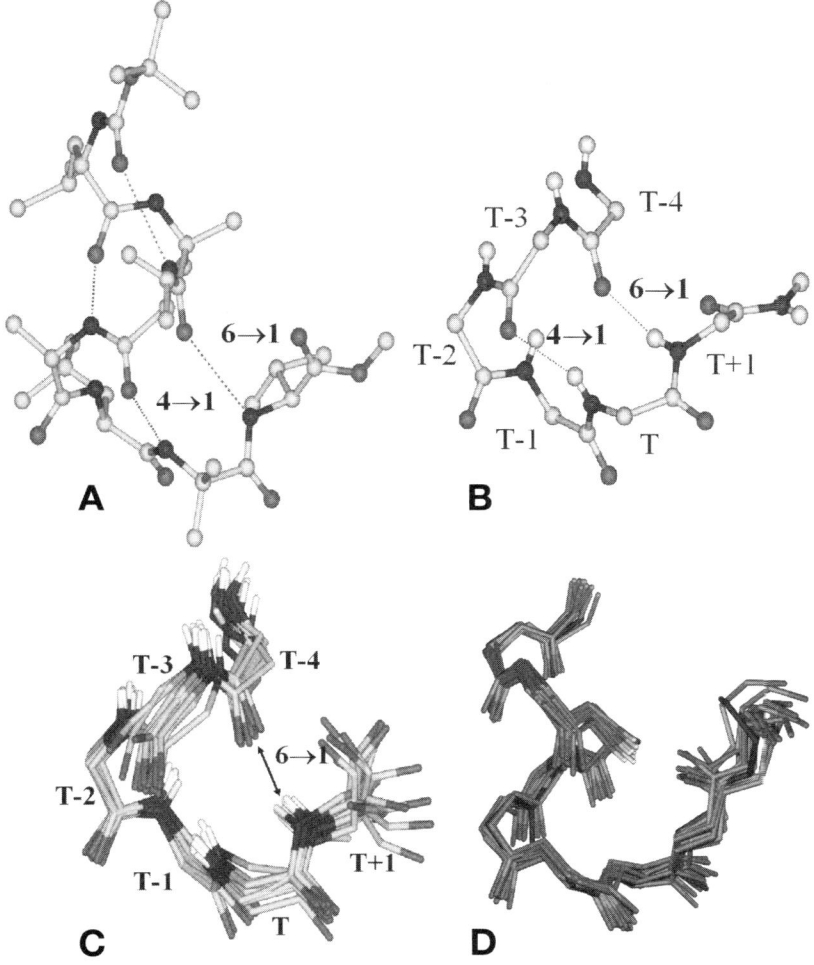

Fig. 6. (**A**) Boc-Leu-Aib-Val-Ala-Leu-Aib-Val-OMe showing the formation of a Schellman motif in crystals. (**B**) Helix-terminating Schellman motif stabilized by 4→1 and 6→1 hydrogen bonds. (**C**) Overlay of Schellman motifs in 18 crystal structures of peptide helices. (**D**) Overlay of 15 helix termination motifs observed in protein crystal structures. Adapted from **ref. 8**.

ment of an Aib residue at the penultimate position from the C-terminus results in the formation of the helix terminating Schellman motif in which Aib adopts the α_L conformation when the preceding helix is right-handed. **Figure 6** shows examples of helix termination involving Aib in both peptides and protein crystal structures (*8,23–25*). The Schellman motif, signals a reversal of chirality of the polypeptide chain. Fusing peptide helices of opposite screw sense can result in the termination of the N-terminus helix in a Schellman motif, with concomitant propagation of a helix of the opposite hand at the C-terminus. Such ambidextrous helix structures have been realized in sequences of the type Boc-D-(Val-Ala-Leu-Aib-Val-Ala-Leu)-L-(Val-Ala-Leu-Aib-Val-Ala-Leu)-OMe as well as Boc-L-(Val-Ala-Leu-Aib-Val-Ala-Leu)-D-(Val-Ala-Leu-Aib-Val-Ala-Leu)-OMe, in which the two distinct helical segments are each nucleated by a single centrally positioned Aib residue *(26,27)*.

Aib and related residues play a dominant role in forcing helical folding. Indeed, the presence of Aib together with the energetic advantage of co-operative hydrogen bond formation can drive the incorporation of D-amino acid residues into right-handed helical structures, as illustrated in **Fig. 7**. Even two contiguous D-residues can be forced into the right-handed helix as illustrated in the structures of the 13- and 19-residue peptides shown in **Fig. 7** *(28,29)*.

2.3. β-*Hairpins*

β-hairpins can be nucleated by a central two-residue segment that adopts a β-turn conformation of the appropriate stereochemistry. In proteins, prime turns (type I'/II') are most frequently observed as nucleating segments. The stereochemistry of these turns facilitates hydrogen bond registry of the antiparallel β-strands. The conformational parameters for these turn types are: type I' ϕ_{i+1} = +60° ψ_{i+1} = +30° ϕ_{i+2} = +90° ψ_{i+2} = 0°; type II' ϕ_{i+1} = +60° ψ_{i+1} = −120° ϕ_{i+2} = −80° ψ_{i+2} = 0°; type III' ϕ_{i+1} = +60° ψ_{i+1} = +30° ϕ_{i+2} = +60° ψ_{i+2} = 30°. Generation of these turn types in designed peptides requires the residue at the i + 1 position to adopt positive values for the dihedral angle ϕ. This is readily accomplished by using [D]Pro, in which ϕ is constrained to approx +60° ± 20°. [D]Pro-Gly segments were first shown to be effective nucleators of hairpin structures in the octapeptide Boc-Leu-Val-Val-[D]Pro-Gly-Leu-Val-Val-OMe *(30,31)*. In this case, both solution NMR and X-ray diffraction studies in crystals provided conclusive evidence for the formation of a hairpin structure nucleated by a type II' β-turn (**Fig. 8**). Subsequently, several apolar peptides containing central [D]Pro-Xxx segments have been shown to form β-hairpin conformations in crystals and in solution (**Fig. 8**) *(32,33)*. [D]Pro-Xxx segments are effective hairpin nucleators even in water soluble peptides *(34,35)* and stabilize folded structures to a greater extent than the corresponding Asn-Gly segments *(36)*. In most of the hairpin crystal structures, the terminal hydrogen bond that links the N- and the C-termini of the peptide is absent as a consequence of strand fraying *(37)*. Crystal structures of model β-hairpins provide insights into packing of β-sheets and cross-strand side chain–side chain interactions. Several reviews on the design and construction of β-sheets have been published *(4,8,38,39)*.

Hairpin nucleating type I' turns can also be generated by the use of Aib-[D]Xxx segments. The propensity of Aib residues to adopt helical ϕ-ψ values

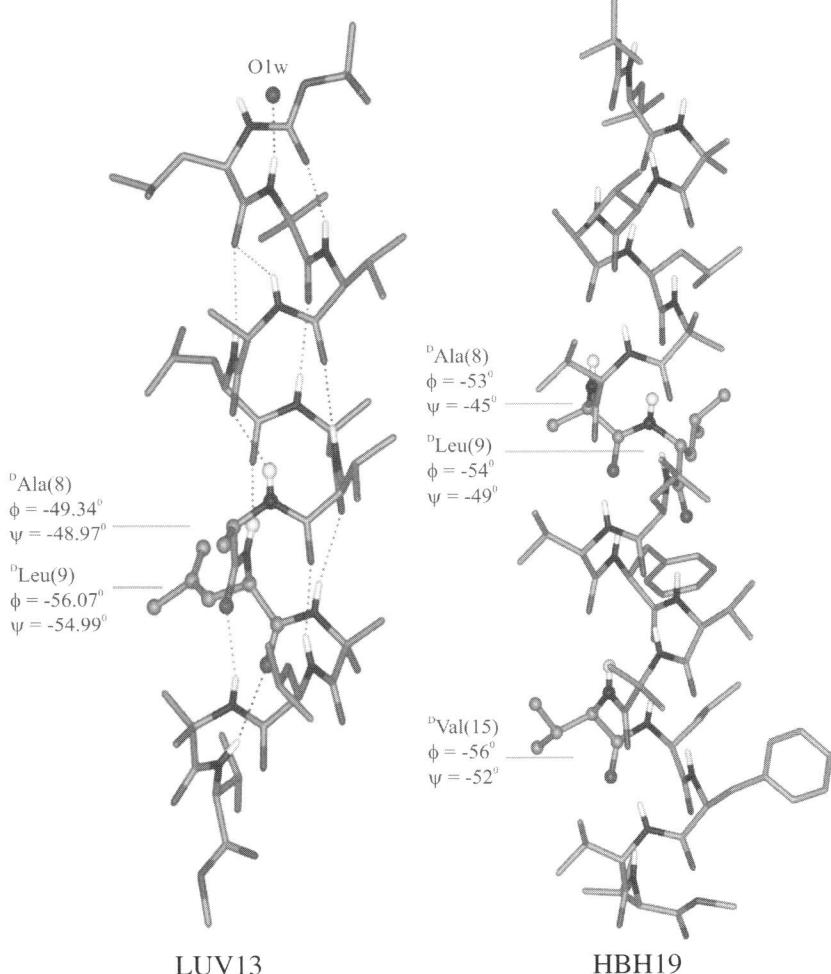

LUV13   HBH19

Fig. 7. Molecular conformation of peptide helices LUV13 (Boc-Leu-Aib-Val-Ala-Leu-Aib-Val-ᴰAla-ᴰLeu-Aib-Leu-Aib-Val-OMe) *(28)* and HBH19 (Boc-Leu-Aib-Val-Ala-Leu-Aib-Val-ᴰAla-ᴰLeu-Leu-Val-Phe-Val-Aib-ᴰVal-Leu-Phe-Val-Val-OMe) *(29)* in crystals. Both peptides accommodate guest ᴅ-amino acids in a continuous right handed helical structure. The ᴅ-amino acids in both the cases take up the α_R conformation suggesting that helical sequences can be designed to harbor residues of opposite chirality along the backbone.

($\pm50°$, $\pm50°$) precludes the formation of type II' β-turns. The crystal structure of the octapeptide Boc-Leu-Phe-Val-Aib-ᴰAla-Leu-Phe-Val-OMe shown in **Fig. 9** illustrates a designed hairpin nucleated by a central Aib-ᴰAla type I' turn

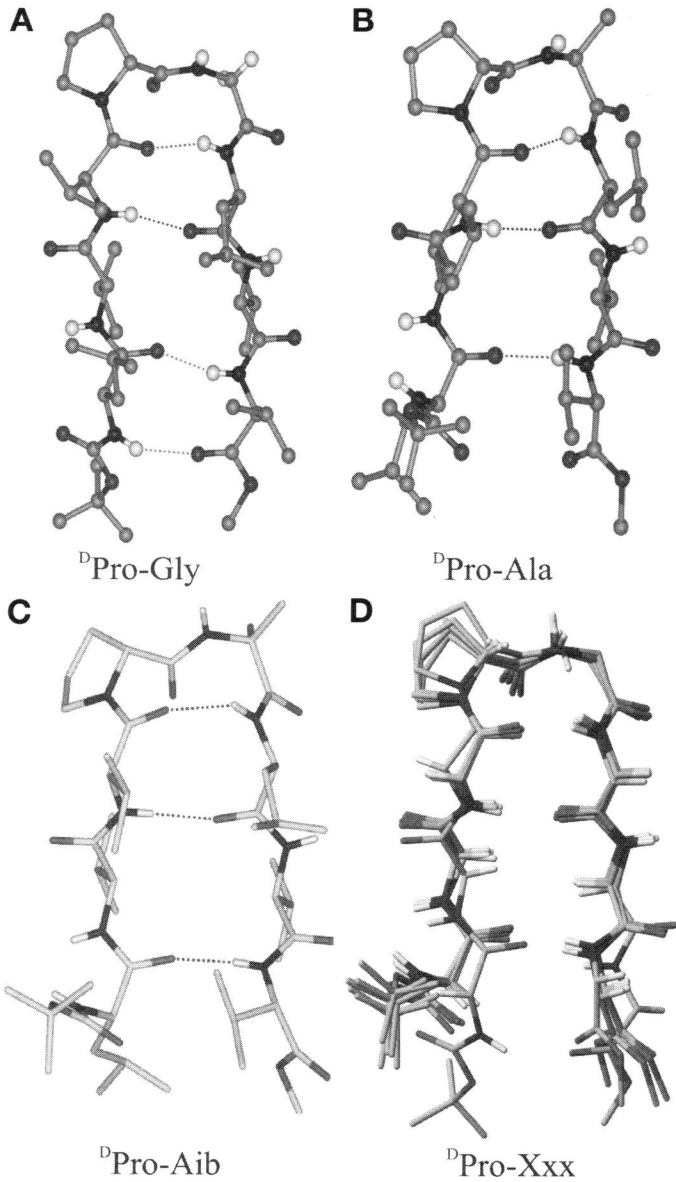

A ^{D}Pro-Gly

B ^{D}Pro-Ala

C ^{D}Pro-Aib

D ^{D}Pro-Xxx

Fig. 8. Crystal structures of peptide hairpins with ^{D}Pro in the turn region. Sequences: (A) Boc-Leu-Val-Val-^{D}Pro-Gly-Leu-Val-Val-OMe *(31)*; (B) Boc-Leu-Val-Val-^{D}Pro-Ala-Leu-Val-Val-OMe *(33)*; (C) Boc-Leu-Val-Val-^{D}Pro-Aib-Leu-Val-Val-OMe. In all three peptides, ^{D}Pro-Xxx segment adopts a type II′ turn. (D) Superposition of crystal structures of six peptide hairpins with ^{D}Pro-Xxx forming II′ turns. (A) and (B) adapted from **ref. 8**.

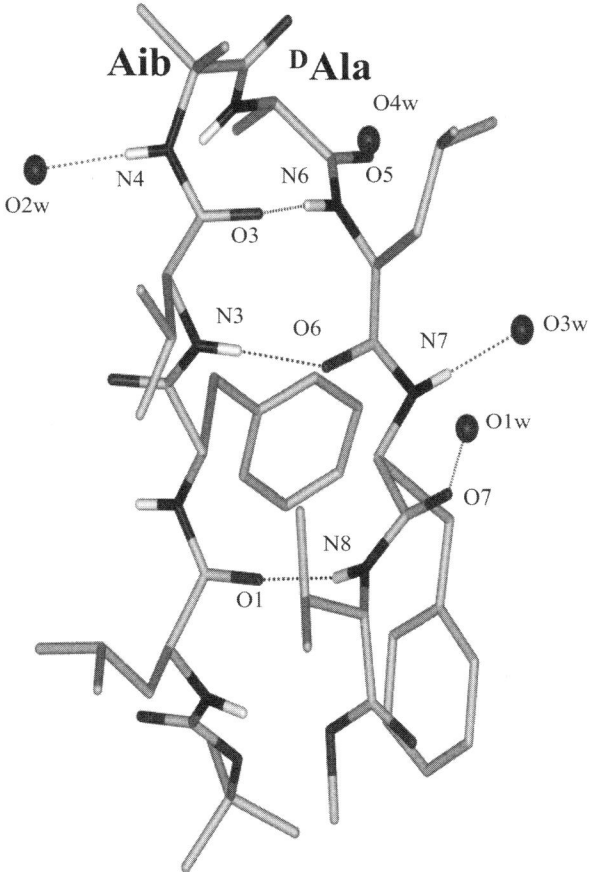

Fig. 9. Crystal structure of a peptide hairpin Boc-Leu-Phe-Val-Aib-DAla-Leu-Phe-Val-OMe with the Aib-DAla segment forming a type I' turn *(40)*.

(40). The generation of hairpins by turns formed by DPro-Ac$_n$c sequences has also been demonstrated by X-ray diffraction studies in crystals *(41)*.

2.4. β-Sheets

The successful formation of hairpins by appropriately positioning DPro-Xxx segments immediately points toward a strategy for the creation of multistranded β-sheet structures. In principle, placement of multiple nucleating DPro-Xxx segments along a polypeptide segment should facilitate the formation of multistranded sheets in which adjacent antiparallel β-strands are connected by a tight turn. A three-stranded β-sheet has been established in the 14-residue peptide (Boc-Leu-Phe-Val-DPro-Gly-Leu-Val-Leu-Ala-DPro-Gly-Phe-Val-

Leu-OMe) (**Fig. 10A**) *(42)*, whereas a four-stranded β-sheet has been demonstrated in the 26-residue sequence (**Fig. 10B**) *(43)*. This approach has been extended to the formation of a single five-stranded β-sheet *(44)*. Preformed β-sheets have been covalently cross-linked using a disulfide bridge to generate an eight-stranded β-sheet structure in a C_2 symmetric 70-residue peptide (**Fig. 10C**) *(45)*. The choice of residues in the strands determines the polarity of the sheet surface. Large residues and β-branched residues (Val, Ile, and Thr) promote strand formation. Small residues such as Gly and Ala appear to destabilize hairpin structures when incorporated into strands. Designed β-hairpins and multistranded sheets constructed with predominantly apolar amino acids display a high solubility in organic solvents and do not show a pronounced tendency to aggregate. This is presumably a consequence of effective antiparallel strand registry, which limits unsatisfied hydrogen bond acceptor and donor sites to the edge strands. The absence of strand registry often promotes aggregation, resulting in broad NMR spectra. Future design targets might involve the closure of multistranded sheet structures to for β-barrels or the creation of designed β-sandwiches by face-to-face association of flat sheets.

2.5. Mixed Secondary Structures

The helix and hairpin structures considered in the preceding sections are conformationally well defined and can be used in a modular fashion to create supersecondary structure motifs. Assembly of modules can be achieved by the use of linking segments that control orientation, or by building favorable interactions between the modules. Thus far, the success in building mixed α/β structures has been limited. **Figure 11** shows an example of a 17-residue helix-hairpin peptide that has been characterized in crystals *(46)*. In this case, an octapeptide hairpin has been linked to a heptapeptide helix using a 2-residue Gly-Gly linking segment. The conformations of both the modules are maintained in the target peptide. **Figure 12** provides a counterexample in which profound changes in the structure of the module may accompany its incorporation in the longer sequence. In this case, a helical module (Boc-Leu-Aib-Val-Ala-Leu-Aib-Val-OMe) and a β-hairpin nucleated by an Aib-DAla turn (Boc-Leu-Phe-Val-Aib-DAla-Leu-Phe-Val-OMe) were connected by a 2-residue linking segment, DAla-DLeu, to yield a target 19-residue sequence. The crystal structure of the 19-residue peptide shown in **Fig. 12** revealed that the molecule has adopted a continuous helical structure despite the presence of as many as three D-amino acids including a centrally positioned double-D segment. Ironically, this is the longest synthetic helix characterized in crystals of peptides so far *(29)*. NMR studies in chloroform solution confirm that the structure observed in crystals is also maintained in solution (unpublished).

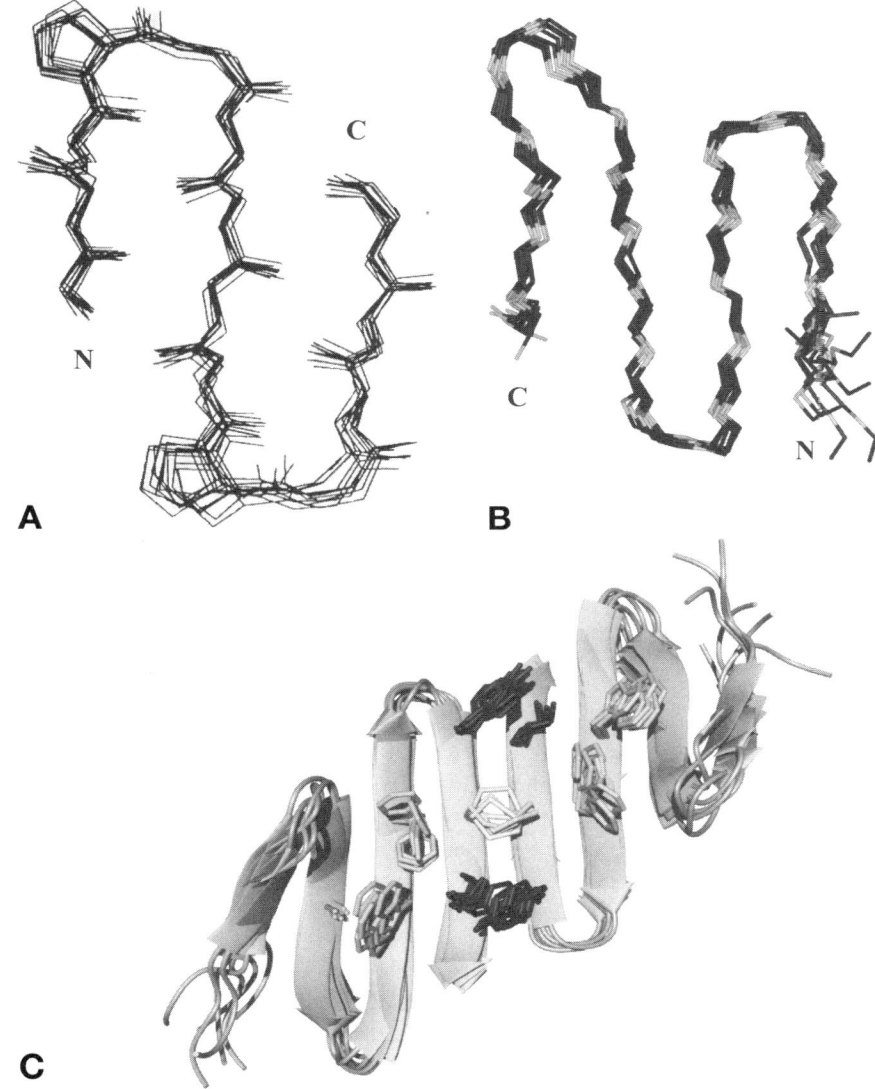

Fig. 10. Multistranded β-sheets obtained using ᴰPro-Xxx sequence for turn nucleation. (**A**) Superposition of 10 structures of a three-stranded peptide hairpin (Boc-Leu-Phe-Val-ᴰPro-Gly-Leu-Val-Leu-Ala-ᴰPro-Gly-Phe-Val-Leu-OMe) *(42)* calculated using NOEs in the solution NMR spectrum recorded in methanol. (**B**) Water-soluble four-stranded 26-residue β-sheet *(43)*. (**C**) NMR derived structure of an open eight-stranded β-sheet in a designed 70-residue peptide with a C_2 axis of symmetry *(45)*. Superposition of 10 best structures calculated for methanol.

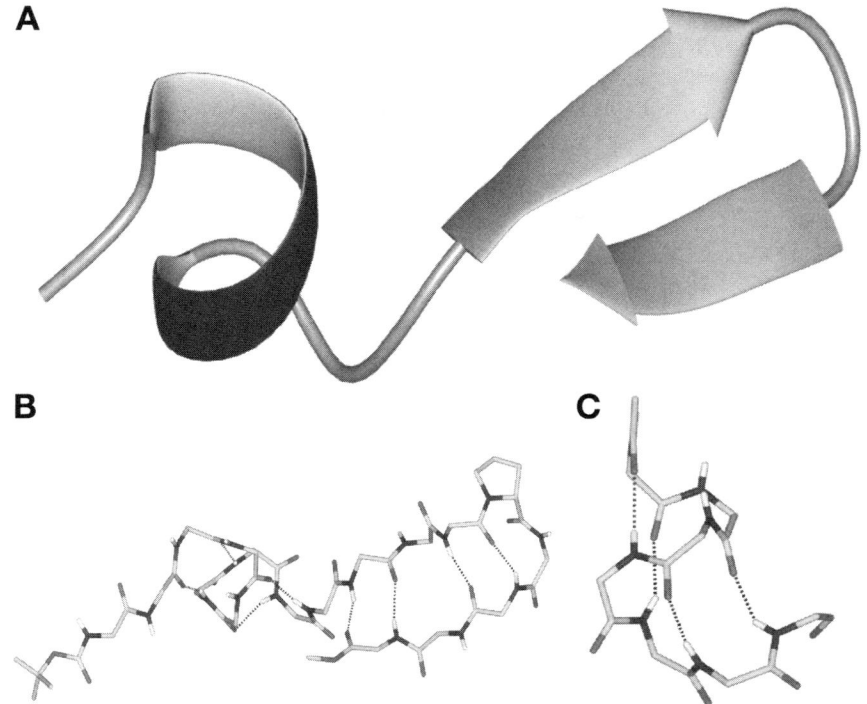

Fig. 11. 17-residue helix-hairpin peptide (Boc-Val-Ala-Leu-Aib-Val-Ala-Leu-Gly-Gly-Leu-Phe-Val-DPro-Gly-Leu-Phe-Val-OMe) *(46)*. (**A**) Ribbon diagram of the peptide. (**B**) Stick representation showing hydrogen-bonding pattern. (**C**) The crystal structure of this peptide revealed the presence of the Schellman motif as the helix-termination signal. Stick representation of the residues forming the Schellman motif. From **ref. *4***.

2.6. Conformational Characterization of Designed Peptides

In any project on secondary structure design, it is important to provide an unambiguous determination of the molecular conformation of the chosen sequence. The methods of choice are X-ray diffraction in single crystals and NMR spectroscopy in solution, with an emphasis on the determination of key nuclear Overhauser effects (NOEs) characteristic of a specific fold. Low-resolution methods like circular dichroism (CD) are of limited use in defining molecular conformation. In particular, design strategists can often be prejudiced in favor of a particular structure, resulting in overly optimistic interpretations of CD data. Recent cautionary reports on the interpretation of the CD spectra of designed helical peptides must be noted *(47,48)*.

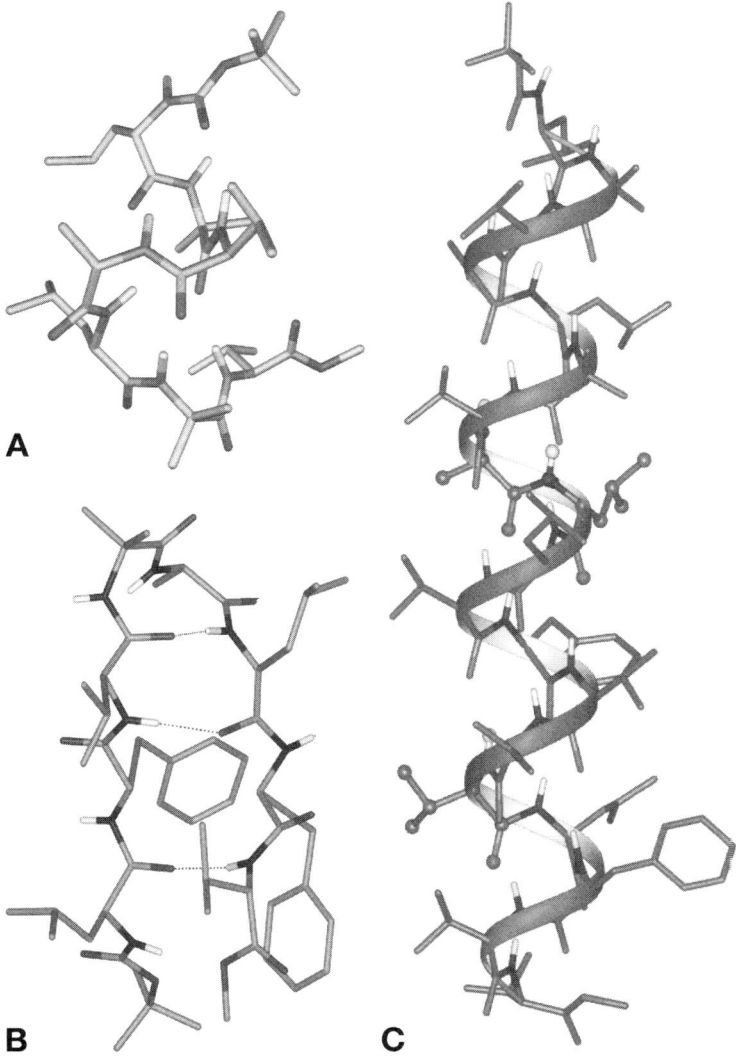

Fig. 12. A 19-residue synthetic peptide intended to produce independent helical and hairpin domains formed, instead, the longest reported right-handed α-helix in crystals with three D-amino acids adopting the α_R conformation *(29)*. The structure of this peptide is an outstanding example of profound structural changes in a module that was expected to be conformationally rigid. (**A**) Boc-Leu-Aib-Val-Ala-Leu-Aib-Val-OMe shows helical structure in the crystal. (**B**) Boc-Leu-Phe-Val-Aib-DAla-Leu-Phe-Val-OMe forms a β-hairpin with a type I' turn. (**C**) Combination of the helical and hairpin segments however lead to the formation of a single long helix with cocrystallized toluene in the crystal.

Structural characterization of helices by NMR methods relies on the observation of key NOEs between backbone protons. Under normal experimental conditions, NOEs are usually observed up to an interproton distance of <4 Å in peptides of sizes up to 20 residues. For peptides of low molecular weight, rotating frame nuclear Overhauser effect spectroscopy (ROESY) is a better choice for structural characterization compared with NOE spectroscopy experiments generally used for proteins. Most simply, the observation of successive $N_iH \leftrightarrow N_{i+1}H$ and C^α_iH 1 $N_{i+3}H$ and C^α_iH 1 $N_{i+4}H$ is characteristic of a helical backbone. For apolar peptides in organic solvents, the number of intramolecularly hydrogen-bonded NH groups can also be estimated by the use of solvent dependence of NH chemical shift in mixtures of hydrogen bonding and non-hydrogen bonding solvents (e.g., $CDCl_3$ + DMSO-d_6, $CDCl_3$ + CD_3OH) and by nitroxide-induced line broadening in inert solvents such as $CDCl_3$. Solvent-exposed and solvent-shielded NH groups can also be differentiated by measurement of temperature dependence of NH chemical shifts in hydrogen bonding solvents. The HD exchange rate is also used as an indicator of exposed vs hydrogen-bonded amide NH groups. Although HD exchange rates can be quite rapid in short designed peptides, experiments carried out at low temperatures can be useful in distinguishing between exposed and hydrogen-bonded amides. Additional parameters that provide information on local conformation are the $^3J_{NH-C^\alpha H}$ coupling constants and the chemical shifts of the individual residues when present in different secondary structure motifs. Characterization of peptide hairpins by NMR is done on the basis of characteristic turn NOEs (**Fig. 13**) accompanied by long-range backbone $C^\alpha H$-$C^\alpha H$ NOEs at the non-hydrogen–bonded positions and the NH-NH NOEs at the hydrogen bonded regions of an antiparallel β-sheet. In contrast to helices, wherein most of the observed NOEs are limited to the backbone, it is possible to observe side chain–side chain interactions of residues distant in sequence in the case of β-hairpins. An octapeptide hairpin with an unusually large number of NOEs in solution is illustrated in **Fig. 14 (49)**. Identification of exposed and hydrogen-bonded NH groups in hairpins is carried out using methods similar to those outlined for helices. In contrast to helices, however, β-hairpins are characterized by large values of $^3J_{NH-C^\alpha H}$ coupling constants and a much greater chemical shift dispersion for the backbone $C^\alpha H$ and NH protons.

The formation of single crystals is, of course, an essential prerequisite for applying X-ray diffraction methods. For apolar peptide sequences, which exhibit a high level of conformational homogeneity, crystals may be obtained by trial-and-error procedures involving a variety of organic solvents. Helical peptides appear to yield crystals more readily than do β-hairpins. Inspection of determined crystal structures can point to favorable packing features, which facilitate crystal formation. In the case of helices, arrangement into helical columns linked by head-to-tail hydrogen bonds is invariably found with adjacent col-

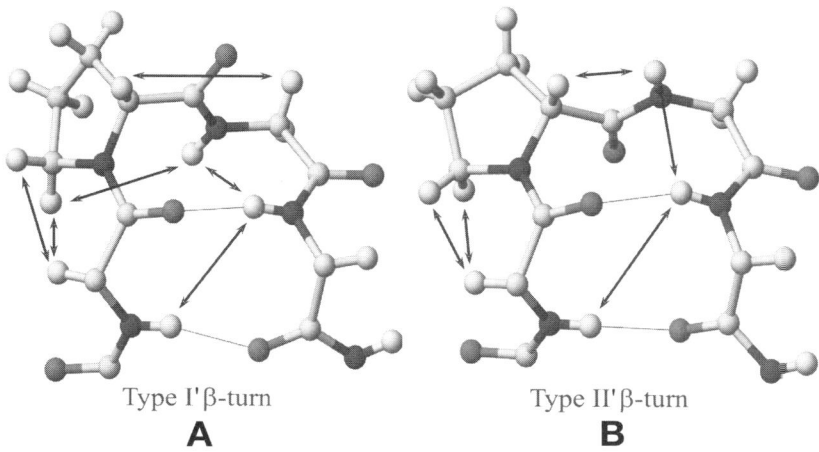

Type I' β-turn   Type II' β-turn
A   **B**

Fig. 13. Short interproton distances (double-edged arrows) that result in observable NOEs in ROESY spectra of peptides adopting type I' (**A**) or type II' (**B**) β-turns. These NOEs are used in demarcating the two turn types in solution. (Only sequential and long-range interresidue NOEs are indicated. All intraresidue NOEs are expected to be present; intensities of the NOEs observed will vary, however.)

umns interacting via nonpolar side chains. Amino acid residues such as Ala, Leu, Val, Phe, and Met are particularly effective in promoting inter-helix interactions. In the case of hairpins, two-dimensional sheet formation by intermolecular hydrogen bond formation is the dominant mode of association. The central peptide unit of the nucleating β-turn is usually pointed approximately perpendicular to the plane of the sheet, providing interactions in the third dimension in crystals. Effective side chain packing in the third dimension is necessary to get a well-ordered crystal. As the available body of literature on crystalline β-hairpins grows, it should be possible to engineer sequences to crystallize. The rapid advancement of direct methods of structure determination *(50)* and the use of single wavelength anomalous dispersion using laboratory X-ray sources *(51)* augurs well for the determination of the structures of designed peptides which fall into the "no man's land" between small molecules and proteins.

2.7. Concluding Remarks

The unfolded state of polypeptides is favored by conformational entropy and solvation effects. In hydrogen-bonding solvents, solvent-solute hydrogen bonding can overwhelm contributions because of intramolecular backbone–backbone hydrogen bonds. Folded peptide structures are invariably stabilized by the formation of multiple intramolecular hydrogen bonds. The approach

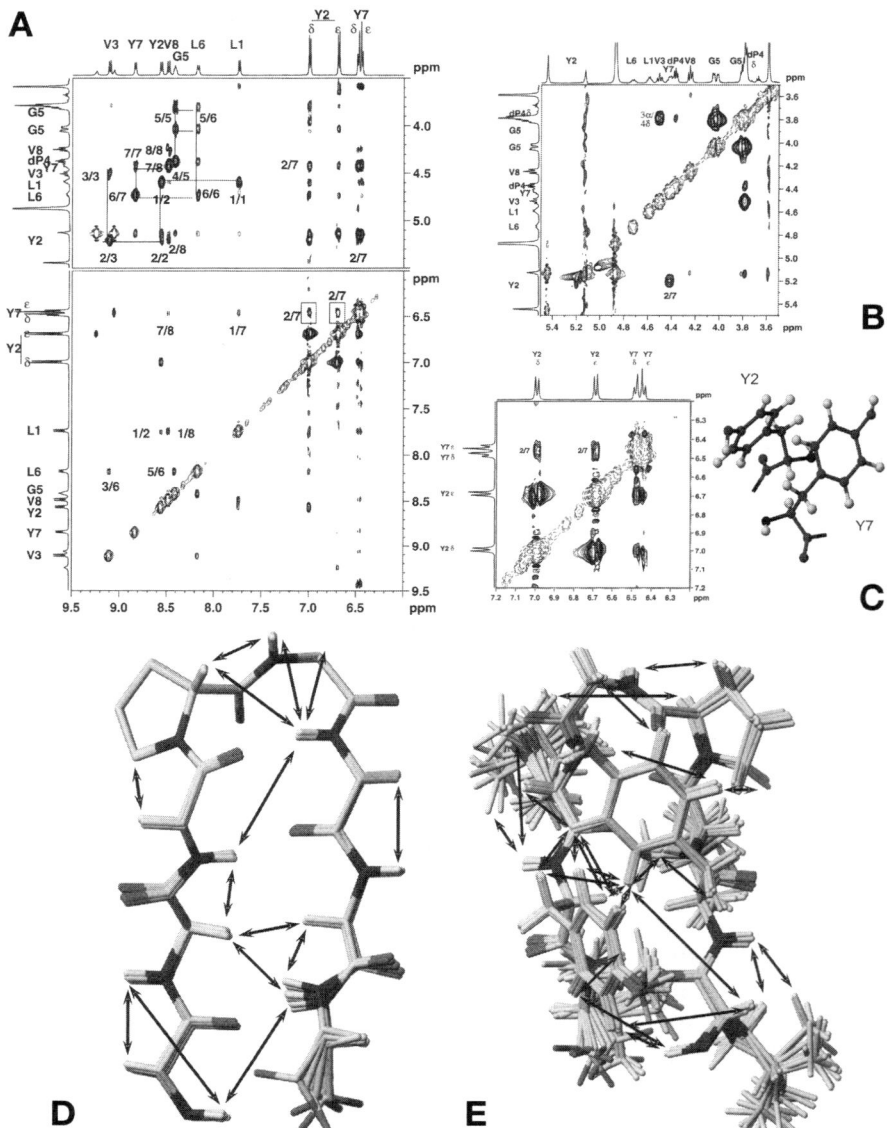

Fig. 14. NMR data of peptide Ac-Leu-Tyr-Val-DPro-Gly-Leu-Tyr-Val-OMe recorded in CD₃OH at 273K. (**A,B**) Partial expansions of the NH-NH and NH-CᵃH region (**A**) and the CᵃH region (**B**) of the ROESY spectrum of the peptide. (**C**) Strong ring-ring ROEs between the phenolic rings of Tyr2 and Tyr7: a clear indication of the close proximity of the two aromatics. (**D,E**) Superposition of 10 best structures of the peptide calculated using DYANA. Backbone (**D**) and side chain (**E**) ROEs observed in solution are indicated using arrows.

described in this chapter is based on limiting specific folding nuclei to a restricted region of conformational space, favoring the formation of intramolecular hydrogen-bonded turn structures. Repetitive turns result in the nucleation of helices, whereas isolated turns can facilitate the formation of β-hairpins. For peptides that are soluble in nonpolar organic solvents, intramolecular hydrogen bond formation can be a major stabilizing factor in generating secondary structures. The design of sequences that are soluble in a variety of organic solvents also facilitates crystallization, permitting unambiguous characterization of designed structures.

Acknowledgments

RM is supported by the award of a Senior Research Fellowship from the Council of Scientific and Industrial Research (CSIR), India. Research in this area has been supported by grants from the Department of Science and Technology (DST) and Department of Biotechnology (DBT).

References

1. Ramachandran, G. N., Ramakrishnan, C., and Sasisekharan, V. (1963) Stereochemistry of polypeptide chain configurations. *J. Mol. Biol.* **7,** 95–99.
2. Ramachandran, G. N. and Ramakrishnan, C. (1965) Stereochemical criteria for polypeptide and protein chain conformation. II. Allowed conformations for a pair of peptide units. *Biophys. J.* **5,** 909–933.
3. Ramachandran, G. N. and Sasisekharan, V. (1968) Conformation of polypeptides and proteins. *Adv. Protein. Chem.* **23,** 283–438.
4. Venkatraman, J., Shankaramma, S. C., and Balaram, P. (2001) Design of folded peptides. *Chem. Rev.* **101,** 3131–3152.
5. Prasad, B. V. V. and Balaram, P. (1984) The stereochemistry of peptides containing α-aminoisobutyric acid. *CRC Crit. Rev. Biochem.* **16,** 307–384.
6. Kaul, R. and Balaram, P. (1999) Stereochemical control of peptide folding. *Bioorg. Med. Chem.* **7,** 105–117.
7. Allen, F. H. (2002) The Cambridge Structural Database: a quarter of a million crystal structures and rising. *Acta Crystallogr.* **B58,** 380–388.
8. Aravinda, S., Shamala, N., Roy, R. S., and Balaram, P. (2003) Non-protein amino acids in peptide design. *Proc. Indian Acad. Sci. (Chem. Sci.)* **115,** 373–400.
9. Smith, J. A. and Pease, L. G. (1980) Reverse turns in peptides and proteins. *CRC Crit. Rev. Biochem.* **8,** 315–399.
10. Rose, G. D., Gierasch, L. M., and Smith, J. A. (1985) Turns in peptides and proteins. *Adv. Protein Chem.* **37,** 1–109.
11. Venkatachalam, C. M. (1968) Stereochemical criteria for polypeptides and proteins. V. Conformation of a system of three linked peptide units. *Biopolymers* **6,** 1425–1436.
12. Wilmot, C. M. and Thornton, J. M. (1988) Analysis and prediction of the different types of β-turn in proteins. *J. Mol. Biol.* **203,** 221–232.

13. Sibanda, B. L. and Thornton, J. M. (1985) β-hairpin families in globular proteins *Nature* **316**, 170–174.

14. Gunasekaran, K., Ramakrishnan, C., and Balaram, P. (1997) β-hairpins in proteins revisited: lessons for *de novo* design. *Protein Eng.* **10**, 1131–1141.

15. Richardson, J. S. and Richardson, D. C. (1989) Principles and patterns of protein conformation, in Prediction of Protein Structure and the Principles of Protein Conformation (Fasman, G. D., ed.), Plenum, New York, pp. 1–98.

16. Srinivasan, N., Anuradha, V. S., Ramakrishnan, C., Sowdhamini, R., and Balaram, P. (1994) Conformational characteristics of asparaginyl residues in proteins. *Int. J. Peptide Protein Res.* **44**, 112–122.

17. Stigers, K. D., Soth, M. J., and Nowick, J. S. (1999) Designed molecules that fold to mimic protein secondary structures. *Curr. Opin. Chem. Biol.* **3**, 714–723.

18. Toniolo, C., Bonora, G. M., Bavoso, A., Benedetti, E., di Blasio, B., Pavone, V., et al. (1983) Preferred conformations of peptides containing α,α-disubstituted α-amino acids. *Biopolymers* **22**, 205–215.

19. Karle, I. L. and Balaram, P. (1990) Structural characteristics of α-helical peptide molecules containing Aib residues. *Biochemistry* **29**, 6747–6756.

20. Karle, I. L., Flippen-Anderson, J. L., Uma, K., and Balaram, P. (1990) Apolar peptide models for conformational heterogeneity, hydration, and packing of polypeptide helices: crystal structure of hepta- and octapeptides containing α-aminoisobutyric acid. *Proteins* **7**, 62–73.

21. Karle, I. L., Flippen-Anderson, J. L., Uma, K., and Balaram, P. (1993) Unfolding of an α-helix in peptide crystals by solvation: conformational fragility in a heptapeptide. *Biopolymers* **33**, 827–837.

22. Datta, S., Kaul, R., Rao, R. B., Shamala, N., and Balaram, P. (1997) Stereochemistry of linking segments in the design of helix-helix motifs in peptides. Crystallographic comparison of a glycyl-dipropylglycyl-glycyl segment in a tripeptide and a 14-residue peptide. *J. Chem. Soc. Perkin Trans.* **2**, 1659–1664.

23. Karle, I. L., Flippen-Anderson, J. L., Uma, K., and Balaram, P. (1993) Peptide mimics for structural features in proteins. Crystal structures of three heptapeptide helices with a C-terminal 6→1 hydrogen bond. *Int. J. Peptide Protein Res.* **42**, 401–410.

24. Datta, S., Shamala, N., Banerjee, A., Pramanik, A., Bhattacharjya, S., and Balaram, P. (1997) Characterization of helix-terminating Schellman motifs in peptides. Crystal structure and nuclear Overhauser effect analysis of a synthetic hexapeptide helix. *J. Am. Chem. Soc.* **119**, 9246–9251.

25. Babu, M. M., Singh, S. K., and Balaram, P. (2002) A C-H...O hydrogen bond stabilized polypeptide chain reversal motif at the C-terminus helices in proteins. *J. Mol. Biol.* **322**, 871–880.

26. Banerjee, A., Raghothama, S., Karle, I. L., and Balaram, P. (1996) Ambidextrous molecules: cylindrical peptide structures formed by fusing left- and right-handed helices. *Biopolymers* **39**, 279–285.

27. Karle, I. L. (2001) Controls exerted by the Aib residue: helix formation and helix reversal. *Biopolymers (Pept. Sci.)* **60**, 351–365.

28. Aravinda, S., Shamala, N., Desiraju, S., and Balaram, P. (2002) A right handed peptide helix containing a central double D-amino acid segment. *Chem. Commun.* 2454–2455.

29. Karle, I. L., Gopi, H. N., and Balaram, P. (2003) Crystal structure of a hydrophobic 19-residue peptide helix containing three centrally located D amino acids. *Proc. Natl. Acad. Sci. USA* **100,** 13946–13951.

30. Awasthi, S. K., Raghothama, S., and Balaram, P. (1995) A designed β-hairpin peptide. *Biochem. Biophys. Res. Commun.* **216,** 375–381.

31. Karle, I. L., Awasthi, S. K., and Balaram, P. (1996) A designed β-hairpin peptide in crystals. *Proc. Natl. Acad. Sci. USA* **93,** 8189–8193.

32. Raghothama, S., Awasthi, S. K., and Balaram, P. (1998) β-hairpin nucleation by Pro-Gly β-turns. Comparison of D-Pro-Gly and L-Pro-Gly sequences in apolar octapeptides. *J. Chem. Soc. Perkin Trans.* **2,** 137–143.

33. Das, C., Naganagowda, G. A., Karle, I. L., and Balaram, P. (2001) Designed β-hairpin peptides with defined tight turn stereochemistry. *Biopolymers* **58,** 335–346.

34. Haque, T. S. and Gellman, S. H. (1997) Insights on β-hairpin stability in aqueous solution from peptides with enforced type I' and type II' β-turns. *J. Am. Chem. Soc.* **119,** 2303–2304.

35. Espinosa, J. F. and Gellman, S. H. (2000) A designed β-hairpin containing a natural hydrophobic cluster. *Angew. Chem. Int. Ed. Engl.* **39,** 2330–33.

36. Stanger, H. E. and Gellman, S. H. (1998) Rules for antiparallel β-sheet design: D-Pro-Gly is superior to L-Asn-Gly for β-hairpin nucleation. *J. Am. Chem. Soc.* **120,** 4236–4237.

37. Aravinda, S., Harini, V. V., Shamala, N., Das, C., and Balaram, P. (2004) Structure and assembly of designed β-hairpin peptides in crystals as models for β-sheet aggregation. *Biochemistry* **43,** 1832–1846.

38. Gellman, S. H. (1998) Minimal model systems for β-sheet secondary structure in proteins. *Curr. Opin. Chem. Biol.* **2,** 717–725.

39. Blanco, F., Ramirez-Alvarado, M., and Serrano, L. (1998) Formation and stability of β-hairpin structures in polypeptides. *Curr. Opin. Struct. Biol.* **8,** 107–111.

40. Aravinda, S., Shamala, N., Rai, R., Gopi, H. N., and Balaram, P. (2002) A crystalline β-hairpin peptide nucleated by a type I' Aib-D-Ala β-turn: evidence for cross-strand aromatic interactions. *Angew. Chem. Int. Ed. Engl.* **41,** 3863–3865.

41. Harini, V. V., Aravinda, S., Rai, R., Shamala, N., and Balaram, P. (2005) Molecular conformation and packing of peptide β-hairpins in crystals. Structures of two synthetic octapeptides containing 1-aminocycloalkane-1-carboxylic acid residues at the i+2 position of the β-turn. *Chem. Eur. J.* In press.

42. Das, C., Raghothama, S., and Balaram, P. (1998) A designed three stranded β-sheet peptide. *J. Am. Chem. Soc.* **120,** 5812–5813.

43. Das, C., Raghothama, S., and Balaram, P. (1999) A four stranded β-sheet structure in a designed synthetic polypeptide. *Chem. Commun.* 967–968.

44. Venkatraman, J., Naganagowda, G. A., Sudha, R., and Balaram, P. (2001) *De novo* design of a five-stranded β-sheet anchoring a metal-ion binding site. *Chem. Commun.* 2660–2661.

45. Venkatraman, J., Naganagowda, G. A., and Balaram, P. (2002) Design and construction of an open multistranded β-sheet polypeptide stabilized by a disulfide bridge. *J. Am. Chem. Soc.* **124,** 4987–4994.
46. Karle, I. L., Das, C., and Balaram, P. (2000) De novo protein design: crystallographic characterization of a synthetic peptide containing independent helical and hairpin domains. *Proc. Natl. Acad. Sci. USA* **97,** 3034–3037.
47. Wallimann, P., Kennedy. R. J., Miller, J. S., Shalongo. W., and Kemp, D. S. (2003) Dual wavelength parametric test of two-state models for circular dichroism spectra of helical polypeptides: anomalous dichroic properties of alanine-rich peptides. *J. Am. Chem. Soc.* **125,** 1203–1220.
48. Aravinda, S., Datta, S., Shamala, N., and Balaram, P. (2004) Hydrogen-bond lengths in polypeptide helices: no evidence for short hydrogen bonds. *Angew. Chem. Int. Ed. Engl.* **43,** 6728–6731.
49. Mahalakshmi, R., Raghothama, S., and Balaram, P. (2006) NMR analysis of aromatic interactions in designed peptide β-haripins. *J. Am. Chem. Soc.* **128,** 1125–1138.
50. Bunkoczi, G., Vertesy, L., and Sheldrick, G. M. (2005) The antiviral antibiotic Feglymycin: first direct-methods solution of a 1000+ equal-atom structure. *Angew. Chem. Int. Ed. Engl.* **44,** 1340–1342.
51. Konnert, J., Karle, J., Karle, I. L., Uma, K., and Balaram, P. (1999) Isomorphous replacement combined with anomalous dispersion in the linear equations: applications to a crystal containing four nonapeptide conformers. *Acta Crystallogr.* **D55,** 448–457.

5

Design and Synthesis of β-Peptides With Biological Activity

Marc J. Koyack and Richard P. Cheng

Summary

β-Peptides have been used as a platform for developing bioactive compounds with various types of bioactivity such as antimicrobial activity, cholesterol absorption inhibition, somatostatin receptor agonist, and hDM2 inhibition. These bioactive β-peptides have been designed based on bioactive α-peptides. Three main strategies have been used to design bioactive β-peptides: direct conversion of α-peptide sequences into β-peptide sequences, placement of side chains to provide desirable distribution of physicochemical properties, and the grafting of proteinaceous side chains critical for bioactivity onto β-peptide structures. This chapter briefly discusses the various strategies employed to design bioactive β-peptides, followed by protocols for the synthesis of N-α-fluorenylmethyloxycarbonyl (Fmoc)-protected β³-amino acids from Fmoc-protected α-amino acids, and synthesis of β-peptides by solid phase methods using Fmoc-based chemistry.

Key Words: β-Amino acids; bioactivity; foldamers; β-peptides; rational design.

1. Introduction

Biologically active compounds have been typically dominated by small molecules *(1,2)* and α-peptides *(3)*, because small molecules are attractive for therapeutic drug development, and libraries of α-peptides are easy to generate and screen. Recently, β-peptides *(4–6)*, or oligo-(β-amino acids), have provided a novel avenue for the discovery of bioactive compounds. β-Peptides are known to adopt local regularly repeating structures reminiscent of protein secondary structures (helices, sheets, and turns) *(7–10)*, despite the insertion of a carbon unit in the backbone of each residue compared to proteins or α-peptides (**Fig. 1**). Although poor oral availability *(11)* may limit the development of β-peptide based therapeutics, β-peptides are stable to proteolytic degradation *(12,13)* with long elimination half-lives *(11,13)*, permeable to cell membranes

From: *Methods in Molecular Biology, vol. 340: Protein Design: Methods and Applications*
Edited by: R. Guerois and M. López de la Paz © Humana Press Inc., Totowa, NJ

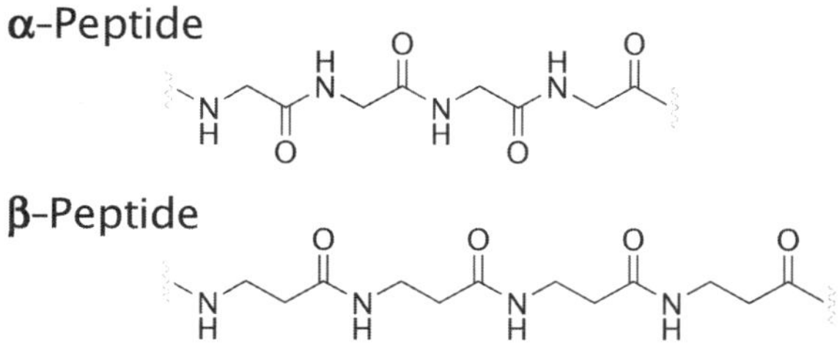

α-Peptide

β-Peptide

Fig. 1. Backbones of α- and β-peptide tetramers. α-Peptide backbones (poly-Gly) have one carbon unit per residue between the amine and carbonyl. β-Peptide backbones (poly-hGly) have two carbon units per residue between the amine and carbonyl.

(14–17), and nonmutagenic by the Ames test *(18)*. Therefore, bioactive β-peptides may serve as potential research tools and perhaps injectable therapeutics.

1.1. Biorelevant Characteristics of β-Peptides

α-Peptides are frequently used in biological studies because of the ease of generation. However, β-peptides are more attractive than α-peptides for developing bioactive compounds because of several desirable characteristics such as high metabolic stability *(12,13,19)*, long elimination half-lives *(11,13)*, cell membrane permeability *(14–17)*, and nonmutagenicity *(18)*. Various robust peptidases (bacterial, fungal, and eukaryotic) were unable to degrade β-peptides after 48 h *(12,13)*, whereas α-peptides were completely degraded within 1 h *(12,13)*. Furthermore, several bacterial species could not grow individually on media supplemented with β-peptides as the only carbon and nitrogen source, but the same bacteria readily grew on media supplemented with α-peptides, showing that bacteria could metabolize α-peptides but not β-peptides *(13)*. Interestingly, multiple bacteria species synergistically metabolized the β-peptides completely in 400 h, significantly longer than the 150 h for α-peptides *(19)*. After intravenous injection into rats, the elimination half lives of β-peptides were 3–30 h *(11,13)*, whereas the elimination half-lives of α-peptides were in the range of minutes *(20)*. However, peroral administration of radiolabeled β-peptides in rats resulted in no bioaccumulation, suggesting poor oral bioavailability *(11)*. Furthermore, the radiolabeled β-peptides were quantitatively eliminated from rats as feces after 96 h, indicating lack of absorption via the digestive tract *(11)*. Oligo-β-homoarginines *(14)* and Tat-derived β-peptides *(15,16)* were shown to cross cell membranes and localize within cell nuclei. Ames tests with a number of β-amino acids (the proteolytic degradation products of β-peptides) were all negative, revealing that β-amino acids are nonmutagenic *(18)*. These biorelevant characteristics set the stage for using β-peptides as scaffolds for developing biologically active compounds.

1.2. β-*Peptide Structures and Monomer Selection*

β-Peptides provide attractive scaffolds for developing bioactive compounds, because they can adopt various local regular repeating structures such as helices, sheets, and turns *(7–10)*. In particular, multiple different helical conformations have been discovered *(6–8,21)*. Importantly, control of the β-peptide helix type can be achieved by changing the constituting residues *(7,21,22)*. β-Peptides with exclusively β^2- or β^3-amino acids adopt the 14-helix motif, whereas alternating β^2- and β^3-amino acids promotes the 10/12-helix *(22)*. β-Amino acids constrained by cyclopentyl rings promote the 12-helix *(21)*, whereas cyclohexyl β-amino acids promote the 14-helix *(7)*. The 14-helix is one of the most well-characterized β-peptide secondary structure with an approximate three residue repeat *(6)*. β^3-Residues with chirality analogous to natural α-amino acids favor the formation of a left-handed 14-helix *(4,5)*, whereas natural α-amino acids favor the formation of a right-handed α-helix (**Fig. 2**). Also, the helix macrodipole of the 14-helix involves a partial positive charge at the C-terminus and partial negative charge at the N-terminus, which is the exact opposite of the helix dipole for an α-helix. Initial structural studies in organic solvents were important for the understanding of β-peptide helices *(7,8)*; however, structural information in aqueous environment is critical for the development of bioactive compounds. Accordingly, several recent studies have focused on the stability of β-peptide helices in water *(23–27)*. The β-peptide 14-helix have been stabilized in water through intramolecular salt bridges *(23,24)*, branched amino acids in the sequence *(25,26)*, or neutralization of the helix macrodipole *(26,27)*, similar to strategies for stabilizing α-helices *(28)*. These studies have led to the development of a number of bioactive β-peptides based on the 14-helix scaffold (*vide infra*).

Short β-peptides (less than six residues) are readily available through solid-phase peptide synthesis using Fmoc-based chemistry *(29)*. Furthermore, numerous protocols have been published for the synthesis of a wide variety of enantiomerically pure β-amino acids *(30)*. Because the bioactivity of proteins and α-peptides is derived from the spatial arrangements and properties of proteinaceous side chains, analogous side chains should be sufficient for developing bioactivity in β-peptide scaffolds. Some of the appropriately protected Fmoc-β^3-amino acid monomers are commercially available, but none of the Fmoc-protected β^2- or β^2,β^3-amino acids can be purchased. Also, these Fmoc-protected β^3-amino acids can be prepared with relative ease compared to β^2- and β^2,β^3-amino acids *(29–32)*; most Fmoc-β^3-amino acids can be synthesized from commercially available Fmoc-α-amino acids *(29)* or Fmoc-α-amino pentafluorophenylesters *(33)* via Arndt-Eistert homologation. Therefore, most of the bioactive β-peptides to date are composed of β^3-amino acids.

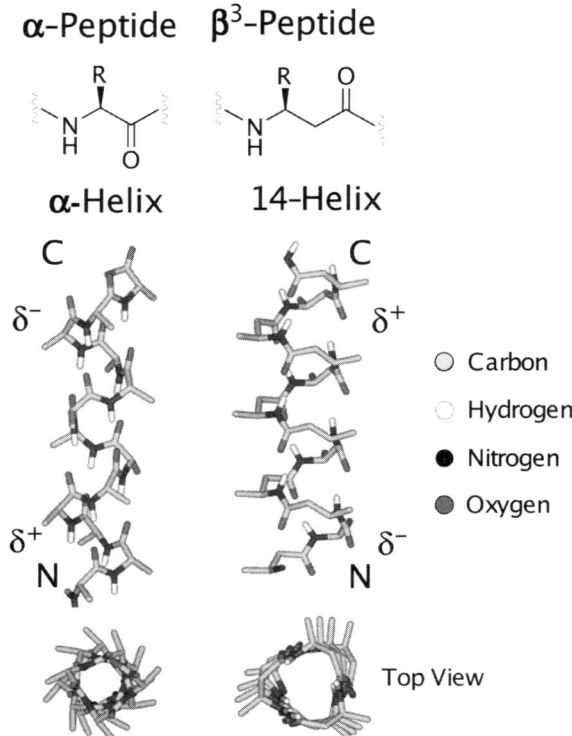

Fig. 2. The α-peptide α-helix and β-peptide 14-helix and the corresponding monomers. α-Peptide α-helices composed of L-amino acids are right-handed; carbonyl
oxygens point toward the C-terminus whereas -NH groups point toward the N-terminus, generating a helix dipole with positive and negative charges at the N– and C–
termini, respectively. β-Peptide 14-helices composed of β^3-amino acids with analogous
stereochemistry as L-amino acids are left-handed; carbonyl oxygens point toward the
N-terminus whereas -NH groups point toward the C-terminus, generating a helix
dipole opposite of the α-helix. Termini are labeled N and C with the respective associated partial charge. For clarity, all hydrogens except for backbone –NH groups have
been omitted.

1.3. Design Strategies

Approaches for designing bioactive β-peptides are all based on bioactive α-
peptides. There are three main strategies: direct conversion of α-peptide sequences
into β-peptide sequences, placement of side chains to provide desirable distribution of physiochemical properties, or the grafting of proteinaceous side
chains critical for bioactivity onto β-peptide structures. All bioactive β-pep-

tides are based on known bioactive α-peptides, which may be somewhat limiting. Therefore, the next major challenge is the *de novo* design of bioactive β-peptides based directly on the molecular properties of the intended targets.

1.3.1. Direct Sequence Conversion

The direct conversion of α-peptide sequences into the homologous β-peptide sequences has been used to generate β-peptides that bind somatostatin receptors *(34–38)*, or cross the cell membrane and localize to the cell nucleus *(14–17)*. Somatostatin receptor-binding linear and cyclic tetra-β-peptides have been generated based on four consecutive residues of the α-peptide octreotide responsible for somatostatin receptor-binding *(34–38)*. These tetra-β-peptides exhibited micromolar *(35,37)* to nanomolar *(34,36,38)* binding to the somatostatin receptor, because the spatial arrangement of bioactive side chains were analogous to the parent α-peptide. Based on the direct sequence conversion strategy, oligo-β³-homoarginines exhibited nuclear localization in mouse fibroblast cells despite the insertion of a carbon unit into the monomer backbone of oligo-Arg peptides *(14)*. Also, β-peptides based on the sequence of residues 47–57 of the HIV-Tat peptide localized to the nucleus of HeLa cells, similar to the natural Tat peptide *(15,16)*; the Tat-based β-peptide selectively bound to TAR RNA with nanomolar affinity *(39)*. The cell membrane–penetrating ability of these β-peptides were influenced by the conformational stability of the 14-helix scaffold and the spatial arrangement of hydrophobic side chains *(17)*. These findings suggest that the distribution of physicochemical properties has a major effect on membrane permeability.

1.3.2. Distribution of Physicochemical Properties

In addition to designing cell-membrane–penetrating β-peptides, the distribution of physicochemical properties has been used to design β-peptides that demonstrate antimicrobial activity *(40–47)* and *in vivo* inhibition of cholesterol uptake *(48)*. β-Peptides that fold into amphiphilic helices (with hydrophobic groups aligned on one helical face and cationic groups on the other) exhibited antimicrobial activity against several classes of bacteria *(40–47)*. Antimicrobial activity has been demonstrated by β-peptides that adopt either 12- *(41,43)*, 10/12- *(45,46)*, or 14-helices *(40,42,44,47)*. β-Peptides with minimal inhibitory concentrations two times less than required for hemolysis have also been demonstrated *(40)*. β-Peptides that inhibit cholesterol absorption in CaCO-2 cell models were also designed to adopt amphiphilic 14-helices *(48)*. Although the *in vivo* inhibition of cholesterol adsorption was achieved in millimolar range, tests of α-peptides in the same system demonstrated no inhibition from proteolytic degradation *(48)*.

1.3.3. Grafting Bioactive Side Chains

The grafting of proteinaceous side chains onto β-peptide scaffolds is the most challenging strategy for designing bioactive β-peptides, because β-peptide scaffolds have different structural details compared with α-peptides and therefore the spatial arrangement of side chains need to be considered in detail. Nonetheless, this grafting strategy was used to design β-peptides that can bind and inhibit hDM2 *(49,50)*. The spatial arrangement of three key hydrophobic residues for binding hDM2 (residues F19, W23, and L26 of p53) was recapitulated on a β-peptide 14-helix *(49)*. Fluorescence polarization studies revealed that the β-peptide design bound to hDM2 with an affinity only 1.6 to 2.5 times lower than an α-peptide sequence derived from p53 *(49)*.

1.4. Strategy Selection and Implementation

Choosing the appropriate strategy for designing bioactive β-peptides depends on the nature of the interaction being targeted. Interactions that depend on local side chain presentation (of an α-peptide) may be targeted by β-peptides with the same sequences. Interactions that rely on the allocation of certain physicochemical properties (hydrophobic and hydrophilic) may be targeted by structured β-peptides with the appropriate distribution of these properties. Finally, if an interaction requires exact placement of side chains on an extended binding interface, grafting side chain functionalities onto a β-peptide structural scaffold would be most appropriate. Current bioactive β-peptides are either helical or cyclic, whereas sheets and turns have not been used as scaffolds for developing bioactive β-peptides. The appropriate structural scaffold should be determined by the nature of the binding interface of the target. Many proteins have pitted and grooved surfaces (i.e., enzymes) that may be targeted using helical motifs. Conversely, flat protein surfaces that cannot be easily recognized by helices may be targeted by sheets, which present a flat surface topology.

After a target and the most suitable strategy for designing the bioactive β-peptide are chosen, some molecular modeling may be performed to provide a first evaluation of achieving the intended bioactivity. Molecular modeling may not be critical for β-peptide sequences based on the direct conversion of α-peptide sequences or the overall distribution of physicochemical properties. However, for grafting bioactive side chains onto a β-peptide structural scaffold, molecular modeling is important for evaluating the spatial orientation of these side chains to potentially exert bioactivity. Various commercially available molecular modeling software packages are capable of performing this task and will not be discussed in detail in this chapter. Also, the experimental evaluation of the bioactivity of the designed β-peptide will depend on the target chosen, and is beyond the scope of this chapter. The remainder of this chapter will be used to describe the synthesis of appropriately protected Fmoc-β³-amino acids and the solid-phase peptide synthesis of β³-peptides.

2. Materials

2.1. Synthesis of Fmoc-β³-Amino Acid via Arndt-Eistert Homologation

1. Fmoc-protected α-amino acids (Nova Biochem). Store at −20°C.
2. For *in situ* generation of mixed anhydrides: tetrahydrofuran (EMD), ethylchloroformate (Aldrich), and N-methylmorpholine (Aldrich).
3. For generation of diazomethane: 2-(2-ethoxyethoxy) ethanol (EMD), potassium hydroxide pellets (Mallinckrodt Baker), anhydrous ethyl ether (Mallinckrodt Baker), and Diazald (Aldrich). Diazald should be stored at 4°C.
4. For workup of Fmoc-α-amino-diazomethylketones: acetic acid (EMD), 1 *N* hydrochloric acid (EMD), saturated sodium bicarbonate (Fisher Scientific), sodium sulfate (Mallinckrodt Baker), hexanes (EMD), and ethyl acetate (EMD).
5. Silver trifluoroacetate (Aldrich).
6. For workup of Fmoc-β³-amino acids: 0.2 *M* citric acid (EMD), saturated sodium chloride (EMD), chloroform (EMD), and methanol (EMD).
7. Diazomethane generation kit with polished glass joints (Aldrich).

2.2. β-Peptide Synthesis via Solid-Phase Methods Using Fmoc-Based Chemistry

1. Fmoc-PAL-PEG-PS resin (Applied Bioscience). Store at 4°C. The final product afforded by this resin is a carboxyamide, other resins may be chosen to afford the desired C-terminal function group depending on the design *(51)*.
2. 20% Piperidine in N,N-dimethylformamide (*see also* **Note 4**).
3. For coupling of Fmoc-β³-amino acids: N,N-dimethylformamide (EMD), diisopropylethylamine (Aldrich), N-hydroxybenzotriazole (HOBt, NovaBiochem), and 2-(1-H-benzotriazole-1-yl)-1,1,3,3-tetramethyluronium hexafluorophosphate (HBTU, NovaBiochem). Store HOBt and HBTU at <8°C.
4. For cleavage of β-peptides from resin: trifluoroacetic acid (TFA, Aldrich), triisopropylsilane (TIPS, Aldrich), and hexanes (EMD).
5. Peptide synthesis vessels (Chemglass) and Mistral multimixer (Lab-Line Instruments).

3. Methods

3.1. Synthesis of Fmoc-β³-Amino Acid via Arndt-Eistert Homologation

Fmoc-β³-amino acids with orthogonal side chain protecting groups serve as building blocks for synthesizing β³-peptides by solid phase peptide synthesis using Fmoc-based chemistry *(51,52)*. The side chain protecting groups are needed for performing solid phase peptide synthesis using Fmoc-based chemistry. Typical side chain–protecting groups include *tert*-butoxycarbonyl group (Lys, Trp), *tert*-butyl group (Asp, Glu, Ser, Thr, Tyr), trityl group (Asn, Cys, Gln, His), and 2,2,4,6,7-pentamethyldihydro-benzofuran-5-sulfonyl group (Arg) *(51)*. Although some appropriately protected Fmoc-β³-amino acid monomers are commercially available from Fluka (Aldrich), most others can be syn-

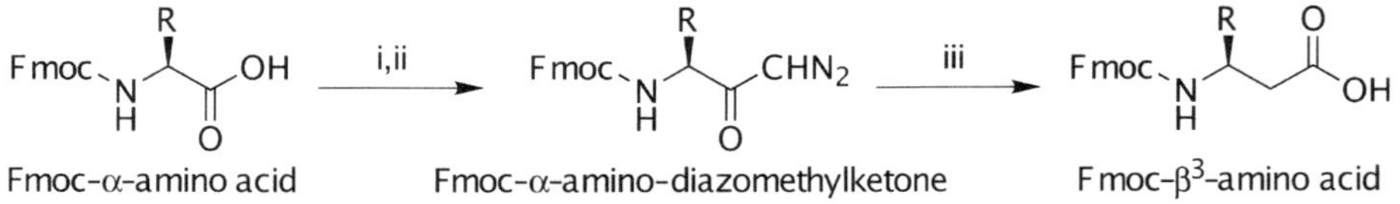

i. Ethylchloroformate, N-methylmorpholine, THF, 0°C.
ii. CH_2N_2/Et_2O, 0°C -> RT.
iii. CF_3COOAg, N-methylmorpholine, H_2O, THF, 40°C -> RT.

Fig. 3. Synthesis of Fmoc-β^3-amino acids by Arndt-Eistert homologation of Fmoc-α-amino acids.

thesized from commercially available Fmoc-α-amino acids *(29)* or Fmoc-α-amino pentafluorophenylesters *(33)* via Arndt-Eistert homologation (**Fig. 3**).

The procedures presented here involve converting the Fmoc-protected α-amino acid into the corresponding Fmoc-α-amino-diazomethylketone via a mixed anhydride *(29)* (*see* **Fig. 3**). Alternatively, the Fmoc-α-amino-diazomethylketones can be obtained by reacting the commercially available Fmoc-protected α-amino pentafluorophenyl ester with diazomethane etherate *(33)*. The Fmoc-α-amino-diazomethylketone is subsequently rearranged to the desired Fmoc-β^3-amino acid via Wolff rearrangement (*see* **Fig. 3**). Typical syntheses are carried out on a 5- to 15-mmol scale. Because diazomethane is highly toxic and explosive, procedures presented here involve the diazomethane etherate due to safety concerns, even though higher yields of β^3-amino acids were reported using methylene chloride solutions of diazomethane *(29,53)*. Diazomethane etherate can be generated from Diazald using the commercially available diazomethane generation kit following standard procedures provided by the vendor (**Fig. 4**). It is critical that polished glassware is used during generation of Fmoc-α-amino-diazomethylketones to avoid explosions and that all procedures be performed in a hood because of the toxicity of diazomethane.

3.1.1. Synthesis of Fmoc-α-Amino-Diazomethylketone

1. Dissolve commercially available N-α-fluorenylmethyloxycarbonyl protected α-amino acids (*see* **Note 1**) in tetrahydrofuran (THF) (3 mmol/mL) and cool to 0°C.
2. Add N-methylmorpholine (1 equivalent) and ethylchloroformate (1 equivalent) to the amino acid solution and stir under nitrogen for 1 h at 0°C to form the mixed anhydride product.
3. Filter the solution into a *smooth glass joint* round bottom flask (from kit) to remove any precipitated salts. *Do not add a stir bar or use glassware that is scratched.* Attach the flask to the receiving end of a *smooth glass joint* diazomethane generation setup (*see* **Fig. 4**; **Note 2**). Cool this flask to 0°C in an ice bath.

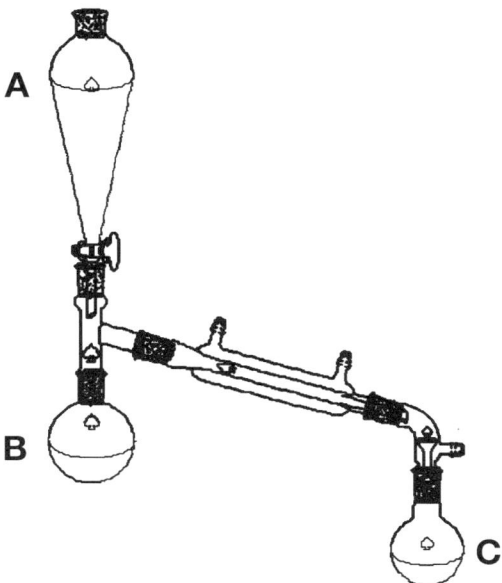

Fig. 4. Glassware setup for diazomethane generation. Separatory funnel (**A**) contains diazald dissolved in anhydrous ethyl ether. Diazomethane generation flask (**B**) contains KOH, H_2O, di(ethylene glycol) ethyl ether, and ethyl ether. Receiving flask (**C**) contains the THF solution of the mixed anhydride which is generated *in situ*.

4. Dissolve 8 Eq of KOH pellets in H_2O (1.5 mL/g KOH), 2-(2-ethoxyethoxy) ethanol (3 mL/g KOH) and ethyl ether (1 mL/g KOH) in a separate *smooth glass joint* round bottom flask (diazomethane generation flask). Attach this round bottom flask to the diazomethane generation set up (*see* **Fig. 4**; **Note 3**). Heat the diazomethane generation flask to 65°C in a water bath.

5. Add dropwise a 0.5 *M* ethereal solution of diazald (4 Eq) to the KOH solution to generate the diazomethane etherate. Add ethyl ether dropwise to the reaction flask until the distillate turns from yellow to clear.

6. Cap the receiving flask with a rubber septum, and leave it to react overnight. Allow to equilibriate from 0°C to room temperature. *Do not stir the solution.*

7. After confirming the absence of starting material by thin layer chromatography, quench residual CH_2N_2 by adding dropwise acetic acid (5.25 Eq).

8. Concentrate the yellow Fmoc-α-amino-diazomethylketone solution under reduced pressure and re-suspend the resulting oil in 100 to 150 mL of ethyl ether.

9. Wash the ether solution with saturated $NaHCO_{3(aq)}$, 1 *N* HCl, and H_2O. Dry the ether solution over Na_2SO_4 and concentrate the ether solution under reduced pressure.

10. Purify the product by flash chromatography using hexanes/ethyl acetate mixtures as the eluent.

3.1.2. Synthesis of Fmoc-β³-Amino Acids via Wolff Rearrangement

1. Dissolve Fmoc-α-amino-diazomethylketones in tetrahydrofuran and H$_2$O under nitrogen in a round bottom flask. Wrap the flask in aluminum foil to exclude light. Cool the solution to –40°C in a dry ice/acetone bath with magnetic stirring.
2. In the dark, suspend CF$_3$COOAg (0.1 Eq) in N-methylmorpholine (2.5 Eq) and add the suspension to the Fmoc-α-amino-diazomethylketone solution under nitrogen.
3. Maintain the bath temperature at –40°C for 3 h. Then, allow the reaction to proceed for a total of 8 to 12 h in the dark as the bath equilibrates to room temperature.
4. Concentrate the reaction mixture under reduced pressure and dissolve the resulting oil in 75 to 150 mL dichloromethane.
5. Wash the dissolved Fmoc-β³-amino acids with 0.2 *M* citric acid, saturated NaCl$_{(aq)}$, and H$_2$O. Dry the organic phase over Na$_2$SO$_4$. Concentrate under reduced pressure.
6. Purify the product by flash chromatography using chloroform/methanol mixtures as the eluent.

3.2. β-Peptide Synthesis Using Solid-Phase Methods by Fmoc-Based Chemistry

Synthesis of β-peptides longer than 6 residues have not been straightforward because of the difficulty of removing the Fmoc-protecting group at the fifth through seventh residues (*44,54*). However, the difficulty in Fmoc removal at these residues appears to be sequence dependent (unpublished results). Several modifications to the standard protocol have been developed to overcome this apparent difficulty including alternative deblock procedures involving both piperidine and DBU (*54*) and elevated temperatures (*44*). These modifications are not reflected in the procedures provided in this section, but may be implemented if necessary as described in the notes (*see* **Note 4**).

3.2.1. β-Peptide Synthesis

1. Swell Fmoc-protected resin in N,N-dimethylformamide (DMF) for 20 min with shaking.
2. Deprotect the resin by adding 20% piperidine solution in DMF (*see* **Note 4**). Wash the resin with DMF.
3. Dissolve Fmoc-protected β-amino acids (3 Eq), HOBt (3 Eq), and HBTU (3 Eq) in DMF to a concentration of 0.05 to 0.1 *M* Fmoc-β³-amino acid. Add diisopropylethylamine (8 Eq; *see* **Note 5**) to the solution and add the solution to the resin. For the first residue, shake the reaction for 8 h; couple subsequent residues for 90 to 120 min. Drain the solution from the resin and wash the resin with DMF.
4. Repeat **steps 2–3** for each residue (*see* **Note 6**).
5. Lyophilize the resin to dryness.

3.2.2. Cleavage of β-Peptide From Resin

1. Cleave the peptide from the resin by adding TIPS (0.5 mL/g lyophilized resin) and TFA (10 mL/g lyophilized resin) (*see* **Note 6**). Run reaction for 2 h with shaking.
2. Separate resin and β-peptide by filtration.
3. Gently blow nitrogen over the filtrate to evaporate TFA to yield an oily product.
4. Triturate the product with hexanes, and dissolve the sample in water.
5. Freeze samples and lyophilize to dryness.
6. β-Peptides can be purified by reverse-phase, high-performance liquid chromatography.

4. Notes

1. If Fmoc-α-amino pentafluorophenyl esters are used to synthesize Fmoc-β³-amino acids, begin at **step 3**. Dissolve the Fmoc-α-amino pentafluorophenyl esters in THF in a *smooth glass joint* round bottom flask and attach to the diazomethane generation setup as the receiving flask and cool to 0°C. The solubility of these esters may differ from carboxylic acids, and therefore the amount of THF necessary to dissolve the esters will be different.
2. Arndt-Eistert homologations use *toxic* and *explosive* diazomethane. All synthetic steps involving diazomethane should be carried out using polished glassware in the hood behind a shield. *Never* use glassware with ground glass joints or scratches. *Never* use magnetic stir bars during the reaction of diazomethane with mixed anhydrides.
3. Increased yields of β³-amino acids have been obtained by using methylene chloride solutions of diazomethane *(29,53)*. However, diazomethane etherates can be generated following standard procedures using the commercially available diazomethane generation kit.
4. Increased yields for solid phase β-peptide synthesis have been reported using alternative deblocking procedures (1:1:48 DBU/piperidine/DMF) *(54)* or elevated temperatures during deblocking (60°C) *(44)*.
5. Although the combination of HOBt and HBTU is effective for coupling Fmoc-amino acids, other reagents may also be used such as 1-hydroxy-7-azabenzotriazole and *O*-(7-azabenzotriazol-1-yl)-1,1,3,3-bis(tetramethylene)uronium hexafluorophosphate *(51,55,56)*.
6. The most suitable scavenger combination to cleave a β-peptide from resin is dependent on the β-peptide sequence and the nature of the resin linker. Sequences containing hCys, hTrp, and hArg may require scavengers other than TIPS *(51)*.
7. After all residues have been coupled to the resin, removal of the Fmoc-protecting group and N-terminal capping may be desirable depending on the design.

References

1. Bemis, G. W. and Murcko, M. A. (1996) The properties of known drugs. *J. Med. Chem.* **39,** 2887–2893.

2. Bemis, G. W. and Murcko, M. A. (1999) Properties of known drugs. 2. Sidechains. *J. Med. Chem.* **42,** 5095–5099.

3. Falciani, C., Lozzi, L., Pini, A., and Bracci, L. (2005) Bioactive peptides from libraries. *Chem. Biol.* **12,** 417–426.

4. Seebach, D. and Matthews, J. L. (1997) β-Peptides: a surprise at every turn. *Chem. Commun.* 2015–2022.

5. Gellman, S. H. (1998) Foldamers: a manifesto. *Acc. Chem. Res.* **31,** 173–180.

6. Cheng, R. P., Gellman, S. H., and DeGrado, W. F. (2001) β-peptides: from structure to function. *Chem. Rev.* **101,** 3219–3232.

7. Appella, D. H., Christianson, L. A., Karle, I. L., Powell, D. R., and Gellman, S. H. (1996) β-Peptide foldamers: robust helix formation in a new family of β-amino acid oligomers. *J. Am. Chem. Soc.* **118,** 13,071–13,072.

8 Seebach, D., Ciceri, P. E., Overhand, M., Juan, B., Rigo, D., Oberer, L., et al. (1996) Probing the helical secondary structure of short-chain β-peptides. *Helv. Chim. Acta* **79,** 2043–2066.

9. Daura, K. X. G., Schaefer, H., Juan, B., Seebach, D., and van Gunsteren, W. F. (2001) The β-peptide hairpin in solution: conformational study of a β-hexapeptide in methanol by NMR spectroscopy and MD simulation. *J. Am. Chem. Soc.* **123,** 2393–2404.

10. Langenhan, J. M., Guzei, I. A., and Gellman, S. H. (2003) Parallel sheet secondary structure in β-peptides. *Angew. Chem. Int. Ed. Engl.* **42,** 2402–2405.

11. Wiegand, H., Wirz, B., Schweitzer, A., Camenisch, G. P., Perez, M. I. R., Gross, G., et al. (2002) The outstanding metabolic stability of a C_{14}-labeled β-nonapeptide in rats—in vitro and in vivo pharmacokinetic studies. *Biopharm. Drug Dispos.* **23,** 251–262.

12. Frackenpohl, J., Arvidsson, P. I., Schreiber, J. V., and Seebach, D. (2001) The outstanding biological stability of β- and γ-peptides toward proteolytic enzymes: an in vitro investigation with fifteen peptidases. *ChemBiochem* **2,** 445–455.

13. Seebach, D., Abele, S., Schreiber, J. V., Martinoni, B., Nussbaum, A. K., Schild, H., et al. (1998) Biological and pharmacokinetic studies with β-peptides. *Chimia* **52,** 734–739.

14. Rueping, M., Mahajan, Y., Sauer, M., and Seebach, D. (2002) Cellular uptake studies with β-peptides. *ChemBiochem* **3,** 257–259.

15. Umezawa, N., Gelman, M. A., Haigis, M. C., Raines, R. T., and Gellman, S. H. (2002) Translocation of a β-peptide across cell membranes. *J. Am. Chem. Soc.* **124,** 368–369.

16. Potocky, T. B., Menon, A. K., and Gellman, S. H. (2003) Cytoplasmic and nuclear delivery of a TAT-derived peptide and a β-peptide after endocytic uptake into HeLa cells. *J. Biol. Chem.* **278,** 50,188–50,194.

17. Potocky, T. B., Menon, A. K., and Gellman, S. H. (2005) Effects of conformational stability and geometry of guanidinium display on cell entry by β-peptides. *J. Am. Chem. Soc.* **127,** 3686–3687.

18. Hintermann, T. and Seebach, D. (1997) The biological stability of β-peptides: no interactions between α- and β-peptidic structures. *Chimia* **51,** 244–247.

19. Schreiber, J. V., Frackenpohl, J., Moser, F., Fleischmann, T., Kohler, H. P. E., and Seebach, D. (2002) On the biodegradation of β-peptides. *ChemBiochem* **3**, 424–432.
20. Fauchere, J. L. and Thurieau, C. (1992) Evaluation of the stability of peptides and pseudopeptides as a tool in peptide drug design. *Adv. Drug Res.* **23**, 127–159.
21. Appella, D. H., Christianson, L. A., Klein, D. A., Powell, D. R., Huang, X. L., Barchi, J. J., et al. (1997) Residue-based control of helix shape in β-peptide oligomers. *Nature* **387**, 381–384.
22. Seebach, D., Overhand, M., Kuhnle, F. N. M., Martinoni, B., Oberer, L., Hommel, U., et al. (1996) β-Peptides: synthesis by Arndt-Eistert homologation with concomitant peptide coupling. Structure determination by NMR and CD spectroscopy and by x-ray crystallography. Helical secondary structure of a β-hexapeptide in solution and its stability towards pepsin. *Helv. Chim. Acta* **79**, 913–941.
23. Cheng, R. P. and DeGrado, W. F. (2001) De novo design of a monomeric helical β-peptide stabilized by electrostatic interactions. *J. Am. Chem. Soc.* **123**, 5162–5163.
24. Arvidsson, P. I., Reuping, M., and Seebach, D. (2001) Design, machine synthesis, and NMR solution structure of a β-heptapeptide forming a salt bridge stabilised 3_{14}-helix in methanol and in water. *Chem. Commun.* 649–650.
25. Raguse, T. L., Lai, J. R., and Gellman, S. H. (2002) Evidence that the β-peptide 14-helix is stabilized by $β^3$-residues with side-chain branching adjacent to the β-carbon atom. *Helv. Chim. Acta* **85**, 4154–4164.
26. Kritzer, J. A., Tirado-Rives, J., Hart, S. A., Lear, J. D., Jorgensen, W. L., and Schepartz, A. (2005) Relationship between side chain structure and 14-helix stability of $β^3$-peptides in water. *J. Am. Chem. Soc.* **127**, 167–178.
27. Hart, S. A., Bahadoor, A. B. F., Matthews, E. E., Qiu, X. Y. J., and Schepartz, A. (2003) Helix macrodipole control of $β^3$-peptide 14-helix stability in water. *J. Am. Chem. Soc.* **125**, 4022–4023.
28. Armstrong, K. M. and Baldwin, R. L. (1993) Charged histidine affects α-helix stability at all positions in the helix by interacting with the backbone charges. *Proc. Natl. Acad. Sci. USA* **90**, 11,337–11,340.
29. Guichard, G., Abele, S., and Seebach, D. (1998) Preparation of N-Fmoc-protected $β^2$- and $β^3$-amino acids and their use as building blocks for the solid phase synthesis of β-peptides. *Helv. Chim. Acta* **81**, 187–206.
30. Juaristi, E. (ed.). (1997) *Enantioselective Synthesis of β-Amino Acids*. Wiley-VCH, New York.
31. Lelais, G. and Seebach, D. (2004) $β^2$-Amino acids—syntheses, occurrence in natural products, and components of β-peptides. *Biopolymers* **76**, 206–243.
32. Juaristi, E., Escalante, J., Lamatsch, B., and Seebach, D. (1992) Enantioselective synthesis of β-amino acids. 2. preparation of the *like* stereoisomers of 2-methyl- and 2-benzyl-3-aminobutanoic acid. *J. Org. Chem.* **57**, 2396–2398.
33. Babu, V. V., Gopi, H. N., and Ananda, K. (1999) Homologation of α-amino acids to β-amino acids using Fmoc-amino acid pentafluorophenyl esters. *J. Pept. Res.* **53**, 308–313.

34. Seebach, D., Rueping, M., Arvidsson, P., Kimmerlin, T., Micuch, P., Noti, C., et al. (2001) Linear, peptidase-resistant β^2/β^3-di- and α/β^3-tetrapeptide derivatives with nanomolar affinities to a human somatostatin receptor—preliminary communication. *Helv. Chim. Acta* **84,** 3503–3510.

35. Gademann, K., Ernst, M., Hoyer, D., and Seebach, D. (1999) Synthesis and biological evaluation of a cyclo-β-tetrapeptide as a somatostatin analogue. *Angew. Chem. Int. Ed. Engl.* **38,** 1223–1226.

36. Nunn, C., Rueping, M., Langenegger, D., Schuepbach, E., Kimmerlin, T., Micuch, P., et al. (2003) β^2/β^3-di- and α/β^3-tetrapeptide derivatives as potent agonists at somatostatin sst_4 receptors. *Naunyn. Schmiedebergs Arch. Pharmacol.* **367,** 95–103.

37. Gademann, K., Kimmerlin, T., Ernst, M., Seebach, D., and Hoyer, D. (2000) The cyclo-β-tetrapeptide (β-HPhe-β-HThr-β-HLys-β-HTrp): synthesis, NMR structure in methanol solution, and affinity for human somatostatin receptors. *Helv. Chim. Acta* **83,** 16–33.

38. Gademann, K., Kimmerlin, T., Hoyer, D., and Seebach, D. (2001) Peptide folding induces high and selective affinity of a linear and small β-peptide to the human somatostatin receptor 4. *J. Med. Chem.* **44,** 2460–2468.

39. Gelman, M. A., Richter, S., Cao, H., Umezawa, N., Gellman, S. H., and Rana, T. M. (2003) Selective binding of TAR RNA by a tat-derived β-peptide. *Org. Lett.* **5,** 3563–3565.

40. Epand, R. F., Raguse, T. L., Gellman, S. H., and Epand, R. M. (2004) Antimicrobial 14-helical β-peptides: potent bilayer disrupting agents. *Biochemistry* **43,** 9527–35.

41. Epand, R. F., Umezawa, N., Porter, E. A., Gellman, S. H., and Epand, R. M. (2003) Interactions of the antimicrobial β-peptide β-17 with phospholipid vesicles differ from membrane interactions of magainins. *Eur. J. Biochem.* **270,** 1240–1248.

42. Liu, D. H. and DeGrado, W. F. (2001) *De novo* design, synthesis, and characterization of antimicrobial β-peptides. *J. Am. Chem. Soc.* **123,** 7553–7559.

43. Porter, E. A., Weisblum, B., and Gellman, S. H. (2002) Mimicry of host-defense peptides by unnatural oligomers: antimicrobial β-peptides. *J. Am. Chem. Soc.* **124,** 7324–7330.

44. Raguse, T. L., Porter, E. A., Weisblum, B., and Gellman, S. H. (2002) Structure-activity studies of 14-helical antimicrobial β-peptides: probing the relationship between conformational stability and antimicrobial potency. *J. Am. Chem. Soc.* **124,** 12,774–12,785.

45. Arvidsson, P. I., Ryder, N. S., Weiss, H. M., Gross, G., Kretz, O., Woessner, R., et al. (2003) Antibiotic and hemolytic activity of a β^2/β^3 peptide capable of folding into a 12/10-helical secondary structure. *ChemBiochem* **4,** 1345–1347.

46. Arvidsson, P. I., Ryder, N. S., Weiss, H. M., Hook, D. F., Escalente, J., and Seebach, D. (2005) Exploring the antibacterial and hemolytic activity of shorter- and longer-chain β, α,β, and γ-peptides, and of β-peptides from β^2-3-Aza- and β^3-2-methylidene-amino acids bearing proteinogenic side chains—a survey. *Chem. Biodiv.* **2,** 401–420.

47. Hamuro, Y., Schneider, J. P., and DeGrado, W. F. (1999) *De novo* design of antibacterial β-peptides. *J. Am. Chem. Soc.* **121,** 12,200–12,201.

48. Werder, M., Hauser, H., Abele, S., and Seebach, D. (1999) β-Peptides as inhibitors of small-intestinal cholesterol and fat absorption. *Helv. Chim. Acta* **82,** 1774–1783.

49. Kritzer, J. A., Lear, J. D., Hodsdon, M. E., and Schepartz, A. (2004) Helical β-peptide inhibitors of the p53-hDM2 interaction. *J. Am. Chem. Soc.* **126,** 9468–9469.

50. Kritzer, J. A., Stephens, O. M., Guarracino, D. A., Reznik, S. K., and Schepartz, A. (2005) β-peptides as inhibitors of protein-protein interactions. *Bioorg. Med. Chem.* **13,** 11–16.

51. White, P., Dorner, B., and Steinauer, R. (eds.) Synthesis notes, in Novabiochem 2004–2005 catalog, pp. 1.1–6.4.

52. Fields, G. B. and Noble, R. L. (1990) Solid-phase peptide synthesis utilizing 9-flurorenylmethoxycarbonyl amino acids. *Int. J. Pept. Protein Res.* **35,** 151-214.

53. Leggio, A., Liguori, A., Procopio, A., and Sindona, G. (1997) Convenient and stereospecific homologation of N-fluorenylmethoxycarbonyl-α-amino acids to their β-homologs. *J. Chem. Soc., Perkin Trans.* **13,** 1969–1971.

54. Schreiber, J. V. and Seebach, D. (2000) Solid-phase synthesis of a β-dodecapeptide with seven functionalized side chains and CD spectroscopic evidence for a dramatic structual switch when going from water to methanol solution. *Helv. Chim. Acta* **83,** 3139–3152.

55. Carpino, L. A. (1993) 1-Hydroxy-7-azabenzotriazole. An efficient peptide coupling additive. *J. Am. Chem. Soc.* **115,** 4397–4398.

56. Carpino, L. A., El-Faham, A., Minor, C. A., and Albericio, F. (1994) Advantageous applications of azabenzotriazole (triazolopyridine)-based coupling reagents to solid phase peptide synthesis. *J. Chem. Soc. Chem. Commun.* 201–203.

II

DESIGN OF ENTIRE PROTEINS AND PROTEIN COMPLEXES

6

Design of Miniproteins by the Transfer of Active Sites Onto Small-Size Scaffolds

François Stricher, Loïc Martin, and Claudio Vita

Summary

Natural miniproteins (e.g., animal toxins, protease inhibitors, defensins) can express specific and powerful biological activities by using a stable and minimal (<80 amino acids) structural motif. Artificial activities have been designed on these miniscaffolds by transferring previously identified protein active sites into regions structurally compatible with the site and permissive for sequence mutations. These newly designed miniproteins, presenting a specific and high activity within a small size and well-defined three-dimensional structure, represent novel tools in biology, biotechnology, and medical sciences, and are also useful intermediates to develop new therapeutic agents. The different steps used to design and characterize new bioactive miniproteins are here described in detail. Two successful examples are here reported. The first one is a metal-binding miniprotein (MBP, 37 residues), which possesses a metal specificity resembling that of natural carbonic anhydrase; the second is a CD4 mimic (CD4M33, 27 residues), which is a powerful inhibitor of HIV-1 entry but also a fully functional substitute of the human receptor CD4 and, hence, a potential component of an AIDS vaccine.

Key Words: Natural scaffold; protein structure; active site transfer; protein–protein interaction; hot spot; disulfide bonds; scorpion toxin; ^{1}H-NMR; crystallography; peptide synthesis; drug design; metal binding; HIV inhibition; AIDS vaccine.

1. Introduction

The ultimate goal of protein design is to obtain a protein with a desired structure and function. From the start of protein design, new proteins have been obtained by two different and complementary approaches: "rational" *de novo* design (*1–3*) and "irrational" combinatorial and evolution methods (*4,5*). The first approach mainly focuses on the design of a protein structure or fold and a function is then subsequently incorporated, whereas the second focuses on a specific function, for which minimal knowledge of the protein structure is

From: *Methods in Molecular Biology, vol. 340: Protein Design: Methods and Applications*
Edited by: R. Guerois and M. López de la Paz © Humana Press Inc., Totowa, NJ

required. A third method has recently met considerable success and is based on the grafting of functional sites on natural scaffolds (*6,7*). In this case, the protein designer, by imitating nature, which uses common scaffolds to obtain proteins with various functions, uses a natural structure as a scaffold and designs a new function by the transfer of an active site on a specific structural region, which is permissive for sequence mutations and structurally compatible with the new function. Although the *de novo* design of an overall scaffold and active site may represent an opportunity to test our knowledge on protein structure and function and relation between them, the design of an active site on a natural scaffold may provide more opportunities to engineer new functional proteins of practical utility in biology, biotechnology, and biomedicine (*6,7*). Furthermore, this approach is not expected to lead to molten globule-like structures, which was the drawback in early *de novo* designs (*8*), but, because of the high stability and extraordinary tolerance for residue substitution of many natural scaffolds, it should directly lead to a specific function on a scaffold that has already been proved by the natural evolution.

Proteins of small size, miniproteins, containing 30 to 80 amino acid residues are particularly attractive scaffolds for protein design. They possess a well-defined 3D structure and biological activity and are usually stabilized by structural constraints like disulfide bridges or metal ions. Examples of natural miniproteins are: animal toxins (from conus shells, scorpions, snakes), protease inhibitors, knottins, defensins, cellulose binding domains, pheromones, zinc fingers, Leu zippers, and epidermal growth factor modules (*see* **Note 1**). All these natural miniproteins, though possessing different structural motifs, represent examples of successful solutions adopted by nature to express specific activities in different biological systems of a small size and within a stable structure. Natural constraints (disulfide bonds or metal ions), together with intermolecular noncovalent forces, effectively stabilize such small architectures from the interior, leaving most of the exterior amino acid side chains accessible to the solvent and to the binding to their biological partners. This structural organization may explain why these small structures are conformationally very stable and also compatible with rather different sequences and biological activities. The structural stability and functional versatility of these miniprotein systems suggest that new tailored artificial functions can be engineered in such small architectures.

The advantage of using small presentation scaffolds in the engineering of novel functional miniproteins is manifold. First, the reproduction of a biological activity in such small systems also corresponds to a significant minimization and simplification of a more complex natural system. Second, structural information can be obtained by nuclear magnetic resonance (NMR) (or crystallography) much more easily because of the small size and the stability of the miniproteins used. Third, engineered miniproteins are expected to be more resistant to proteolytic degradation in different biological applications because of their reduced flexibility. Fourth, engineered miniproteins can be also produced by chemical synthesis easily and rapidly, thus allowing the introduction of useful chemical probes, such as unnatural or isotopically labeled amino acids, photoactivable groups, appropriate radioactive labels, polymers, carbohydrates, fatty acids, cofactors, fluorescent tags, FRET dye pairs, affinity tags, and other chemical moieties, which can be incorporated in specific structural positions, not disturbing the function of the engineered miniprotein. Finally, as a consequence of its higher biological stability and chosen scaffold, the miniprotein also can be obtained in large amount by low cost biological expression systems for industrial applications.

In this chapter, we will describe two examples of the design of miniproteins obtained by the transfer of a well-defined protein active site onto stable, permissive, and structurally compatible miniprotein scaffolds. We will first describe the strategy and methods used in their design, then will review the structural and functional properties of these artificial miniproteins.

2. Materials

2.1. Design Strategy

The design strategy can be decomposed in the following fundamental steps.

2.1.1. Choice of the Active Site

The choice of the target site depends on the specific interest of the protein designer and is usually defined by the availability of the functional test to be used to characterize the newly designed miniprotein. The exact definition of a protein active site and the role played by each residue are also important because this serves to exactly define the size of the structure that one wants to transfer and reproduce on the scaffold. This is usually performed by extensive mutagenesis and structure-function relationship studies. The "alanine scanning" method is the most widely used approach to identify a protein active site or the "hot spot" (*9*) of a protein–protein interaction surface, which is the site defined by the residues that contribute the most to the binding energy. However, the exact definition of an active site or "hot spot" is not always unambiguous and may vary from one study to another, most often depending on the functional test used. Thus, some degree of subjectivity may be included in the definition of the hot spot or active site.

2.1.2. Selection of the Appropriate Miniprotein Scaffold

This step serves to identify an appropriate miniprotein scaffold possessing a backbone motif structurally similar to the target active site (*see* **Note 2**). The miniprotein structure to be selected as a scaffold has to contain the following characteristics: (1) a 3D structure known to high resolution and the atomic coordinates available; (2) a region that is structurally similar to the target active site, as defined by the superposition of the host and guest active site; (3) a stability and permissiveness for sequence mutations; and (4) obtainable in high yields by chemical synthesis or biological expression systems.

2.1.3. Transfer of the Active Site to the Selected Scaffold

After a miniprotein scaffold with a backbone structural region similar to that of the target active site has been found, this structure receives the appropriate residues of the active site. To do this, the active site is transferred on the scaffold region, by mutating the sequence of the scaffold site with that of the target protein site, using an appropriate molecular display system, for example Sybyl (Tripos). In doing this, care should be taken in not perturbing elements important for the correct folding of the scaffold. In practice, given the close similarity between the two backbone structures, this step simply corresponds to the transfer of the target side chains to a scaffold backbone very similar to the original one. Thus, chances are quite high that most of the transferred side chains maintain the conformation of the target site.

After that, a model of the mutated miniprotein is obtained by a simple and rapid energy minimization step (using for example the Tripos force field of Sybyl). Analysis of this model may be useful to verify if additional mutations are needed also in nearby or distant regions of the scaffold to adapt the inserted active site in the new scaffold. The exact structure to be synthesized is exactly defined at this moment.

2.1.4. Synthesis and Testing

The sequence defined in the previous step is synthesized here by either chemical synthesis or bacterial expression. If the stability and folding properties of the scaffold have not been perturbed because of the multiple mutations introduced, the chimeric miniprotein should be obtained in a pure and well-structured form in good yields. If the newly designed miniprotein contained disulfide bridges, peptide mapping may be required to ascertain the formation of correct cysteine pairings. Furthermore, to verify the correct folding of the designed miniprotein, a simple and quick conformational analysis by circular dichroism may be sufficient. More detailed ^{1}H-NMR (or crystallography) analysis may be conducted later.

At this step, the biological activity of the chimera has to be tested. Testing will be performed by appropriate spectroscopic techniques (absorption, fluorescence spectroscopy), enzyme activity, surface plasmon resonance (Biacore), ELISA or cell receptor binding. This will represent the real first check of the extent of success of the entire design process. However, because of the small size or possible lower biological activity of the first designed miniprotein, the conditions for testing may have to be somehow changed from those used in the native system.

2.1.5. Final Optimization

Functional assays may reveal that the designed miniprotein presents low activity. This occurs because the designed miniprotein does not possess the desired structure or that, in spite of the presence of a native-like structure, the potency is too low because the active site selected is not sufficient to express the expected activity.

In this case, a systematic mutational analysis, such as the "alanine scanning" approach, may be useful to evaluate the contribution of each residues of the active site to function.

Furthermore, because the purpose of the protein design is the reproduction of the structure and function of the target site, a structural analysis is here required to verify the degree of nativelikeness of the transferred site in the new miniprotein context. Given the small size of the scaffold used, structure determination in solution by ¹H-NMR may be quick and effective to ascertain this point; it may eventually provide useful information for additional modifications which may improve the nativelikeness of the designed miniprotein active site.

Ultimately, crystallization and structure resolution of the miniprotein in complex with its receptor may provide precious information not only on the extent of success of protein design but also precious data on how to further improve the design in a structure-based approach.

2.2. Experimental Procedures

2.2.1. Peptide Synthesis and Purification

Chemical synthesis by automated solid phase method represents a useful and an efficient alternative to biological expression of manipulated genes in producing miniproteins of about 30–80 residue in length for engineering studies. Indeed, this method allows the synthesis and purification of tens of milligrams of entirely new sequences in about 1 wk *(10)*.

We used an Applied Biosystems 433A Peptide Synthesizer and fluorenylmethyloxycarbonyl (Fmoc)-protected amino acids, providing monitoring of Fmoc-deprotection steps at each cycle with feedback on the deprotection and coupling time to optimize deprotection and coupling yields. Syntheses use polyethylene glycol-polystyrene resin (PAL-PEG-PS resin, Applied Biosystems) and single 2-(1-H-benzotriazol-1-yl)-1,1,3,3-tetramethyluronium hexafluorophosphate (HBTU) coupling (ABI FastMoc protocol on 0.1 mmol scale, with 10 molar excess of Fmoc amino acids derivatives suggested by the Synthesizer manufacturer), and systematic capping with acetic anhydride. Peptide was cleaved from the resin with simultaneous removal of side-chain protecting groups by treatment with reagent K' (81.5% trifluoroacetic acid, 5% water, 5% phenol, 5% thioanisole, 2.5% ethanedithiol, and 1% triisopropylsilane) for 2.5 h at room temperature.

Disulfide formation is a critical step and was performed by directly dissolving the crude reduced peptide (at 0.1 mg/mL) in 0.1 M Tris-HCl buffer, pH 7.8, in the presence of glutathione. Different ratios of reduced and oxidized glutathione were tested to find optimal folding conditions, because it is known that both the kinetics and yields of disulfide bond formation are affected by varying the oxidation potential. Folding in reducing conditions, 5 mM reduced and 0.5 mM oxidized glutathione, usually produced the simplest chromatographic profile and highest yields in the corrected folded material. The folded material was then purified directly, without prior purification of the reduced material (*see* **Note 3**), by loading a large solution (100 mg/L) on a semipreparative reverse-phase high-performance liquid chromatography (RP-HPLC) column, thus allowing concentration and efficient purification even from a large volume. Peptide homogeneity of eluted fractions was assessed by analytical HPLC, pure fractions were pooled then lyophilized, and identity of purified material was finally verified by amino acid analysis and electrospray mass spectrometry.

2.2.2. Metal Binding Miniprotein

Metal binding in the newly designed metal-binding miniprotein (MBP) was studied by exploiting the quenching of the single tryptophan of MBP *(11)*, produced by Cu^{2+} addition. Excitation was performed at 280 nm and emission was monitored at 352 nm. The protein was dissolved at 1 µM in 2.5 mL of 50 mM sodium acetate buffer, pH 6.5. For copper binding studies, $CuCl_2$ (10^{-4}–10^{-3} M) was added in 5–10 µL aliquots and fluorescence measured after 5 min from metal addition and mixing of the sample. For inhibition of copper quenching of Trp fluorescence, Cu^{2+} was added to the protein solution in the presence of varying concentration of the spectroscopically silent metals, Zn^{2+}, Cd^{2+}, Ni^{2+}, or Mn^{2+}. Data were corrected for background copper quenching by comparison with identical experiments conducted on charybdotoxin. Experimental data were fitted with both standard binding equations and Scatchard transformation. In the calculations, the ratio between the fluorescence quenching after each metal addition and the maximum fluorescence quenching was assumed proportional to the bound metal; free metal was obtained as the difference between the total metal added and the bound metal.

2.2.3. Structure Calculations

NMR experiments were collected on a 600 MHz spectrometer (Bruker AMX600) with 4 mM miniprotein (e.g., CD4M33) sample (pH 3.5, 25°C).

Standard pulse sequences and parameters were employed to record 2D homonuclear DQF-COSY, TOCSY (isotropic mixing period: 80 ms) and NOESY (mixing times: 80 and 200 ms) spectra. The standard strategy was used for assignment *(12)*. Residual overlaps were resolved from analogous spectra recorded at 15°C and 35°C. The data sets were processed with the GIFA software *(13)*. NOE intensities used for the structure calculations were obtained from the NOESY spectra recorded with a 80 ms mixing time on the fully protonated sample. NH-C$^{\alpha}$H and C$^{\alpha}$H-C$^{\beta}$H coupling constants used to define angular restraints were measured in gradient enhanced DBF-COSY *(14)* and DQF-COSY spectra, respectively.

Collected NOEs (about 300) were partitioned into three categories of intensities that were converted into distance upper limits of 2.8 Å, 3.6 Å, and 4.8 Å. Additionally, some cross-peaks present only in 200 ms 2D NOESY maps were converted to distance constraints of 5.5 Å. When no stereospecific assignment was possible, pseudo-atoms were defined and corrections added as described by Wüthrich et al. *(12)*. Angular restraints (about 27–31; ϕ and χ_1), obtained from J coupling constants analysis, were added, in addition to the usual distance restraints used to enforce the three disulfide bridges.

Final structure calculations using the torsion angle dynamics protocol of DYANA *(15)* were started from 999 randomized conformers. The 20–30 conformers with the lowest final target function value (<0.02 Å^2) were retained and energy-minimized with the AMBER 5 program *(16)*, using the AMBER all-atom force field with parameters from Cornell et al. *(17)*, including improved torsional angle parameters. All calculations were performed on a Silicon Graphics Origin 200 workstation.

2.2.4. Structure Analysis and Modeling With Sybyl

Structures were displayed, analyzed, and compared on a Silicon Graphics Octane 2 station using the Sybyl software (Tripos Associates, Inc.). Mimic models were obtained by Sybyl software, using the atomic coordinates of the selected scaffold, introducing the mutations envisaged, and finally energy minimizing the resulted model by Tripos force field using the minimize routine of Sybyl, and the following parameters: for compute-minimize: Powell method, 500 interactions; for energy setup: Tripos force field, Gasteiger-Marsili charges, distance dielectric function, one dielectric constant, no constraints, 8 Å cutoff.

Molecular graphics were produced using the open-source molecular visualization software MolMol and PyMol.

2.2.5. Molecular Modeling of the Complex gp120:CD4M9

The structure of the complex gp120:CD4M9 was determined by molecular modeling. The crystallographic structure of the complex gp120$_{HXB2}$:CD4:17b

(PDB code 1g9m) was used as a template to dock the CD4M9 mimic structure on gp120, after 17 antibody was removed. All molecular mechanics and molecular dynamic calculations were performed with the AMBER 5 package *(16)*. The CD4M9 NMR structure with the lowest energy was used to replace the CD4 on the gp120:CD4 complex structure by superimposing the backbone atoms of CD4M9 (residues 19–28) to those of CD4 (residues 39–48). A total of 5000 cycles of energy minimization were carried out on all CD4M9 side chains and on the gp120 side chains at a maximum of 30 Å around CD4M9 mass center to reduce any bad contacts. A 25 ps long simulated annealing procedure followed, in which the temperature was raised to 700K for 15 ps and gradually lowered to 300K. During this stage, the relative weights for the nonbonded energy terms and for the torsion energy terms were gradually increased from 0.1 to 1.0 to allow side-chain rearrangement. Finally, the full miniprotein and gp120 side chains were minimized for 5000 cycles. This procedure was repeated 10 times, and the complex with the minimal energy was retained.

2.2.6. ELISA

Binding to gp120 was measured by a competition of CD4 binding to gp120 in an ELISA in 96-well plates (Maxisorb, Nunc, Denmark), by using two different formats: the CD4-coated method and the gp120 captured method, as originally described by J. Moore *(18)*.

2.2.6.1. CD4-COATED METHOD

A total of 250 ng/well of recombinant human CD4 (Progenics) was coated overnight at 4°C in 96-well plates; 80 ng/well of gp120 from HIV-1LAI (Progenics) or other isolates, were then added, followed by addition of different concentrations of soluble competitors (CD4, peptides, miniproteins), anti-gp120 NEA mAb (NEN-DuPont), a goat anti-mouse peroxidase-conjugated antibody (Jackson ImmunoResearch), and the 3,3',5,5'-tetramethylbenzidine substrate (Sigma) for revelation. After acidification with 25 µL of 1 M H_2SO_4, inhibition of binding was calculated from the 450 nm OD using the formula: % inhibition = $100 \times (OD_{gp120} - OD_{gp120+comp})/OD_{gp120}$. Results are means of duplicate experiments repeated twice.

2.2.6.2. GP120 CAPTURED METHOD

A total of 50 ng/well of the D7324 antibody (Aalto Bio Reagents), which is specific for the C-terminal sequence of gp120, was coated overnight at 4°C in 96-well plates, and 15 ng/well of gp120 from HIV-1LAI, JRFL (Progenics), or other isolates were then added, followed by the addition of 1 ng/well four-domain soluble CD4 and different concentrations of soluble competitors, the anti-CD4 monoclonal antibody L120.3 (specific for CD4 domain 3 and 4,

CFAR, NIBSC, UK - EU Programme EVA/MRC), a goat anti-mouse peroxidase-conjugated monoclonal antibody (Jackson ImmunoResearch), and the 3,3',5,5'-tetramethylbenzidine substrate (Sigma) for revelation (at 450 nm). The following formula was used to calculate percent inhibition of CD4 binding: $100 \times (O.D._{450\ nm} - O.D._{450\ nm} + competitor)/O.D._{450\ nm}$. Results are means of duplicate experiments repeated twice.

2.2.7. Surface Plasmon Resonance Biosensor Analysis

Experiments were performed at 25°C with 20 μL/min flow rate in HBS (HEPES-buffer-saline, 3 mM EDTA, 0.05% Biacore surfactant, pH 7.4) with a BIACORE 3000 instrument (Biacore AB). For direct binding of gp120 glycoproteins to CD4 miniprotein, a B1 sensor chip, precoated with 700 RU of streptavidin, was used to immobilize biotinylated-CD4M33 (50 RU). For binding of gp120 to 48d, 17b, and X5 monoclonal antibodies, they were immobilized at approx 1500 RU by the amine coupling kit, provided by the manufacturer. The gp120$_{HXB2}$ (125 nM) and CD4 or CD4 miniproteins (375 nM) were premixed at 37°C for 60 min before injection at 20 μL/min. All the sensorgrams were corrected by subtracting the signal from reference flow cell and, in the case of CD4, normalized to the miniprotein mass. Kinetics were recorded with different flow-rate to show the absence of mass transport and rebinding effects. The association and dissociation phase data were fitted to a Langmuir 1:1 global fitting model.

2.2.8. HIV-1 Infection Assays

A one-round infection assay was used with HeLa P4 or HeLa-P4 CCR5 target cells *(19)* that constitutively express CD4, coreceptors, and β-galactosidase under the control of HIV-1-LTR. Five hundred TCID$_{50}$ of virus were mixed with CD4-based molecules for 1 h at 25°C in the absence of serum. Subsequently, 2.5×10^4 target cells were incubated for an additional hour with the CD4-virus complexes, followed by extensive washes. Then cells were incubated in culture DMEM medium, 10% fetal calf serum (Gibco-BRL) in 5% CO_2 atmosphere at 37°C for 24 h, and β-galactosidase activity was measured by chemiluminescent assay according to the manufacturer's instructions (Roche Molecular Biochemicals). Experiments were performed in triplicate and repeated twice. The 50% inhibitory concentration (IC$_{50}$) was calculated by GraphPad3.

The acute infection assay was performed using the PM1 cell line *(20)*, an immortalized CD4$^+$CCR5$^+$CXCR4$^+$ T-cell clone susceptible to a wide variety of HIV-1 strains. The cells were seeded at 2.5×10^4 cells per well into flat-bottom 96-well plates in 200 μL of complete RPMI medium and then exposed to the viral stocks (~50 TCID/well), pretreated or not with serial dilutions of

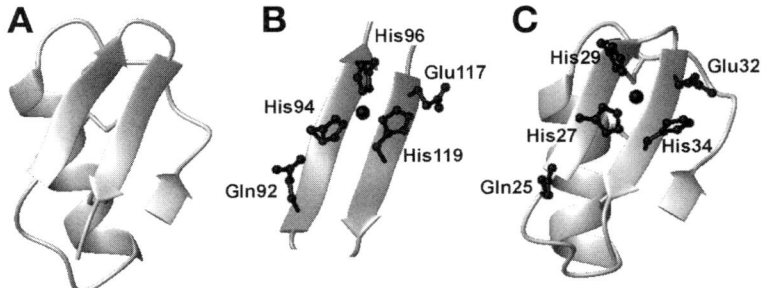

Fig. 1. Design of the metal-binding miniprotein (MBP). Three-dimensional structure of (**A**) the charybdotoxin scaffold (2crd), (**B**) the metal-binding site of carbonic anhydrase B (2cab), and (**C**) the designed MBP. Residues involved in metal binding are in sticks and balls.

CD4-based molecules. After overnight incubation at 37°C, the cells were washed twice and recultured in complete RPMI medium supplemented with the appropriate inhibitors. After 3, 5, and 7 d, the culture supernatant was removed for p24 antigen testing and replaced with fresh medium containing the appropriate inhibitors. The extracellular p24 concentrations were measured by ELISA using commercial antibodies (Aalto Bio Reagents).

3. Methods

In this section we will describe two examples of miniprotein design. The first example reports on the design of a metal-binding miniprotein that expresses metal specificity for different metals and has been characterized using different spectroscopic techniques to verify its structure and function. The second reports on the design of a mimic of human CD4, which was ultimately optimized to a native-like activity by two different steps, the first used to increase its nativelikeness (after the miniprotein 3D structure was solved by solution ^{1}H-NMR) and the second used to increase the fitness of the miniprotein active surface for the receptor binding surface (after the structure of the mimic:receptor complex was modeled).

3.1. Design of a Metal-Binding Site

3.1.1. Choice of the Scaffold

The feasibility of designing a functional miniprotein by the transfer of an appropriate active site on a miniprotein scaffold was first tested on the structure of charybdotoxin, a 37 amino acid toxin from *Leiurus quinquestriatus hebraeus* scorpion, whose 3D structure *(21)* (**Fig. 1A**), function *(22)*, and folding properties *(23)* have been well-characterized. This small toxin contains an antiparallel β-sheet which is linked in the interior core to an α-helix and an extended segment by three disulfide bridges, leaving the solvent exposed face of the β-sheet available to protein design (*see* **Note 4**).

3.1.2. Choice of the Active Site

Although the forces that stabilize a metal-protein interaction are not yet fully understood and cannot be treated rigorously yet by ordinary force field algorithms, the structures of metal-binding proteins are numerous in the PDB, metal-binding sites have been described in several reviews *(24,25)* and metal-binding site engineered. Indeed, the introduction of a metal-binding activity in proteins of known structure *(26,27)* and in *de novo* design structures *(28,29)* has been a classical exercise in protein engineering. An advantage of designing a site binding heavy metal ions may reside in the existence of absorption spectroscopy techniques that can be effectively used to characterize the binding affinity and structure of the metal coordination sphere, without the need to determine the 3D structure of the designed miniprotein.

We chose to design a metal-binding site mimicking the Zn^{2+} binding site of carbonic anhydrase B (2cab) which includes three histidine residues—His 94, His 96, His 119—on the two antiparallel strands, 92–96 and 117–121, of a β-sheet and bind the metal by the imidazole ε-N, ε-N, and δ-N atoms, respectively (**Fig. 1B**).

3.1.3. Design of the Metal-Binding Site

First, we superimposed the backbone atoms of the two charybdotoxin strands 25–29 and 32–36 to the carbonic anhydrase strands 92–96 and 117–121, respectively. This superimposition yielded a 0.82 Å root mean square (RMS) deviation, which was considered quite low and encouraging for the transfer of the three-histidine metal-binding site. Second, the three histidine residues His 94, His 96, and His 119, coordinating the metal in carbonic anhydrase, were incorporated in the corresponding positions, His 27, His 29, and His 34, in the two antiparallel strands of charybdotoxin (**Fig. 1C**). Then, we attempted to stabilize the metal-binding site by incorporating other residues of the second coordination sphere of the metal. In carbonic anhydrase, the side chains Gln 92 and Glu 117 appear to hydrogen-bond His 94 and His 119, respectively. Thus, we incorporated Gln 25 and Glu 32, respectively. Finally, we introduced four additional mutations in the charybdotoxin structure, either to avoid unfavorable contacts with the introduced metal site (Phe2Ser, Lys31Asp, Tyr36Thr) or to prevent alternative metal-binding mode (His21Phe). Nine residues altogether (of 37) were therefore mutated in the original toxin sequence, eight located at the surface of the β-sheet. Importantly, the three cysteine residues holding the sheet from the interior were unchanged. A model incorporating all residue mutations and representing the newly designed MBP was then obtained by using Sybyl software (**Fig. 1C**).

3.1.4. Synthesis and Purification

We synthesized the 37-residue MBP by solid-phase synthesis. The polypeptide chain was assembled in less than 2 d, the deprotected peptide folded efficiently in the presence of glutathione, and it was finally purified by RP-HPLC in 10.2% overall yields in about 1 wk. We also verified that the corrected disulfide bonds were formed by peptide mapping analysis *(30)*.

3.1.5. Structural Characterization and Stability

CD spectrum in the far-UV region, i.e., 180 to 250 nm, provides information about the peptide bond asymmetric environment and reflects the secondary structure content of the designed miniprotein. However, a significant contribution from aromatic residues and disulfide bonds can significantly perturb the spectrum in this region, thus making its interpretation in terms of secondary structure uncertain. Thus, we limited our analysis to a simple comparison between the CD spectra of the synthetic MBP and charybdotoxin. Because these spectra were superimposable, we concluded that similar ordered secondary structures were present in both proteins *(30)* and that the mutations introduced in the scaffold did not significantly perturb its overall architecture. Furthermore, we also recorded the far-UV CD spectrum at 93°C or in the presence of 5.0 *M* guanidine hydrochloride. Because we did not observe any significant change in the spectrum *(30)*, we concluded that the new protein maintained a remarkable conformational stability even after incorporation of nine mutations.

Because the newly designed protein was soluble in water at millimolar concentration, we performed a ¹H-NMR characterization and we recorded DQF-COSY, Z-TOCSY and NOESY spectra. We also observed that the backbone proton chemical shifts were close to those of charybdotoxin, indicating that the global folding of the protein was similar to that of charybdotoxin. The secondary structure of the protein was established by the analysis of the sequential and non-sequential backbone proton NOEs and of $^3J_{HN-H\alpha}$ coupling constants *(11)*. Strong sequential dNN connectivity for residues 10–20 indicates the presence of a helical conformation in this region; strong sequential $d_{\alpha N}$ NOEs, the typical pattern of long-range $d_{\alpha\alpha}$, d_{NN}, and $d_{\alpha N}$ connectivities, and the large values of $^3J_{HN-H\alpha}$ coupling constants indicate that residues 1–2, 25–29, and 32–36 are involved in a short triple-stranded antiparallel β-sheet *(11)*. Thus, the designed MBP protein contained the expected secondary structure elements of the scorpion toxin fold and they were present at the expected location. The presence of specific NOEs between secondary structure elements and preliminary data on 3D structure resolution indicate that the tertiary structures of the MBP protein and charybdotoxin were highly similar.

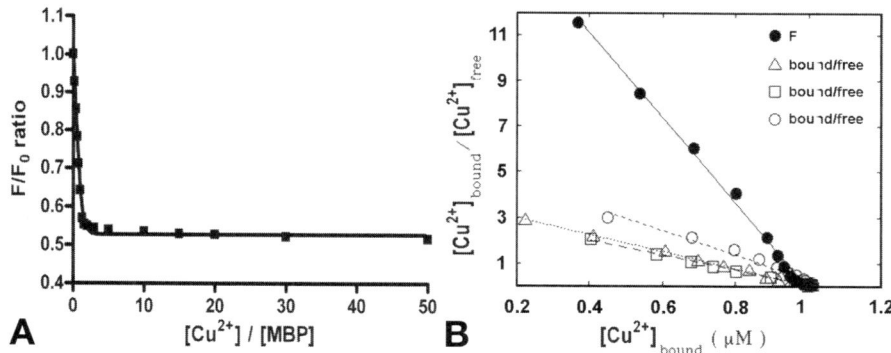

Fig. 2. Metal binding activity of the newly designed MBP miniprotein. (**A**) Relative emission fluorescence at 352 nm as a function of Cu^{2+} ions added. (**B**) Scatchard transformations of the fluorescence emission intensities at 352 nm in the absence (●), or in the presence of Zn^{2+} ions added at 30 μM (○), 60 μM (△) and 100 μM (□) to a 1 μM miniprotein solution.

3.1.6. Metal Binding

Metal binding was first analyzed by addition of Cu^{2+} ions to the miniprotein to observe metal absorption bands in the visible region. Indeed, we recorded electronic absorption spectra with maxima at 330 nm and 600 nm, which are indicative of a Cu^{2+}-histidine charge-transfer transition (*11*). We also performed a CD spectrum in the same region and recorded a corresponding negative spectrum indicating the existence of an asymmetric (structured) environment about the copper ion (*30*). Because a unique Trp 14 is present in the scaffold, we also performed Trp-fluorescence measurement and observed that addition of Cu^{2+} quenched the Trp fluorescence (**Fig. 2A**). Fluorescence quenching was metal concentration dependent, was saturable and reached a maximum after addition of equimolar amount of Cu^{2+} ions, suggesting the presence of an energy transfer mechanism of quenching from the Trp residue to the bound Cu^{2+} ion and one metal-binding site per molecule of protein. As a control, we verified that native charybdotoxin titrated with Cu^{2+} ions did not show the 330 nm and 600 nm absorption bands and that Trp 14 fluorescence was only marginally affected. Thus, we concluded that native charybdotoxin did not bind Cu^{2+} ions. We took advantage of the copper fluorescence quenching to determine the Cu^{2+} binding affinity for MBP; thus, we fitted the relative fluorescence intensity data as a function of Cu^{2+} ion concentration with the standard binding equation and calculated a copper dissociation constant of 4.2 (± 0.2) $\times 10^{-8}$ M. Thus, copper binds to the newly designed miniprotein with a strikingly high affinity. We performed the Scatchard analysis of the fluorescence titration curves and could confirm the high affinity for the Cu^{2+} ion and the presence of only one binding site per molecule (**Fig. 2B**). Then we analyzed Zn^{2+} binding. Because Zn^{2+} does not give transitions in the near-UV or visible region useful for binding studies, we performed competition experiments with Cu^{2+}. We thus repeated the quenching experiments in the presence of fixed amounts of Zn^{2+} and we calculated an apparent dissociation constant of 5.3 (± 0.4) $\times$ 10^{-6} M. We also confirmed Zn^{2+} binding by gel filtration experiments on a Sephadex G15 column by comparing the protein and metal concentrations in the eluted fractions by 280 nm absorbance and atomic absorption, respectively *(30)*. In each protein fraction an equimolar amount of zinc was associated with the protein. We then analyzed other metal ions, Cd^{2+}, Ni^{2+}, and Mn^{2+}. All these ions that are spectroscopically inactive inhibited copper fluorescence quenching, but to a lesser extent than zinc ions. Using the same competition experiments performed with Zn^{2+} and Scatchard analyses, we could calculate apparent dissociation constants, 3.6 (± 0.3) $\times$ 10^{-5} M for Cd^{2+}, 4.6 (± 0.4) $\times$ 10^{-5} M for Ni^{2+}, and 8.7 (± 0.4) $\times$ 10^{-5} M for Mn^{2+}. Thus, the newly designed binding site shows specificity for metals in an order resembling that exhibited by carbonic anhydrase B *(31)*.

Then, using ^{1}H-NMR spectroscopy, we verified that all three histidines were involved in metal ion coordination. By recording ^{1}H-NMR spectra in the downfield region, we could observe that the two resonances associated with the δ (6.6–7.2 ppm) and ϵ (7.5–7.9 ppm) imidazole C-H protons of the three His residues were clearly separated in the absence of metal ions but were shifted and broader when one Eq of Zn^{2+} ions was added *(11)*. Irrespective of the origin of the broadening, probably because of intermediate exchange between the Zn^{2+} saturated and free protein, it was clear that all three His imidazole rings bound zinc, as predicted. We could also confirm the involvement of His ligands in metal binding by ESR analysis of the Cu^{2+}-MBP complex and the presence of a water molecule as the fourth metal ligand.

Finally, we also verified that this newly designed miniprotein did not present the toxic activity of the native scaffold charybdotoxin, which is a powerful K^+-channel blocker. Thus, we tested MBP on single high-conductance Ca^{2+} activated K^+-channels from rat skeletal muscle inserted into planar lipid bilayer and demonstrated that MBP was completely inactive on these K^+-channels *(11)*. Indeed, the positively charged residues that are believed to contact the K^+-channel, as identified by mutagenesis, namely Lys 27, Lys 31, and Lys 32, were modified in MBP. Thus, this protein design completely transformed a toxin that presented nM affinity for K^+-channel into a miniprotein able to bind heavy metals with high affinity and with specificity reflecting that shown by the targeted carbonic anhydrase metal-binding site.

This example clearly demonstrated the potential of the protein design strategy, based on the transfer of critical residues of an active site to a structurally compatible region of a stable and permissive miniprotein scaffold.

3.2. Design of a CD4 Mimic

In **Subheading 3.2.1.** following, we describe the protein design of a miniprotein that binds to the envelope protein gp120 of HIV-1 with native-like affinity and reproduces the "hot spot" of the surface of the cellular CD4 receptor, engaged by the virus in the entry process.

3.2.1. Choice of the Active Site: Protein Design Hypothesis

The binding of the gp120 glycoprotein of the HIV-1 exterior envelope subunit to cellular receptor CD4 represents the first step in virus entry mechanism. This interaction triggers a conformational change in the envelope glycoprotein gp120 that favors its subsequent binding to the CCR5 or CXCR4 chemokine receptor, essential coreceptor for entry. This binding, in turn, leads to further conformational changes in gp120 and exposure of the fusogenic domain of gp41 (the other protein subunit of the HIV-1 envelope), which then fixes the host cell membrane, thus facilitating fusion of the viral and cell membranes, and finally cell infection. Crystallography has provided important clues on the structures involved in the process of HIV-1 entry. Indeed, the structure of gp120 was determined in complex with CD4 and the antigen binding fragment (Fab) of the 17b monoclonal antibody *(32)*, and it shows that a large surface (742 Å²) of the domain D1 of CD4 binds to a large (800 Å²) depression on gp120 (**Fig. 3**). The CD4 protein, which possesses an immunoglobulin fold, uses three regions, residues 31–35, 40–48, and 58–64, for binding. Residues 40–48, which represent the C-terminal β-strand of the β-hairpin 35–47 and correspond to the CDR2 motif (in analogy with antibody structure), are central in gp120 binding, and thus may represent the "hot spot" of the CD4 binding surface. Indeed, this strand H-bonds to the β-15 strand of gp120 via the backbone atoms in a β-sheet–like interaction, and projects the Phe 43 side chain into a deep pocket of gp120, hence called "Phe 23 cavity," thus plugging its entrance. Furthermore, the Arg 59 side chain, located just behind Phe 43, forms a double H-bond with the Asp 368 side chain of gp120, which is the C-terminal residue of the gp120 β-15 strand, and appears to stabilize the β-sheet interaction. The structure of the CD4:gp120 complex justifies the critical functional role played by most of the residues of the CDR2, suggested by previous mutagenesis experiments *(33,34)*.

We hypothesized that the design of a molecule that could reproduce the hot spot of the CDR2 loop should result in an inhibitor of CD4-gp120 interaction and, consequently, an inhibitor of virus attachment and infection. Thus, we planned to design a molecule that could present the CDR2 loop in a native-like conformation.

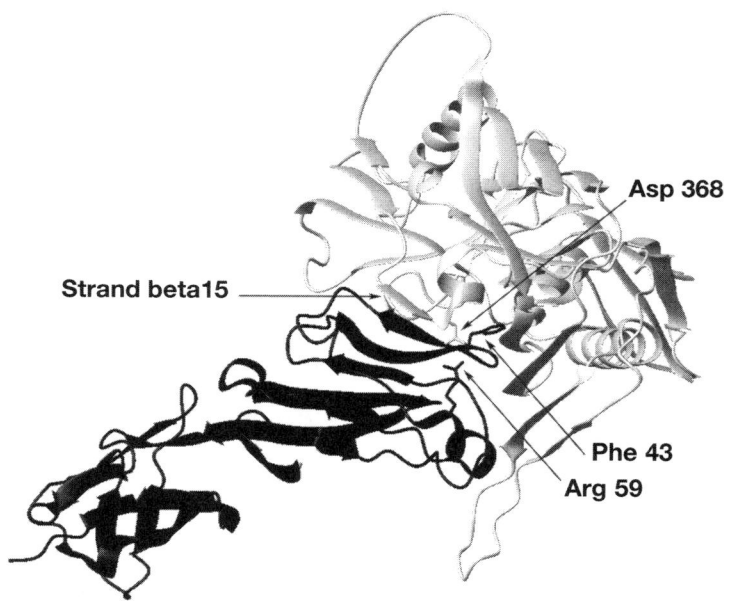

Fig. 3. Three-dimensional structure of the CD4:gp120 complex. For sake of simplicity, the Fab 17b present in the original structure (1g9m) was deleted. Indicated are the gp120 β-15 strand, which interacts with the C-terminal strand of CDR2; the CD4 Phe 43, which engages the entrance of the "Phe 43" cavity of gp120; and the Arg 59 (CD4)–Asp 368 (gp120) interaction.

3.2.2. Experimental Protein Design

3.2.2.1. EARLY DESIGNS

To reproduce the structural features of the β-strand 40–48, the "hot spot" of CD4-gp120 interaction, we though of reproducing the entire CDR2 loop that corresponds to the β-hairpin 35–47. We first synthesized several peptides that contained the CD4 35–47 sequence and attempted to stabilize the native β-hairpin conformation by insertion of a critically disulfide bond or D-amino acids. However, none of the synthesized peptides exhibited any gp120 binding affinity in ELISA. This suggested that a different strategy was required to stabilize the "active" conformation and to produce an effective CD4 mimic. We decided to design a CD4 mimic, by transferring its "hot spot" onto a miniprotein scaffold (*see* **Note 5**).

3.2.2.2. CHARYBDOTOXIN-BASED DESIGN

Charybdotoxin, already used in previous protein designs *(11)* and with its solvent-exposed β-hairpin 25–37, appeared as a first appealing scaffold candi-

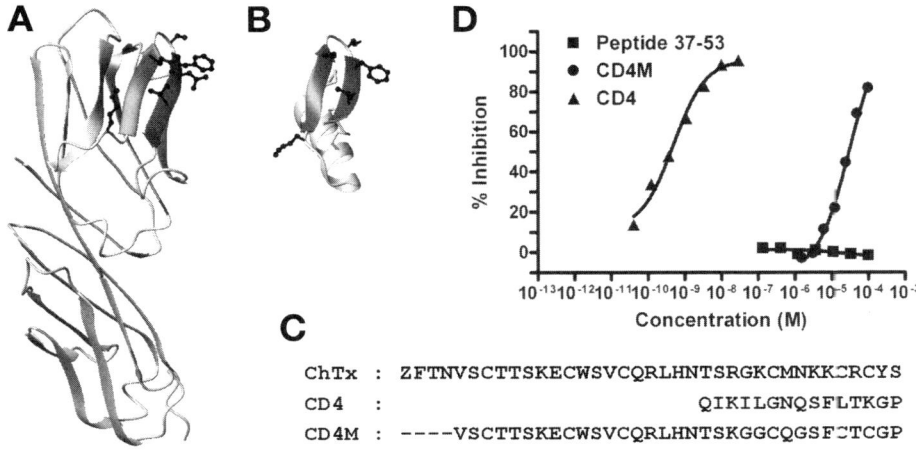

Fig. 4. Structure, sequence, and activity of the target CD4 receptor and the designed mimic. (**A**) Domain 1–2 of CD4 (1cdh) and (**B**) the CD4M mimic designed on charybdo-toxin scaffold. Functional side chains transferred are highlighted. (**C**) Sequence align-ment of charybdotoxin (ChTx), CD4, and CD4M mimic. (**D**) Gp120 binding activity in ELISA of CD4 (▲), CD4M (●), and the linear peptide 37–53 from CD4 (■).

date to reproduce the CDR2 site. Thus, we superposed the backbone atoms of the charybdotoxin β-hairpin 25–37 with those of the CDR2 loop 36–48 of human CD4 (**Figs. 4A, 4B**) and we calculated an RMS deviation of 1.3 Å for this fitting. However, charybdotoxin contains an N-terminus that forms a third β-strand antiparallel to the scaffold C-terminal strand and could interfere with gp120 binding. Thus, because the N-terminal segment of charybdotoxin struc-ture is not required for structure stability *(23)*, we deleted this strand and trun-cated four residues from the N-terminus. Therefore, the 5–37 sequence of charybdotoxin was taken as structural scaffold and on this structure we trans-ferred the CDR2 residues pointing to the solvent exposed side. Thus, Lys 35 and Gly 38 were transferred in the positions 25, 26 of the first β-strand; Gln 40, Gly 41, Ser 42, Phe 43 in the turn 29–32; and Thr 45, Gly 47, Pro 48 in the position 34, 36, 37 of the second β-strand, respectively (**Fig. 4A**). Of course, the three cysteine residues of the scaffold in position 28, 33, and 35, which pointed toward the interior to form disulfide bonds, which are essential to the stability of the β-hairpin, were unchanged. We modeled the resulting 33-residue miniprotein that contained nine CD4 residues in structurally equivalent regions of the scaffold, using the Sybyl software (**Fig. 4B**). After we superposed the transferred CDR2 residues of the modified charybdotoxin on the respective residues of the CD4 structure, it appeared that the designed miniprotein nicely reproduced the CDR2 motif and was thus named CD4M(imic).

3.2.2.2.1. Synthesis and Purification

The 33-residue mimic (CD4M, **Fig. 4C**) was obtained by solid-phase synthesis in a pure form in 9% yields, and in 1 wk. The purified peptide, analyzed by CD spectroscopy in the far-UV region, showed a spectrum indicative of a significant content of mixed α/β secondary structure as expected, and similar to that shown by charybdotoxin *(35)*.

3.2.2.2.2. Binding Assays

The ability of this miniprotein to bind gp120 and to inhibit CD4-gp120 interaction was evaluated by competition ELISA. In this system, CD4M was able to specifically inhibit the binding of soluble recombinant gp120 to coated recombinant CD4, with 2×10^{-5} M IC$_{50}$ (**Fig. 4D**). As a comparison, the linear CD4 peptide 37–53 showed no inhibition. This demonstrates that, by the transfer of the CD4 site to the charybdotoxin scaffold, we succeeded in designing a miniprotein able to bind gp120 with low but specific activity. This result suggested that the CD4 structural and functional properties could be reproduced in a miniprotein scaffold but that further improvement is needed to produce an effective gp120 ligand and HIV-1 inhibitor.

3.2.3. Design of Improved CD4 Mimics

3.2.3.1. PROTEIN DESIGN ON NEW SCAFFOLDS

We looked for more appropriate scaffold in SCOP, Structural Classification of Proteins (scop.mrc-lmb.cam.ac.uk/scop). In the 1999 release of SCOP, about 10,000 structures were classified in 11 classes. The class "peptides" listed 61 folds and the "antimicrobial beta-hairpin" fold contained 4 superfamilies; within this fold, protegrin 1 (1pg1) appeared a very interesting scaffold because it contained a β-hairpin of 18 residues and was stabilized by 2 disulfides. This scaffold was retained for its small size and simple architecture (**Fig. 5A**). Interestingly, the SCOP "small proteins" class included 44 folds, with two families containing a β-hairpin and a disulfide-bonded structure: (1) the proteinase inhibitor PMP-C fold, with PMP-C itself that contained three antiparallel β-strands and three disulfide bridges (**Fig. 5B**), and (2) the knottins fold, which contained 15 superfamilies with at least 3 disulfides and a β-hairpin; in this fold, the short-chain scorpion toxins family contained 21 disulfide-stabilized domains.

In the first attempts to design a CD4 mimic by the transfer of the CD4 binding site, we used the protegrin (*see* **Note 6**) and PMP-C (*see* **Note 7**) scaffolds. However, these mimics were inactive.

Given the relative success of the charybdotoxin-based design, the short-chain scorpion toxins family was examined more in detail. Four structures, scyllatoxin, P05, P01, and Pitx-ka toxins appeared interesting candidates (**Fig. 6**), because they contained the same β-hairpin of charybdotoxin but no

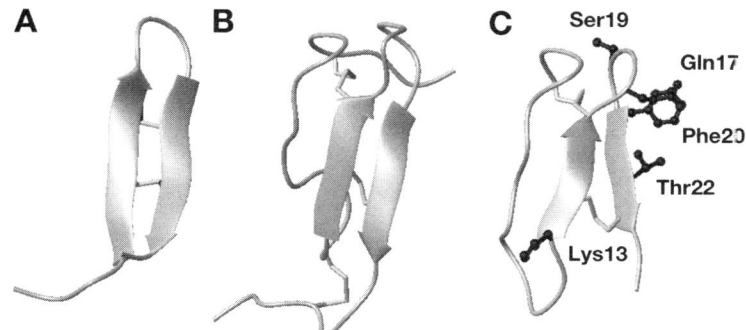

Fig. 5. Scaffolds for CD4 mimicry and a designed miniprotein. (**A**) Protegrin (1pg1); (**B**) PMP-C inhibitor (1mpc); (**C**) and the CD4 mimic designed on PMP-C scaffold. Residues transferred from CD4 are indicated in sticks and balls.

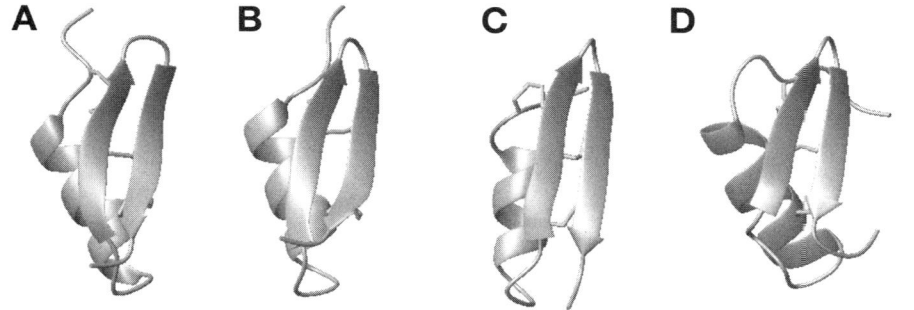

Fig. 6. Scaffolds from the short-chain scorpion toxins family. (**A**) Scyllatoxin from the scorpion *Leiurus quinquestriatus hebraeus* (1scy); (**B**) P05 toxin from the scorpion *Androctonus mauretanicus mauretanicus* (1pnh); (**C**) P01 toxin from the Scorpion *Androctonus mauretanicus mauretanicus* (1acw); (**D**) Pitx-ka toxin from the scorpion *Pandinus imperator* (2pta).

charybdotoxin–like N-terminal strand, which could encumber gp120 binding. Furthermore, scyllatoxin (31 residues only), and the quasi-identical P05, presented a shorter loop connecting the helix to the first β-strand (**Fig. 6**), presumingly allowing a sterically more favorable gp120 binding. Furthermore, the CD4 and scyllatoxin β-hairpins presented a similar spatial orientation in their solvent-exposed side chains, and the scyllatoxin Cys 21, Cys 26, and Cys 28 side chains, which are engaged in disulfide bonds, were oriented similarly to the buried CD4 side chains Asn 39, Lys 44, and Lys 46, respectively. Hence, the backbone atoms of scyllatoxin β-hairpin, sequence 18–29, were superimposed to those of the CDR2 loop of CD4, sequence 36–47, and we calculated

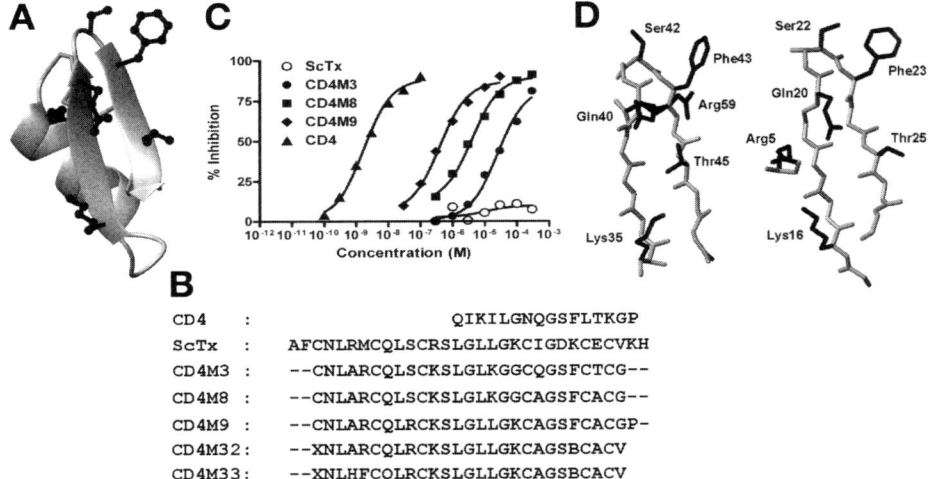

Fig. 7. Mimicry of CD4 based on the scyllatoxin scaffold. (**A**) Model of the CD4M3 mimic, designed on the scyllatoxin scaffold. (**B**) Sequence alignment of CD4, scyllatoxin (ScTx), and mimics CD4M3, CD4M8, CD4M9, CD4M32, and CD4M33. X is thiopropionic acid (Tpa) and B biphenylalanine (Bip). (**C**) Gp120 binding activity in ELISA of ScTx (○), CD4 (▲), CD4M3 (●), CD4M8 (■), and CD4M9 (◆). (**D**) Comparison between the β-hairpins 35–48 of CD4 (1cdh) and 15–28 of CD4M3 (1d5q). Transferred residues are in sticks.

an RMS deviation of only 1.1 Å. Based on these structural results, scyllatoxin was selected as new scaffold to host the CDR2 critical residues (*see* **Note 8**).

3.2.3.2. DESIGN OF A SCYLLATOXIN-BASED CD4M3 MIMIC

The protein design started with the superposition of the scyllatoxin (sequence 18–29) and CD4 (sequence 36–47) structures, followed by the mutation of the solvent exposed residues, which were structurally equivalent to those of CD4 (on the basis of the two β-hairpin superposition) and the preservation of the important cysteines and few other interior residues. Redundant residues, two at the N-terminus and two at the C-terminus, were deleted and the scaffold was reduced to 27 amino acid residues. Then, the residues, Gly 38(18), Gln 40(20), Gly 41(21), Ser 42(22), Phe 43(23), Thr 45(25), and Gly 46(26) (in parentheses are the scaffold host positions) were transferred on the structurally equivalent regions of scyllatoxin (**Fig. 7A**). Furthermore, to increase the structural mimicry with CD4, we included the functional Arg 59 and Lys 35 of CD4 into the topologically equivalent position 5 and 16, respectively. To abolish the original K^+ channel binding activity of scyllatoxin, which is responsible for its toxic activity, Arg 6 and Arg 13 were mutated into Ala and Lys (*see* **Note 9**), respectively. The resulting miniprotein, named CD4M3, contained 27 residues (**Fig. 7B**), of which 9 were topologically positioned as in CD4, and only 17 of the 31 amino acids (55%) of the scyllatoxin sequence subsisted.

3.2.3.3. SYNTHESIS AND FUNCTIONAL CHARACTERIZATION

The new CD4 mimic, named CD4M3, was chemically synthesized, using the same efficient FastMoc protocol, already used for previous miniproteins, folded efficiently as native scyllatoxin, and also presented a CD spectrum similar to that of scyllatoxin, in spite of many mutations in the native sequence. In competitive ELISA, the mimic inhibited CD4 binding to $gp120_{LAI}$ with 4.0×10^{-5} M IC_{50}, which is identical to that shown by CD4M but four orders of magnitude higher than that shown by CD4 (**Fig. 7C**). CD4M3 also bound to gp120 proteins from different CXCR4- and CCR5-using HIV-1 isolates, with affinities in the 20 to 350 μ*M* range *(36)*. Neither the native toxin nor control linear 37–53, nor cyclic 37–46 CD4 peptides bound gp120, suggesting that the scaffold used could stabilize the CD4 active site, at least in part.

3.2.3.4. STRUCTURAL CHARACTERIZATION

To better evaluate the extent of success for the reproduction of the CD4 site in the CD4M3 mimic, we determined its 3D structure by ¹H-NMR spectroscopy and molecular modeling. The final low-energy 20 structures possessed the scyllatoxin expected α/β fold, with a helix in the region 2–13, an antiparallel β-sheet in the region 16–26. The structure of the site transferred from CD4 was also well defined and superimposed on the native CD4 site with striking precision (**Fig. 7D**): the RMS deviation between the backbone atoms of the 17–26 sequence of CD4M3 average structure and the 37–46 sequence of CD4 was only 0.61 Å. The orientations of the side chains of Gln 20, Ser 22, Phe 23, and Thr 25 were similar to those of the corresponding side chains in CD4. In particular, the Phe 23 side chain, which was so well defined in the 20 structures because of many long-range contacts, protrudes into the solvent in a conformation that is unusual for a hydrophobic moiety, but is reminiscent of that of Phe 43 of CD4, which, in the crystal structure of the CD4-gp120 complex, plugs the entrance of the gp120 "Phe43 cavity" *(32)*. However, the position of the Arg 5 and Lys 16 side chains diverged from that of the corresponding Arg 59 and Lys 35 of CD4 by 8.9, 7.7 Å on their C_δ and C_ϵ, respectively; the C-terminus appeared quite flexible, suggesting that all these regions should be further modified to improve the structural reproduction of the CD4 active site.

3.2.3.5. SYSTEMATIC MUTATIONAL ANALYSIS

To more precisely probe the functional role of the transferred residues, we individually substituted the solvent-exposed side chains of the β-hairpin region by an alanine residue and the glycine residues by a bulkier valine. Mutations

Gly21Val and Phe23Ala (equivalent to Gly 41 and Phe 43 of CD4) produced the most dramatic effect and abolished binding, whereas mutation of Ser 22 decreased the apparent affinity by 2.5-fold; substitution of Lys 16, Gly 18, and Gly 27 had no noticeable effect on CD4M3 binding properties *(36)*. In contrast, two substitutions, Gln20Ala and Thr25Ala, increased the apparent binding affinity by about 5-fold. Interestingly, the corresponding two mutations in CD4 also increased its binding affinity for gp120 *(34)*. Altogether, these results suggested that the miniprotein may bind to gp120 in a manner similar to that of the CDR2 region of CD4 and that four positions, Gln 20, Gly 21, Phe 23, and Thr 25, were critical to reproduce a substantial and specific gp120 binding activity in the miniprotein.

3.2.4. From CD4M3 to CD4M9 Mimic

3.2.4.1. IMPROVEMENT OF THE CD4 LIKENESS

The performed structural and functional analysis of CD4M3 pointed to a remarkable structural and functional resemblance between the CD4 and the miniprotein binding regions, but, more importantly, also suggested some changes that could increase the mimic structural and functional mimicry with the CD4 site. The two mutations Q20A and T25A that in the alanine/valine mutational analysis enhanced binding affinity were thus incorporated in a double mutant, denominated CD4M8. This miniprotein was synthesized, purified and, tested in gp120 binding, presented a $2.3 \times 10^{-6} M$ IC$_{50}$ (**Fig. 7C**), which, compared with CD4M3, corresponds to almost one log affinity increase.

Additional mutations were suggested by the structural analysis and the comparison of the two hairpins. Hence, to better mimic the side chain of Lys 35 of CD4, Lys 16 was moved to position 18; this mutant showed increased binding affinity. To reduce the flexibility of the miniprotein C-terminus, as evidenced by the structural analysis, an additional residue, Pro 28, corresponding to Pro 48 of CD4, was added at the C-terminus: this sequence extension also produced some increase in binding affinity.

Given the important role of CD4 Arg 59 in gp120 binding, we incorporated an arginine in position 9 of the miniprotein structure and the corresponding mutant showed some increase in binding affinity. Subsequent designs (*see* **Note 10**), aiming to better reproduce the otherwise important functional role of the Arg 59 residue in gp120 binding failed.

3.2.4.2. DESIGN OF THE CD4M9 MIMIC

At the end, we modeled a new miniprotein, based on the CD4M3 structure and denominated CD4M9, which included all the mutations described to enhance affinity, Ser9Arg, Lys16Leu, Gly18Lys, Gln20Ala, Thr25Ala, and Pro28. The new 28-residue peptide, CD4M9 (**Fig. 7B**), was obtained by solid-phase peptide synthesis in good yields and folded efficiently as shown by HPLC analysis. Furthermore, folding efficiency was insensitive to the presence of 4.0 *M* guanidine-hydrochloride and only slightly perturbed by 6M concentration of this potent protein denaturant, emphasizing the high folding efficiency and conformational stability of the engineered miniprotein *(10)*. The purified miniprotein exhibited a far-UV CD spectrum very similar to that presented by native scyllatoxin. The conformational stability of the CD4M9 miniprotein was also remarkable, as shown by its almost unperturbed CD spectrum obtained at 90°C *(10)*, emphasizing the great sequence permissiveness and stability of the scaffold chosen, even after multiple substitutions.

3.2.4.3. FUNCTIONAL CHARACTERIZATION OF CD4M9

When tested in ELISA in the inhibition of CD4 binding to gp120, the CD4M9 miniprotein exhibited a 4.0×10^{-7} *M* IC_{50}: this represents a remarkable 100-fold increase in apparent gp120 binding affinity, as compared with the CD4M3 miniprotein. Both CD4M3 and the improved CD4M8 and CD4M9 miniproteins were also examined for their ability to prevent infection of HeLa cells and peripheral blood lymphocytes. The CD4M9 miniprotein inhibited infection by HIV-1$_{LAI}$ and HIV-1$_{BaL}$, HIV-1$_{ADA}$ using CXCR4 and CCR5, respectively, as coreceptors for entry, with 10^{-5} to 10^{-6} *M* IC_{50} *(36)*. Protection from infection was effective only when the miniprotein was added before but not after the virus, suggesting that it prevented virus binding to CD4, the primary receptor of virus entry.

3.2.5. From CD4M9 to CD4M33

3.2.5.1. IMPROVEMENT OF THE GP120-BINDING INTERFACE

A 100-fold increase in the binding affinity of CD4 mimicry (from CD4M3 to CD4M9) was obtained uniquely by improving the structural resemblance with CD4. We then planned to further increase binding affinity by improving the mimic binding surface in a way that could better interact with the gp120 interface, by exploiting the availability of the structure of CD4 in complex with gp120 *(32)*.

3.2.5.2. DETERMINATION OF CD4M9 STRUCTURE AND MODELING OF THE GP120:CD4M9 COMPLEX

As a first step, we solved the 3D structure of the CD4M9 miniprotein by ¹H-NMR. Structures were determined from standard 2D ¹H-NMR experiments (TOCSY, DQF-COSY, and NOESY) and molecular modeling. From 297 NOEs and 27 angular restraints 999 3D structures were generated. The 30 conformers with the lowest final target function value were energy-minimized. The final 30 NMR low-energy structures showed that the transferred func-

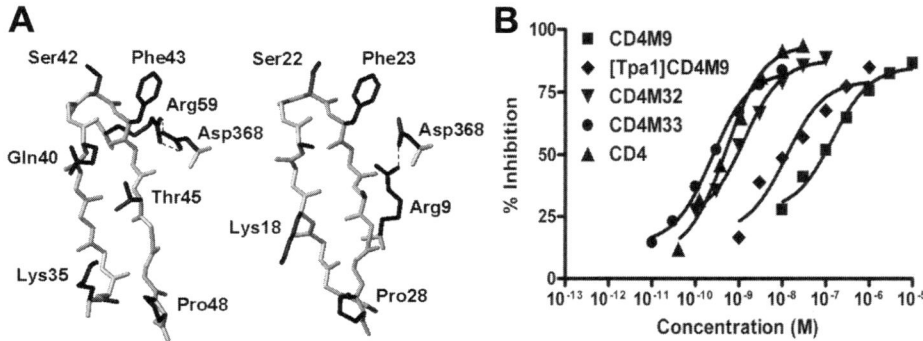

Fig. 8. Mimicry improvement. (**A**) Comparison between the structure of the β-hairpins of CD4(35:48) and CD4M9(15:28). Transferred residues, in sticks, are highlighted. (**B**) Gp120 binding activity in ELISA of CD4 (▲), CD4M9 (■), [Tpa1]CD4M9 (◆), CD4M32, (▼) and CD4M33 (●).

tional site was well defined and the 19–28 β-hairpin of the CD4M9 average structure closely mimicked the corresponding CD4 39–48 region, which is central in gp120 binding, with only 0.7 Å RMS deviation between the two backbones (**Fig. 8A**). On the basis of this structural similarity, the CD4M9 structure was used to model its complex with gp120, which was computed by using the crystallographic structure of the gp120:CD4 complex (1g9m) as a template (*see* **Subheading 2.**). Analysis of this model (**Fig. 9A**) revealed that the CD4M9 binds to the same depression where CD4 also binds and that many interactions that are critical for CD4 to bind gp120 are also present in the modeled CD4M9:gp120 complex. In particular, the Phe 23 of the mimic plugs the entrance of the gp120 "Phe 43 cavity," the mimic β-strand 23–28 forms an antiparallel β-sheet with gp120 β15, and the Arg 9 side chain forms a salt bridge with the carboxyl side chain of gp120 Asp 368. Interestingly, after complex formation, the miniprotein N-terminus (including its N-acetyl group, *see* **Note 11**) moved by more than 1 Å and the side chain of Val 430 of gp120 rotates from 170.1° to –76.5°.

3.2.5.3. DESIGN OF THE OPTIMIZED CD4M33 MIMIC

When binding to gp120 in a CD4-like conformation, CD4M9 encounters unfavorable steric contacts at its N-terminus. Hence, we shortened the N-terminal cysteine and substituted it with a des-α-amino-cysteine residue, i.e., a thiopropionic acid (Tpa), which posed minimal steric hindrance and still allowed formation of a disulfide bond. Furthermore, because the Phe 23 phenyl ring, just like Phe 43 in the CD4:gp120 complex, only engages the entrance of the gp120 "Phe 43 cavity," but leaves further volume free inside the cavity,

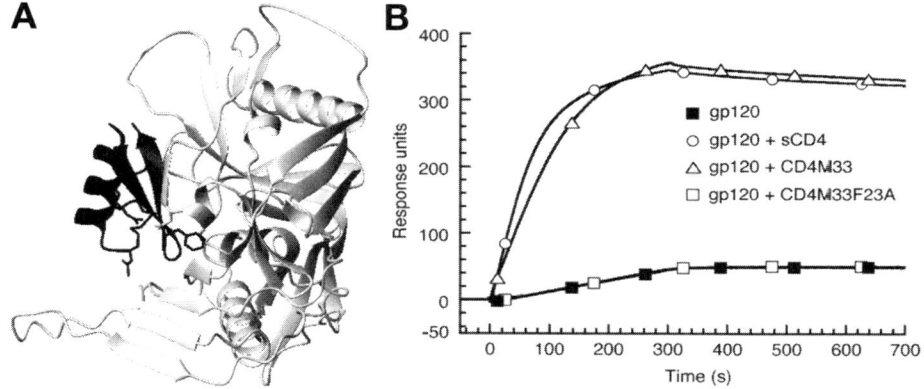

Fig. 9. Mimicry improvement. (**A**) Model of the structure of the CD4M9:gp120 complex, calculated according to the gp120$_{HXB2}$:CD4 complex crystallographic structure (1g9m). Critical interactions are highlighted. (**B**) Stabilization of the CD4-bound conformation of gp120 studied by surface plasmon resonance technology. Sensorgrams were obtained by injecting gp120$_{HXB2}$ (125 nM) to immobilized 48d mAb in the absence (■) or in the presence of threefold molar excess CD4 (○), CD4M33 (△), and inactive [A23]CD4M33 (□).

we thought to fill this free space by adding an additional phenyl ring to the *para* position of the Phe 23 residue, and we incorporated a biphenylalanine (Bip) residue. Finally, to stabilize the C-terminal β-strand and to favor the formation of a β-sheet interaction with the β15 strand of gp120, the C-terminal Gly 27–Pro 28 sequence was replaced by a β-sheet-stabilizing valine residue (Val 27). As a final modification, we introduced a couple of large hydrophobic residues, His 4 and Phe 5, which could shield the disulfide 1–19, which resulted more solvent exposed after the incorporation of the thiopropionic acid, and also stabilize the helical region. The mimic, incorporating all the described mutations, was designated CD4M33 *(37)* (**Fig. 7B**).

3.2.5.4. BINDING TESTS

Each of the modifications, Tpa 1, Bip 23, and Val 27, introduced in a single mutant and tested in competition ELISA, produced an increase in gp120-binding affinity. The simultaneous incorporation of the three mutations into the triple mutant [Tpa 1, Bip23, Val 27]CD4M9 (named CD4M32) resulted in about 40- to 80-fold enhancement in the gp120-binding affinity (depending on the isolate) as compared with CD4M9. At last, CD4M33 incorporating all five mutations, exhibited a 1000-fold increase in gp120-binding affinity (**Fig. 8B**), because it could bind to both gp120$_{HXB2}$, from a CXCR4-using strain, and gp120$_{JRFL}$, from a CCR5-using strain, with 7.5 nM and 4.0 nM IC$_{50}$, respectively. These values, comparable with those obtained with CD4, indicate that the engineered CD4M33 mimic binds to gp120 with a CD4 native-like affinity.

CD4M33, despite maintaining 67% of the original scyllatoxin sequence, but just like the CD4M9 mimic, did not show any significant binding affinity to the potassium channel, which is the cellular target of scyllatoxin.

Direct binding of CD4M33 to different HIV-1 recombinant envelope gp120 glycoproteins was demonstrated by using surface plasmon resonance technology *(37)*. Real-time monitoring of $gp120_{HXB2}$ (X4) or $gp120_{W61D}$ (X4R5) binding to CD4M33 yielded the association, dissociation rate constants, which allowed to calculate the equilibrium dissociation constants $K_d = 8.15 \pm 0.08$ nM for $gp120_{HXB2}$, and $K_d = 12.3 \pm 0.1$ nM for $gp120_{W61D}$. CD4M33 was able to bind gp120 from a variety of isolates, such as SF2 (X4), IIIB (X4), JRFL (R5), and BaL (R5), with K_d values within the 1 to 20 nM range, which are comparable to those determined for native CD4 *(38)*.

Thus, the transfer of critical residues of the CD4 surface binding gp120 onto the scyllatoxin scaffold, and the following structure-guided optimization of the putative binding interface, resulted into a CD4 mimic, CD4M33, presenting a native-like gp120 binding affinity.

3.2.5.5. HIV-1 NEUTRALIZATION

CD4M33 effectively inhibited cell-free infection by different HIV-1 strains, including $HIV-1_{HXB2}$ (X4), $HIV-1_{IIIB}$ (X4), $HIV-1_{YU2}$ (R5), and $HIV-1_{BaL}$ (R5) both in continuous cell lines, such as PM1 and coreceptor-transfected HeLa, and in primary blood peripheral monocyte cells, with IC_{50} in the nanomolar (ng/mL) range *(37)*. Importantly, CD4M33 also inhibited infection by primary HIV-1 isolates, including a pure CCR5-using isolate (224–18), two dual tropic isolates (204–2, 193–21) that use CCR5 and CXCR4 coreceptors, and a promiscuous isolate (196–1) that uses CCR5, CCR3 and CXCR4 coreceptors for entry, with similar efficacy *(37)*. As a specificity control, we tested the variant [A23]CD4M33, in which the critical residue Bip23 was substituted by an alanine residue. This sole substitution was sufficient to completely abolish both the binding to gp120 and the antiviral activity *(37)*. Therefore, the CD4M33 antiviral activity seems to be closely related to its high gp120-binding affinity. The HIV-1 (R5) primary isolate 224–18 is consistently the most resistant to neutralization, requiring 69 nM (202 ng/mL) in PM1 and 1.4 μM (4.3 μg/mL) in blood peripheral monocyte cells to attain 50% virus neutralization by CD4M33.

3.2.5.6. CD4M33 IS A FULLY FUNCTIONAL SUBSTITUTE OF CD4

We then investigated whether the CD4M33 mimic, such as CD4, can stabilize gp120 into the CD4-bound conformation, which is formed as an intermediate structure during cell entry and induces the exposure of antigenic epitopes

(CD4-induced epitopes) that are the target of a class of neutralizing antibodies, represented by the human monoclonal antibodies 48d, 17b, and X5. As shown by surface plasmon resonance experiments, gp120$_{HXB2}$ was found to bind efficiently to the 48d antibody in the presence of either CD4M33 or CD4 (**Fig. 9B**), with calculated K$_d$ values of 3.2 ± 0.5 n*M* and 1.3 ± 0.1 n*M*, respectively. In contrast, in the presence of the inactive [A23]CD4M33 variant or alone, this gp120 showed very little, if any, binding to 48d (**Fig. 9B**). Similarly, CD4M33 was found to increase the binding affinity of other envelope glycoprotein, from isolate SF2 (X4), W61D (R5) and JR-FL (R5) to 48d, as well as to immobilized 17b and X5 MAb. This finding suggests that CD4M33 binds to gp120 proteins and, furthermore, induces gp120 conformational changes that expose CD4-i viral epitopes. Thus, CD4M33 can be used, instead of CD4, in complexes to stabilize the gp120 conformation in an intermediate conformation thats exposes neutralization epitopes.

This strongly suggests that these complexes, stabilized by chemical cross-links, can be used as immunogens to induce neutralizing antibodies. The advantage of using the CD4M33 is that these envelope:mimic complexes will not induce any anti-CD4 immune response, thus will avoid any potential autoimmune response. Thus, CD4M33 in the form of a stable (covalent) complex with a gp120 protein, may represent an essential component of an AIDS vaccine preparation.

3.2.5.7. DETERMINATION OF THE CRYSTAL STRUCTURE OF THE CD4M33-GP120 COMPLEX

To define the CD4M33 mimicry at the atomic level, its structure was first determined in the free state, then in the complex with gp120.

The CD4M33 structure was determined from standard 2D ¹H-NMR in solution and molecular modeling. Complete ¹H resonance assignments of CD4M33 were obtained by standard 2D NMR techniques (TOCSY, DQF-COSY, and NOESY). Analysis of the final 30 lowest energy structures showed that they possessed the expected scyllatoxin αβ-fold, with a helix in the region 2–13, an antiparallel β-sheet in the region 16–26 *(39)*. The structure of the site transferred from CD4 was well defined and superimposed on the native CD4 site with striking precision (*vide infra*). In particular, the Bip 23 side chain was well defined because of many long-range contacts.

Then, the structure of CD4M33, in complex with gp120YU2 and the antigen-binding fragment (Fab) of the 17b antibody, was solved by crystallography *(40)*. The overall structures of gp120 in the Fab 17b ternary complexes with CD4M33 and CD4 were similar (**Fig. 10**). Superpositions, excluding few residues at the N-terminus and at the base of V1/V2 and V4 loop, which are disordered, showed overall Cα RMS deviations of 0.892/0.884 Å (the two numbers are from the two independent CD4M33 complexes in the asymmetric unit).

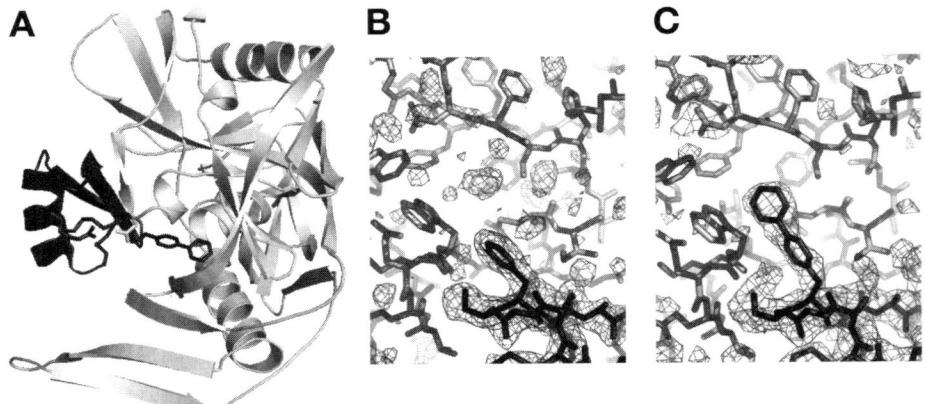

Fig. 10. Three-dimensional structure of CD4M33 in complex with gp120YU2, as determined by crystallography (1yyl). (**A**) Ribbon representation of the CD4M33:gp120 complex. Close-up view of the Phe 43 of CD4 (**B**) and of the Bip 23 of CD4M33 (**C**) interacting with the "Phe 43 cavity" of gp120.

In the complex, the C-terminal β-strand of CD4M33 establishes hydrogen bonds to the β15-strand of gp120 in a manner similar to the CDR2-like loop of CD4 (**Figs. 10B** and **10C**). Superposition of the main chain atoms of residues 22–26 of CD4M33 and residues 42–46 of CD4 produced RMS deviations of 0.432/0.490 Å and of 0.784/0.808 Å for the CDR2-like region (residues 13–27 for CD4M33 and residues 33–47 for CD4). Arg 9 of CD4M33, meanwhile, only partially mimics the important Arg 59 contact of CD4; it adopts two different conformations, only one making a salt bridge with Asp 368 of gp120 *(40)* (**Fig. 10A**). The contact surface between CD4M33 and gp120 is 449/475 Å^2 for CD4M33, approx 55% of the CD4 interface. In CD4M33, the Bip 23 residue was designed to not only mimic Phe 43, but also to fill the CD4-gp120 cavity at the intersection of the inner, outer, and bridging sheet domains of gp120 (**Figs. 10B** and **10C**). This biphenyl group contributes 168/170 Å^2 (37%) of the total area, with roughly 80 Å^2 contributed by its second ring, which reaches the heart of gp120.

We attributed the high affinity of CD4M33, despite making only half the contacts as CD4, to filling the "Phe 43 cavity" in gp120 with its protruding unnatural biphenyl side chain. However, biphenyl is relatively rigid with limited flexibility along its long axis and dihedral constraints between its two rings. The observed ring-ring dihedrals are 38°/40° for CD4M33 in the CD4M33-gp120 complexes and favorably compare with the 10 to 53° range commonly observed in other structures. However, the observed orientation of the first ring of the biphenyl moiety was more than 10° different from those of the Phe 43 of

CD4 (**Fig. 10C**), suggesting that to accommodate the biphenyl moiety, the biphenyl side chain and neighboring atoms adopt an orientation that sacrifices mimicry.

To better evaluate the contribution of the Bip 23 in gp120 binding, we synthesized the reverted mutant [F23]M33 miniprotein, which, like CD4, possesses a natural Phe 23 side chain. Tested in ELISA competition experiments, CD4M33 could bind gp120YU2 protein with only fivefold higher affinity than [F23]M33 *(39)*, which is at odds with the significant binding affinity increase produced by Bip 23 when it was incorporated in the CD4M9 miniprotein. This limited difference in gp120 binding affinity suggests that, in spite of the fact that the Bip 23 side chain more deeply penetrates within the "Phe 43 cavity" than Phe 23, this larger surface of interaction is not translated into a large increase in binding affinity. Indeed, increasing binding surface between an apolar ligand and a hydrophobic cavity is expected to provide a significant increase of binding energy *(41)*. Thus, it is possible that the biphenyl moiety is too rigid and long, thus may not fit appropriately inside the envelope cavity and some adjustments occurred both on the biphenyl side chain and the internal envelope cavity to fit this long side chain in the viral glycoprotein cavity. These adjustments may be of limited entity, but energetically costly, thus may affect binding energy.

These results may suggest that although a biphenyl moiety effectively reaches the gp120 interfacial cavity, other side chains may better fulfill this task and may provide a substantial increase of binding affinity.

3.2.5.8. THREE-DIMENSIONAL STRUCTURE OF [F23]M33 IN COMPLEX WITH GP120

[F23]M33 structure in complex with gp120YU2 was also crystallized and solved *(40)*. Superposition of the gp120 molecules from the two asymmetric units of [F23]M33 with the gp120YU2 in the CD4 ternary complex gave RMS deviations of 0.869/0.850 Å. In [F23]M33 structure, we observed an increased degree of CD4 mimicry especially at the Phe 23 residue. However, increased mimicry was apparent only when the RMS deviations were calculated for the individual side chains. Thus, the same superposition of [F23]M33 and CD4 produced RMS deviations for the side-chain atoms of residue 23 of 0.441/0.233 Å, while CD4M33 superposition resulted in approx twofold higher RMS deviations of 0.784/0.633 Å for the first biphenyl ring. Difference distance matrices *(42)* restricted to a 10 Å radius from the Cα of the biphenyl to phenyl change in the CD4M33 to [F23]M33 showed [F23]M33 to emulate the structural change induced by CD4 significantly better then CD4M33. Also by examining the residue-specific contacts of gp120, comparing the interactive surface of gp120 with CD4, CD4M33 and [F23]M33, we evidenced closer structural similarity between [F23]M33 and CD4 complexes. Thus, although overall RMS deviations were not very revealing, localized measures of analysis indicated that [F23]M33 had a higher degree of CD4 mimicry than CD4M33.

This structure confirms the conclusions reached by the analysis of the CD4M33:gp120 complex structure that further work is needed to design a new side chains that may reach the interior "Phe 43 cavity" and may produce a substantial increase of binding affinity in a new CD4 miniprotein.

3.3. Conclusions

These two examples demonstrate the success obtained by a careful modeling, a structural homology search, and the transfer of a protein active site to stable small presentation scaffolds. However, the design of the CD4 mimic also emphasizes that the genuine reproduction of a protein function on a miniprotein system to a native-like potency may require that the transfer of the active site is followed by optimization steps to adapt the transferred site to the new structural context. Indeed, this was performed by a structure-based approach after the miniprotein 3D structure in the free state and in complex with gp120 was determined by ¹H-NMR and crystallography, respectively. Two intrinsic properties of the chosen scaffold, stability and small size, made this optimization possible. We can predict that in other cases where structural information are available for the protein mimicked in the free state and in complex with its ligand, structure-function relationship studies may effectively optimized the first designed miniprotein and make it a powerful mimic.

Because of its well-defined 3D structure, specific activity, and the ease with which it may be chemically manipulated, the designed CD4 miniprotein represents a useful tool to study the complex biology of the HIV-1 entry process, but also a ligand useful in affinity purification of envelope proteins, and more importantly a therapeutically useful molecule both as an inhibitor of HIV infection (as a microbicide) and as an essential component of an AIDS vaccine.

The examples described here demonstrate that the design of functional miniproteins may provide new exquisite tools in biology, biotechnology, and medicine. They may also lead to practical applications in diagnostics (as new tracers) but also in therapy. Finally, they may also represent structural intermediate useful to determine the bioactive conformation of pharmacophoric groups and to facilitate the development of new drug candidates. Thus the design of functional miniprotein may become not only a useful strategy in biology, but also a convenient step in medicinal chemistry accelerating the development of new drugs.

4. Notes

1. The structure database SCOP, Structural Classification Of Proteins, at scop.mrc-lmb.cam.ac.uk/scop, contains a class designated small proteins that lists 75 different folds.

2. The selection of candidate scaffolds possessing a region structurally similar to a target site remains a difficult task. Indeed, the Protein Data Bank (www.rcsb.org/pdb) today lists more than 30,000 structures and other structure databases, such as SCOP (scop.mrc-lmb.cam.ac.uk/scop/), CATH (www.biochem.ucl.ac.uk/bsm/cath/cath.html) or DALI (www.ebi.ac.uk/dali/index.html) provide an organization of these structures in folds and families. However, in these databases, no tools are available to the protein designer to dig this wealth of structural information to find an appropriate scaffold for a target "hot spot" or epitope. Few software, either freely or commercially available, such as Sybyl (Tripos), Spasm (xray.bmc.uu.se/usf/spasm.html), or What if (swift.cmbi.kun.nl/whatif/) may provide some solution to this task. These software can search a database both by sequence and secondary structure, but they heavily rely on the input structure query and small changes in the query can lead to very different results. Moreover, there is no possibility to filter the results for solvent exposition or compatibility of the selected scaffolds to receptor binding. Thus, a great deal of manipulation, scripting, or the use of other software are required to sort out a candidate scaffolds appropriate for a given target active site. Indeed, no automatic search for structure scaffold is yet available.

3. Purification is facilitated by the efficient folding properties of most scorpion derived miniproteins, because the correctly folded product eluted earlier enough, in a chromatographic region devoid of other incorrectly folded or deletion peptides. The formation of the native disulfide bonds represents a formidable tool in the chromatographic purification of the folded protein. In fact, the correctly folded protein, with most of its hydrophobic side chains buried and fixed in this conformation by disulfide bridges, is expected to interact with the chromatographic matrices of the reverse phase column with a lower affinity than misfolded and incorrect sequences. Thus, as a consequence of the folding, the correctly folded molecule is eluted earlier by the organic solvent (acetonitrile) far from impurities, thereby facilitating enormously the purification task. Starting from a relatively homogeneous crude material one chromatographic step was usually enough to obtain most of the correct sequence in a pure form.

4. Miniproteins containing multiple disulfides are particularly interesting as structural scaffolds, because these covalent bridges represent the major determinants of structural stability, allowing the design of new functions by mutating solvent exposed side chains without perturbing folding efficiency. Within disulfide stabilized natural scaffolds, the scorpion toxin fold is one of the most attractive for protein engineering. This motif, exemplified by charybdotoxin, contains two canonical secondary structures, an antiparallel β-sheet and a short α-helix, which are joined by three disulfide bridges in the interior of the structure that remarkably stabilize such structure to strong denaturant conditions such as high guanidine hydrochloride concentrations or boiling water *(10)*. All known scorpion toxins contain this structural motif, irrespective of their size, amino acid sequence, or function *(21)*. Interestingly, the same structural motif is also present in defensins from insect *(21)*, plant *(43)*, mussel *(44)*, in a sweet-tasting protein *(45)*, and in a family of protease inhibitors *(46)*. Thus, this simple, compact, and

well-organized structural motif seems to have been naturally selected for its high sequence permissiveness (as shown by its compatibility with hundreds of different sequences) and functional versatility (as shown by its compatibility with the Na^+, K^+ channel blockage activity in scorpion toxins, the antimicrobial activity in defensins and the interaction with different proteins and receptors). Furthermore, folding studies of scorpion toxins have revealed that a particular spacing between cysteine residues play a role in facilitating formation of specific native disulfides, avoiding non-native ones (47): this may explain the persistent folding efficiency of some engineered constructions based on this fold even after multiple substitutions. For these reasons, scorpion toxins have been frequently used in our miniprotein design.

5. The CD4 miniprotein design work was performed before the structure of gp120-CD4 complex was solved; thus, in the first part of this protein design work, we could only rely on the 3D structure of CD4, available since 1990, and on the mutagenesis work performed in a number of different laboratories that pointed to the CDR2 as an important element of gp120 interaction, although other CDRs were considered important. Furthermore, the functional role of Arg 59 was not unequivocally demonstrated. However, that in the CD4 structure the CDR2 loop appeared as a β-hairpin well-exposed to solvent, hence well accessible to protein (gp120) binding, was the element that let us to focus on the CDR2 as the "hot spot" and the critical element to be reproduced by protein design.

6. Design of this CD4 mimic started with the transfer of the residues 36–48 of the CD4 CDR2 loop on the 5–17 β-hairpin of the antibacterial peptide protegrin 1(1pg1, 18 residues), leaving the four cysteine residues unchanged. The corresponding chimeric peptide was synthesized. Disulfide bonds were formed chemically, using two different pairs of protecting groups for cysteine residues during synthesis, and by deprotecting and oxidizing two specific pairs, one by one. The folded peptide was purified and tested in ELISA for inhibition of the CD4-gp120 interaction. However, no inhibitory activity was detected, up to 10^{-4} M peptide concentration.

7. The protease inhibitor PMP-C (1pmc, 35 residues) was adopted as scaffold, and the following mutations were introduced: 1, deletion of three residues at the N-terminus and six at the C-terminus (including Cys 32 which is bonded to Cys 14) thus reducing to 25 residues its size; 2, mutation Cys14Gly, because its paired Cys 32 was deleted in **step 1**; 3, replacement of the 5-residue turn 20–24 by the 4-residue turn Gln 40, Gly 41, Ser 42, and Phe 43 of the CDR2; 4, insertion of the CDR2 β-strands residues Lys 35, Gly 39, Thr 45, and Gly 47 in the structurally similar positions 13, 15, 22, and 24 of the 25-residue scaffold (*see* **Fig. 5C**). This molecule was synthesized, the disulfide bonds were formed chemically, and the chimeric peptide was finally purified. It was then tested in the ELISA in competition with CD4 for gp120 binding, but no inhibition of CD4-gp120 interaction was detected.

8. We tried in a quick search to retrieve new small proteins (<50 residues) that match the CDR2 of the CD4 by using Spasm (xray.bmc.uu.se/usf/spasm.html). We looked for candidate proteins presenting residues that could superimpose their α-carbon atoms to those of residues 36–46 of the CDR2. Allowing a maximum

distance of 1.8 Å and an RMS deviation between α-carbon atoms less than 1.8 Å, we found 54 hits in 40 proteins. The pdb files of each hit were then downloaded and, one by one, fitted to the CDR2 of the CD4 in complex with gp120 (1g9m), using software of molecular visualization, like SwissPdbViewer (www.expasy.org/spdbv/) or Sybyl (Tripos). Then, the superimposed hits presenting visible steric clashes with gp120 structure were discarded and four small proteins remained (three epidermal growth factor–like folds: two modules from blood coagulation factors, 1apo and 1f7e, and a fragment of the betacellulin-2; one cysteine knot from the cocaine and amphetamine regulated transcript, 1hy9) that represented potential candidates for grafting the CDR2 site. This fast and simple search shows that finding new scaffolds is feasible. However, this method is mostly manual, because the filtering of the selected hits compatible with gp120 receptor binding is performed manually. However, if one wants to search a large database (without for example restricting the size of the target searched) the hits selected will be too many and would not be possible to screen them manually any more. Thus, one needs to use other software or develop specific scripting, or finally develop a specific software to perform this task.

9. Lys11 residue, located in the helical region, opposite to the CD4 binding epitope, was inserted to provide an appropriate position for labeling. Thus, a fluorescein or biotin probe attached to this position is predicted not to interfere with gp120 binding. Specific and quantitative incorporation of a probe at the side chain of Lys11 is performed by peptide synthesis by using the orthogonal ivDde, (4,4-dimethyl-2,6-dioxocyclohex-1-ylidene-3-methylbutyl) protecting group during polypeptide chain assembly, selective cleavage of this protection by four treatments of 3 min with 2% hydrazine after polypeptide chain completion, and finally coupling of the probe, as a succinimidyl ester. The labeled peptide is then deprotected, folded, and purified as for the unlabeled peptide.

10. Analysis of charybdotoxin-based CD4M mimic suggested that its truncation until residue Cys 7 and insertion of a N-terminal guanidinium group could better mimic the Arg 59 side chain. Therefore, we designed and then produced a CD4M mutant, containing either a guanidinium-acetyl or guanidinium-propionyl group at the N-terminus. Unfortunately, the presence of the positively charged guanidinium group seemed to destabilize the scaffold significantly, as shown by a significant loss in folding efficiency. Furthermore, this derivative did not present any improved gp120-binding affinity. We also used a small scorpion toxin, leiuropeptide II from *Leiurus quinquestriatus hebraeus* (*48*), a structural homologous of scyllatoxin but presenting a shorter sequence (29 residues) and, more importantly, an N-terminus positioned more favorably to present a guanidinium group better mimicking the Arg. 59. Unfortunately, insertion of such (positively charged) group at the N-terminus compromised folding efficiency and did not allow obtaining a homogeneous folded product.

11. The N-terminal cysteine residue of CD4M9 was acetylated to avoid the presence of a free positive charge at the N-terminus that could affect the pK of the thiol group and consequently the folding efficiency. This is a common practice in peptide synthesis.

Acknowledgments

We thank the EU 6th Framework Program EMPRO and the French "Agence Nationale de Recherches sur le SIDA" (ANRS) for financial support. We thank P. Kwong for his comments. We would like to dedicate this work to the memory of Claudio Vita.

References

1. Bryson, J. W., Betz, S. F., Lu, H. S., Suich, D. J., Zhou, H. X., O'Neil, K. T., et al. (1995) Protein design: a hierarchic approach. *Science* **270,** 935–941.

2. Hill, R. B., Raleigh, D. P., Lombardi, A., and DeGrado, W. F. (2000) De novo design of helical bundles as models for understanding protein folding and function. *Acc. Chem. Res.* **33,** 745–754.

3. Baltzer, L. and Nilsson, J. (2001) Emerging principles of de novo catalyst design. *Curr. Opin. Biotechnol.* **12,** 355–360.

4. Petrounia, I. P. and Arnold, F. H. (2000) Designed evolution of enzymatic properties. *Curr. Opin. Biotechnol.* **11,** 325–330.

5. Tobin, M. B., Gustafsson, C., and Huisman, G. W. (2000) Directed evolution: the 'rational' basis for 'irrational' design. *Curr. Opin. Struct. Biol.* **10,** 421–427.

6. Martin, L. and Vita, C. (2000) Engineering novel bioactive miniproteins from small size natural and de novo designed scaffolds. *Curr. Protein. Pept. Sci.* **1,** 403–430.

7. Mathonet, P. and Fastrez, J. (2004) Engineering of non-natural receptors. *Curr. Opin. Struct. Biol.* **14,** 505–511.

8. Betz, S. F., Raleigh, D. P., DeGrado, W. F., Lovejoy, B., Anderson, D., Ogihara, N., et al. (1996) Crystallization of a designed peptide from a molten globule ensemble. *Fold. Des.* **1,** 57–64.

9. Clackson, T. and Wells, J. A. (1995) A hot spot of binding energy in a hormone-receptor interface. *Science* **267,** 383–386.

10. Martin, L., Barthe, P., Combes, O., Roumestand, C. and Vita, C. (2000) Engineering novel bioactive mini-proteins on natural scaffolds. *Tetrahedron* **56,** 9451–9460.

11. Vita, C., Roumestand, C., Toma, F., and Menez, A. (1995) Scorpion toxins as natural scaffolds for protein engineering. *Proc. Natl. Acad. Sci. USA* **92,** 6404–6408.

12. Wütrich, K. (1986). *NMR of Proteins and Nucleic Acids,* Wiley, New York.

13. Pons, J. L., Mallliavin, T. E., and Delsuc, M. A. (1996) Gifa V.4: a complete package for NMR data set processing. *J. Biomol. NMR* **8,** 445–452.

14. Roumestand, C., Delay, C., Gavin, J. A., and Canet, D. (1999) A practical approach to the implementation of selectivity in homonuclear multidimensional NMR with frequency selective-filtering techniques. Application to the chemical structure elucidation of complex oligosaccharides. *Magn. Reson. Chem.* **37,** 451–478.

15. Guntert, P., Mumenthaler, C., and Wuthrich, K. (1997) Torsion angle dynamics for NMR structure calculation with the new program DYANA. *J. Mol. Biol.* **273,** 283–298.

16. Pearlman, D. A., Case, D. A., Caldwell, J. W., Ross, W. S., Cheatham III, T. E., DeBolt, S., et al. (1995) AMBER, a package of computer programs for applying molecular mechanics, normal mode analysis, molecular dynamics and free energy calculations to simulate the structural and energetic properties of molecules. *Comp. Phys. Commun.* **91,** 1–41.
17. Cornell, W. D., Cieplak, P., Bayly, C. I., Gould, I. R., Merz, K. M., Ferguson, D. R. J., et al. (1995) A second generation force field for the simulation of proteins and nucelic acids. *J. Am. Chem. Soc.* **117,** 5179–5197.
18. Moore, J. P., McKeating, J. A., Weiss, R. A., and Sattentau, Q. J. (1990) Dissociation of gp120 from HIV-1 virions induced by soluble CD4. *Science* **250,** 1139–1142.
19. Charneau, P., Mirambeau, G., Roux, P., Paulous, S., Buc, H., and Clavel, F. (1994) HIV-1 reverse transcription. A termination step at the center of the genome. *J. Mol. Biol.* **241,** 651–662.
20. Lusso, P., Cocchi, F., Balotta, C., Markham, P. D., Louie, A., Farci, P., et al. (1995) Growth of macrophage-tropic and primary human immunodeficiency virus type 1 (HIV-1) isolates in a unique CD4+ T-cell clone (PM1): failure to downregulate CD4 and to interfere with cell-line-tropic HIV-1. *J. Virol.* **69,** 3712–3720.
21. Bontems, F., Roumestand, C., Gilquin, B., Menez, A., and Toma, F. (1991) Refined structure of charybdotoxin: common motifs in scorpion toxins and insect defensins. *Science* **254,** 1521–1523.
22. Miller, C., Moczydlowski, E., Latorre, R., and Phillips, M. (1985) Charybdotoxin, a protein inhibitor of single Ca2+-activated K+ channels from mammalian skeletal muscle. *Nature* **313,** 316–318.
23. Vita, C., Bontems, F., Bouet, F., Tauc, M., Poujeol, P., Vatanpour, H., et al. (1993) Synthesis of charybdotoxin and of two N-terminal truncated analogues. Structural and functional characterisation. *Eur. J. Biochem.* **217,** 157–169.
24. Tainer, J. A., Roberts, V. A., and Getzoff, E. D. (1991) Metal-binding sites in proteins. *Curr. Opin. Biotechnol.* **2,** 582–591.
25. Barondeau, D. P. and Getzoff, E. D. (2004) Structural insights into protein-metal ion partnerships. *Curr. Opin. Biotechnol.* **14,** 765–774.
26. Hellinga, H. W. (1998) The construction of metal centers in proteins by rational design. *Fold. Des.* **3,** 1–8.
27. Lu, Y., Berry, S. M., and Pfister, T. D. (2001) Engineering novel metalloproteins: design of metal-binding sites into native protein scaffolds. *Chem. Rev.* **101,** 3047–3080.
28. Lombardi, A., Summa, C. M., Geremia, S., Randaccio, L., Pavone, V., and DeGrado, W. F. (2000) Inaugural article: retrostructural analysis of metalloproteins: application to the design of a minimal model for diiron proteins. *Proc. Natl. Acad. Sci. USA* **97,** 6298–6305.
29. Klemba, M., Gardner, K. H., Marino, S., Clarke, N. D., and Regan, L. (1995) Novel metal-binding proteins by design. *Nat. Struct. Biol.* **2,** 368–373.
30. Pierret, B., Virelizier, H., and Vita, C. (1995) Synthesis of a metal binding protein designed on the alpha/beta scaffold of charybdotoxin. *Int. J. Pept. Protein Res.* **46,** 471–479.

31. Lindskog, S. and Nyman, P. O. (1964) Metal-binding properties of human erythrocyte carbonic anhydrases. *Biochim. Biophys. Acta* **85**, 462–474.

32. Kwong, P. D., Wyatt, R., Robinson, J., Sweet, R. W., Sodroski, J., and Hendrickson, W. A. (1998) Structure of an HIV gp120 envelope glycoprotein in complex with the CD4 receptor and a neutralizing human antibody. *Nature* **393**, 648–659.

33. Sweet, R. W., Truneh, A., and Hendrickson, W. A. (1991) CD4: its structure, role in immune function and AIDS pathogenesis, and potential as a pharmacological target. *Curr. Opin. Biotechnol.* **2**, 622–633.

34. Moebius, U., Clayton, L. K., Abraham, S., Harrison, S. C., and Reinherz, E. L. (1992) The human immunodeficiency virus gp120 binding site on CD4: delineation by quantitative equilibrium and kinetic binding studies of mutants in conjunction with a high-resolution CD4 atomic structure. *J. Exp. Med.* **176**, 507–517.

35. Drakopoulou, E., Vizzavona, J., Neyton, J., Aniort, V., Bouet, F., Virelizier, H., et al. (1998) Consequence of the removal of evolutionary conserved disulfide bridges on the structure and function of charybdotoxin and evidence that particular cysteine spacings govern specific disulfide bond formation. *Biochemistry* **37**, 1292–1301.

36. Vita, C., Drakopoulou, E., Vizzavona, J., Rochette, S., Martin, L., Menez, A., et al. (1999) Rational engineering of a mini-protein that reproduces the core of the CD4 site interacting with HIV-1 envelope glycoprotein. *Proc. Natl. Acad. Sci. USA* **96**, 13091–13096.

37. Martin, L., Stricher, F., Misse, D., Sironi, F., Pugniere, M., Barthe, P., et al. (2003) Rational design of a CD4 mimic that inhibits HIV-1 entry and exposes cryptic neutralization epitopes. *Nat. Biotechnol.* **21**, 71–76.

38. Myszka, D. G., Sweet, R. W., Hensley, P., Brigham-Burke, M., Kwong, P. D., Hendrickson, W. A., et al. (2000) Energetics of the HIV gp120-CD4 binding reaction. *Proc. Natl. Acad. Sci. USA* **97**, 9026–9031.

39. Stricher, F., Martin, L., Barthe, P., Pogenberg, V., Mechulam, A., Menez, A., et al. (2005) A high-throughput fluorescence polarization assay specific to the CD4 binding site of HIV-1 glycoproteins based on a fluorescein-labeled CD4 mimic. *Biochem. J.* **390**, 29–39.

40. Huang, C. C., Stricher, F., Martin, L., Decker, J. M., Majeed, S., Barthe, P., et al. (2005) Scorpion-toxin mimics of CD4 in complex with human immunodeficiency virus gp120 crystal structures, molecular mimicry, and neutralization breadth. *Structure* **13**, 755–768.

41. Sundberg, E. J., Urrutia, M., Braden, B. C., Isern, J., Tsuchiya, D., Fields, B. A., et al. (2000) Estimation of the hydrophobic effect in an antigen-antibody protein-protein interface. *Biochemistry* **39**, 15375–15387.

42. Richards, F. M. and Kundrot, C. E. (1988) Identification of structural motifs from protein coordinate data: secondary structure and first-level supersecondary structure. *Proteins* **3**, 71–84.

43. Bruix, M., Jimenez, M. A., Santoro, J., Gonzalez, C., Colilla, F. J., Mendez, E., et al. (1993) Solution structure of gamma 1-H and gamma 1-P thionins from barley and wheat endosperm determined by 1H-NMR: a structural motif common to toxic arthropod proteins. *Biochemistry* **32,** 715–724.
44. Yang, Y. S., Mitta, G., Chavanieu, A., Calas, B., Sanchez, J. F., Roch, P., et al. (2000) Solution structure and activity of the synthetic four-disulfide bond Mediterranean mussel defensin (MGD-1). *Biochemistry* **39,** 14436–14447.
45. Caldwell, J. E., Abildgaard, F., Dzakula, Z., Ming, D., Hellekant, G., and Markley, J. L. (1998) Solution structure of the thermostable sweet-tasting protein brazzein. *Nat. Struct. Biol.* **5,** 427–431.
46. Ceciliani, F., Bortolotti, F., Menegatti, E., Ronchi, S., Ascenzi, P., and Palmieri, S. (1994) Purification, inhibitory properties, amino acid sequence and identification of the reactive site of a new serine proteinase inhibitor from oil-rape (Brassica napus) seed. *FEBS Lett.* **342,** 221–224.
47. Zhu, Q., Liang, S., Martin, L., Gasparini, S., Menez, A., and Vita, C. (2002) Role of disulfide bonds in folding and activity of leiurotoxin I: just two disulfides suffice. *Biochemistry* **41,** 11488–11494.
48. Buisine, E., Wieruszeski, J. M., Lippens, G., Wouters, D., Tartar, A., and Sautiere, P. (1997) Characterization of a new family of toxin-like peptides from the venom of the scorpion Leiurus quinquestriatus hebraeus. 1H-NMR structure of leiuropeptide II. *J. Pept. Res.* **49,** 545–555.

7

Consensus Design as a Tool for Engineering Repeat Proteins

Tommi Kajander, Aitziber L. Cortajarena, and Lynne Regan

Summary

Repeat proteins were first identified because of their unusual primary structure, in which short amino acid sequences, typically between 20 and 40 residues, are repeated in tandem, often many times. After identification at the sequence level, the three-dimensional structures of representatives from several classes (e.g., ankyrin, tetratricopeptide, leucine rich repeat) have been solved. The structures indeed reveal unusual, nonglobular structures, a linear "string" of the tandem motifs. Perhaps because of the large surface area that is presented as a consequence of such elongated structures, repeat domains are often involved in mediating protein–protein interactions. Here we describe methods of consensus-based design and engineering of repeat proteins. We pay particular attention to the attributes of repeat proteins that make them well-suited to such approaches. In addition, we discuss practical issues related to producing and characterizing such designed proteins. We use the tetratricopeptide repeat, which is well-studied in our group, to illustrate many ideas, but also draw comparisons to other work on repeat proteins, where relevant.

Key Words: Repeat protein; tetratricopeptide repeat; TPR; protein engineering; protein design; binding domain; consensus sequence.

1. Introduction

Protein design initially concentrated on understanding the structural determinants and folding of small, globular all-α helical or all-β sheet proteins (1–4). More recently, various modeling strategies, which incorporate exhaustive conformational searches have been applied to a variety of folds (e.g., **refs. 5–8**).

Another approach to structural design, the design of proteins based on the amino acid consensus from multiple sequence alignments, emerged in the 1990s. The notion was to choose the most probable amino acid for each position by analyzing the statistics of occurrence of different amino acids at each position in a sequence alignment of as many related proteins as possible. The

From: *Methods in Molecular Biology, vol. 340: Protein Design: Methods and Applications*
Edited by: R. Guerois and M. López de la Paz © Humana Press Inc., Totowa, NJ

first successful application of this approach was the design of a consensus zinc-finger peptide by Berg and colleagues. The peptide they designed bound metal ions and adopted the typical zinc finger fold *(9)*. The underlying hypothesis of this approach was that stabilizing amino acids should be present in the sequence data with high frequency, and thus this strategy should not only specify the protein fold, but also yield more stable proteins. This was indeed the case for their zinc-finger peptide designs, and has proven true for many subsequent consensus-based designs. Variations on consensus-based engineering have since been applied in several systems as a method to enhance stability; for example, in immunoglobulin variable domains *(10)*, phytases *(11)*, and SH3-domains *(12)*.

More recently, the consensus design approach has been applied to nonglobular repeat motifs such as the tetratricopeptide repeat (TPR), ankyrin, and leucine-rich repeat (LLR), with slightly differing strategies, in several laboratories *(13–16)*. Such motifs contain a repeating unit of amino acid sequence (34 amino acids in TPR proteins and 33 amino acids in ankyrins, for example), which can occur in tandem arrays of varying length in different proteins. Most characteristically the repeat proteins are "open-ended" structures, which can have varying numbers of repeat units stacked in tandem arrays, forming a domain (**Fig. 1**).

Different repeat motifs form distinct three-dimensional structures. In most cases, tandem repetition of the motif gives rise to elongated structures *(17)*, but with different twists and curvatures *(18)*. The TPRs, for example, form right-handed superhelical structures. Repeat proteins are common, particularly in the higher organisms; comprising 5.3% of the metazoan proteins annotated in the Swiss-Prot database *(17)*. Also, LLR, ankyrin, and TPR repeats are among the 20 most common motifs found in PFAM (i.e., protein families) (http://www.sanger.ac.uk/Software/Pfam/browse/top_twenty.shtml). Thus far, in all known examples, the nonglobular repeat domains appear to be involved in macromolecular interactions, although the exact function and identity of the binding partner is not known for the majority of repeat proteins. In this chapter, we describe methods used in our laboratory to design synthetic consensus TPR proteins and discuss other developments in protein engineering studies of repeat proteins, where applicable.

2. Materials

2.1. Computer Calculations and Software

1. Personal computer (PC) with Windows 98 or 2000 (Microsoft) or equivalent.
2. A UNIX workstation or a PC with Linux operating system (*see* **Note 1**).
3. Microsoft Excel (Microsoft).
4. ClustalX (obtained from ftp://ftp-igbmc.u-strasbg.fr/pub/ClustalX/).

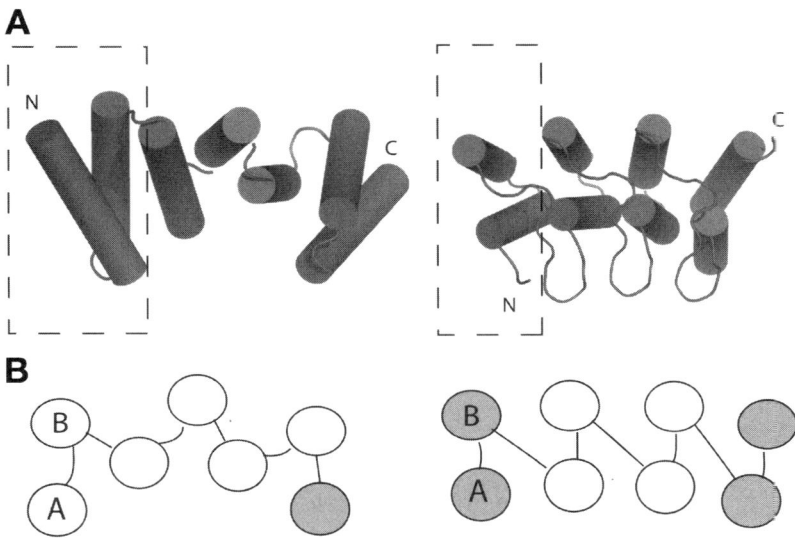

Fig. 1. Structural features of repeat proteins. (**A**) A cartoon representation of the crystal structures of a consensus three-repeat TPR protein(left) *(13)* and four-repeat consensus sequence ankyrin repeat protein (right) *(15)*. Helices are presented as cylinders; one repeat (two helices) is boxed with dashed lines, N- and C-termini are marked with "N" and "C," respectively. (**B**) A schematic two-dimensional presentation of the packing interactions between helices and repeats in both proteins. Capping units are shaded in gray. One repeat in both cases comprises of two helices marked A and B. Note the different capping requirements: in TPRs, the B-helices do not interact with each other, and therefore only one C-terminal "A" helix is required for capping, whereas in the case of ankyrins and other repeat proteins with similar helical arrangements, a whole repeat with one hydrophilic face is needed for capping at both N- and C-terminus.

5. PSI-BLAST, downloaded from the National Center for Biotechnology Information (NCBI) web site (http://www.ncbi.nlm.nih.gov/BLAST).
6. The TPR-containing sequences downloaded from SMART database (http://smart.embl-heidelberg.de/).

2.2. Gene Synthesis and Cloning

1. ThermoPol, EcoPol, T4 ligase buffers, and restriction enzyme buffers (New England Biolabs).
2. Enzymes: Klenow fragment, DeepVent DNA polymerase, T4 ligase, calf intestinal alkaline phosphatase, and restriction enzymes (New England Biolabs).
3. DNA oligonucleotides.
4. Whatman glass fiber paper.
5. Dialysis membrane with low mol wt-cutoff.

6. Ultrafree-MC centrifugal filter devices (low binding Durapore membrane; Amicon).
7. The *Escherichia coli* strain DH10B.
8. Expression vector, preferably with a cleavable His-tag encoded.

2.3. Protein Expression and Purification

1. Ni-NTA agarose (Qiagen).
2. Econo-columns (Biorad).
3. Luria Bertani (LB) medium.
4. IPTG (isopropyl-β-D-thio-galactopyranoside).
5. The *E. coli* strain BL21 (DE3).
6. Lysozyme.
7. DNAse I.
8. Complete EDTA Free-protease inhibitor cocktail tablets (Roche).
9. HiLoad Superdex HR-75 gel filtration column (Amersham Biosciences).
10. A fast performance liquid chromatography (FPLC) system.

3. Methods
3.1. Design for Structure of Repeat Proteins

Here we describe a strategy to obtain fully synthetic consensus repeat proteins, using the TPR repeat as an example, starting with (1) description of retrieving or generating the multiple sequence alignment and generating the consensus, (2) considerations for the final consensus design, and (3) proceeding to the actual gene synthesis and protein purification (**Fig. 2**).

3.1.1. Multiple Sequence Alignment Generation and Analysis

Initially one must obtain the consensus sequence by either: (1) generating the alignment manually or (2) using existing alignment in a domain sequence database such as PFAM or SMART *(19,20)* (http://www.sanger.ac.uk/Software/Pfam/ and http://smart.embl-heidelberg.de/), if suitable.

3.1.1.1. GENERATING MULTIPLE SEQUENCE ALIGNMENTS MANUALLY

One is obliged to generate the sequence alignment manually, if it is not available, or if one wants the alignment of particular combination of domains (in this context, selecting repeat domains that have the same number of tandem repeats) for analysis. This could be desirable, for instance, if one wishes to take into account possible differences between "internal" and "external" (capping) repeats, or if one wishes to work on a functionally related set of proteins, or is interested in covariation or evolutionary analysis of a set of proteins with certain domain or repeat combinations (**Fig. 2**). Also, in general, one must be extremely cautious with the alignments created automatically, because they have not been verified manually, and, for example, the exact definition of domain boundaries can be a problem *(21,22)*.

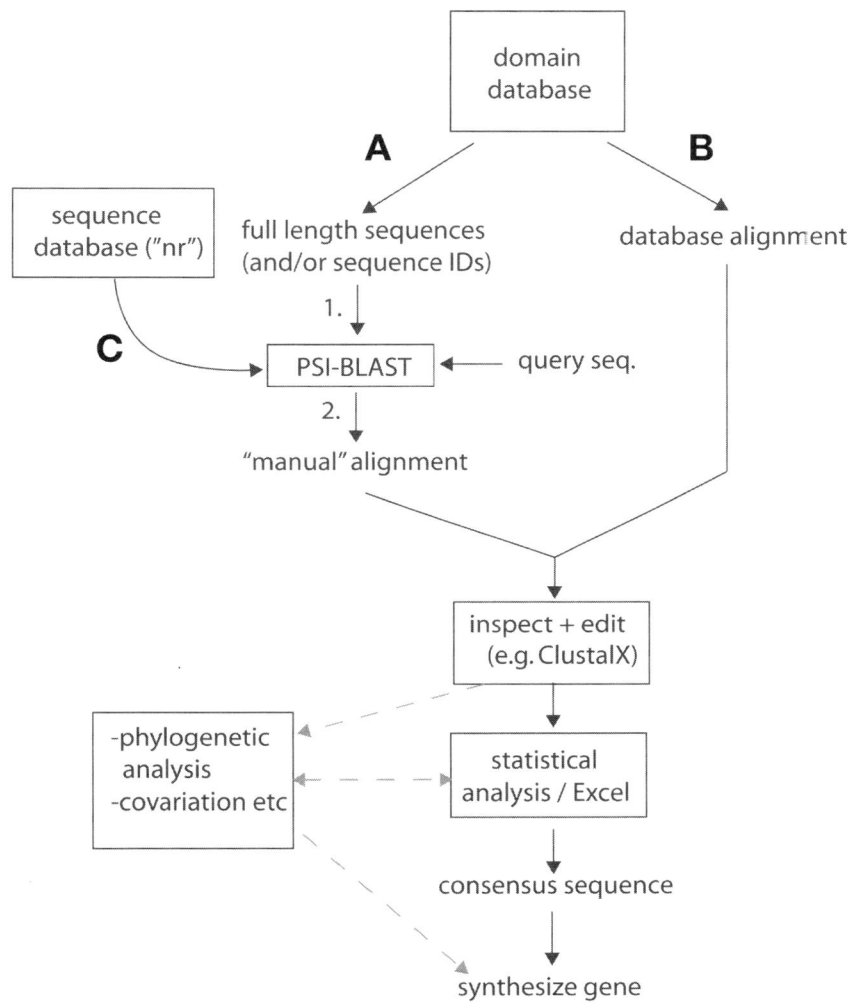

Fig. 2. A flowchart illustrating the protocols suggested for obtaining the consensus sequence. Initially the data for the multiple sequence alignment is retrieved. either from an automatically generated domain (motif) alignment database, here SMART, or from a sequence database (such as the nonredundant database "nr") with command line PSI-BLAST (Route C). Route A describes how an alignment is generated in the case described in the text for three-repeat TPRs (where actually we wish to analyze a family of sequences containing three repeats in tandem). Route B is the one used with an automatically created alignment (e.g., PFAM or SMART). Here step 1 includes the sorting of the data, as described in the text. If an alignment is generated with PSI-BLAST without the preexisting knowledge of targets from, for example, SMART of PFAM, some sorting could be done at step 2 (e.g., removal of multiple hits). Dashed lines indicate additional analysis that could be done on the dataset.

We use PSI-BLAST to generate the multiple sequence alignments, because it allows control over the search criteria and many options for output format. PSI-BLAST generates the multiple sequence alignments with a position-specific scoring matrix *(23)* (*see* **Note 2**). An alternative is to use a program which uses a profile Hidden Markow Model (HMM) method *(24)* in the sequence searches to score the results (*see* **Note 2**). We will provide an example of how to generate the alignment with PSI-BLAST (although we use the data preselected for the SMART database and merely sort it using simple UNIX commands and align it with PSI-BLAST). The example given here is used to generate a three-repeat TPR alignment, with command line PSI-BLAST.

PSI-BLAST can be downloaded from the NCBI web site (ftp:// ftp.ncbi.nlm.nih.gov/blast/) (download the BLAST package and see the README file for instructions on installation). BLAST can be run on essentially any computer operating system (see the README file for details, and see www.ncbi.nlm.nih.gov/BLAST/docs/blastpgp.html for documentation on BLAST usage). The program to use for PSI_BLAST is called "blastpgp" (the gapped BLAST executable).

An example on the command line usage of BLAST (in UNIX or Linux) would be:

```
blastpgp -d swissprot -o 3TPR_search_blast.out -G 32767 -e
0.001 -E 32767 -P 0 -i pp5.seq -m 4 -j 3
```

where -d swissprot defines the database and -o defines the output alignment file, –G and –E define the gap opening and extension penalties, which here are set to very restrictive values (in PSI-BLAST, one has to choose from preset selection of values, hence these particular numbers), -i is the actual search sequence (here protein phosphatase 5 [PP5] TPR domain sequence), -e defines the e-value threshold, –m defines sequence alignment format, and –P 1 means one hit only (0 = multiple) to each homolog found, and -j 3 defines number of iterations for PSI-BLAST (note that if –j flag is not given no PSI-BLAST profile search is performed, but a single round of gapped BLAST). By typing on the command line:

```
blastpgp -
```

you get all the options for the command.

Also download the database against which the searches will be performed; for example, the nonredundant sequence dataset (uniprot or "nr", or other dataset of interest) from ExPaSy (http://www.expasy.ch/) or the NCBI web site (ftp://ftp.ncbi.nlm.nih.gov/blast/). Note, that for use with PSI-BLAST, the downloaded databases must be first formatted with the "formatdb" program provided with the BLAST software package.

Here, we use the SMART database to generate an alignment of three-repeat TPR domains, because single repeat alignments of TPRs, and the information on sequences containing them, already exist (both in PFAM and SMART). From SMART the full sequences containing identified TPR repeats can also be downloaded. Here, we will extract three-repeat sequences from these. First, we retrieve information about the sequences that contain three repeats and then we fetch those and align them. This is done as follows:

1. Obtain list of sequences by downloading the (single-repeat) TPR alignment from SMART in FASTA format (below is a fragment):

```
>smart|TPR-ENSP00000316699/460-493 TETRATRICOPEPTIDE REPEAT PROTEIN 7
(TPR REPEAT PROTEIN 7) (FRAGMENT). [Homo sapiens]

AHALHLLALLFSAQKHHQHALDVVNMAITEHPEN

>smart|TPR-ENSP00000316699/639-672 TETRATRICOPEPTIDE REPEAT PROTEIN 7
(TPR REPEAT PROTEIN 7) (FRAGMENT). [Homo sapiens]

HSVLYMRGRLAEVKGNLEEAKQLYKEALTVNPDG

>smart|TPR-ENSP00000316699/707-740 TETRATRICOPEPTIDE REPEAT PROTEIN 7
(TPR REPEAT PROTEIN 7) (FRAGMENT). [Homo sapiens]

HEAWQGLGEVLQAQGQNEAAVDCFLTALELEASS

>smart|TPR-ENSP00000316740/208-241 NYD-SP14 PROTEIN. [Homo sapiens]

AEAHMHMGLLYEEDGQLLEAAEHYEAFHQLTQGR

>smart|TPR-ENSP00000316740/300-333 NYD-SP14 PROTEIN. [Homo sapiens]

AEASYYLGLAHLAAEEYETALTVLDTYCKISTDL

>smart|TPR-ENSP00000316740/340-373 NYD-SP14 PROTEIN. [Homo sapiens]

GRGYEAIAKVLQSQGEMTEAIKYLKKFVKIARNN

>smart|TPR-ENSP00000316740/380-413 NYD-SP14 PROTEIN. [Homo sapiens]

VRASTMLGDIYNEKGYYNKASECFQQAFDTTVEL

>smart|TPR-ENSP00000317097/3-36 I-KAPPA-B-RELATED PROTEIN.[Homo sapiens]
TRLYLNLGLTFESLQQTALCNDYFRKSIFLAEQN

>smart|TPR-ENSP00000317097/43-76 I-KAPPA-B-RELATED PROTEIN. [Homo sapiens]
FRARYNLGTIHWRAGQHSQAMRCLEGARECAHTM

>smart|TPR-ENSP00000317097/152-185 I-KAPPA-B-RELATED PROTEIN. [Homo sapiens]
MVICEQLGDLFSKAGDFPRAAEAYQKQLRFAELL

>smart|TPR-ENSP00000317097/193-226 I-KAPPA-B-RELATED PROTEIN. [Homo sapiens]
AIIHVSLATTLGDMKDHHGAVRHYEEELRLRSGN
```

   Here one notices several hits for the same sequence (for example, for NYD-SP14, above), because there are several repeats (TPRs) in each sequence.
2. Next count the number of repeats in each sequence in the downloaded alignment (file called tprlist.pl):

```
grep 'smart' tprlist.pl | cut -b1-31 | uniq -c > countlist.txt
```

This chooses lines with "smart" in them, then take first 31 characters of these, and with "uniq -c" counts them ("uniq -c", prints the number of times each line occurred along with the line, type "man uniq" on the command prompt to see manual pages in UNIX or Linux). Contents of the output file "countlist.txt" will look like this:

```
3  >smart|TPR-sptrembl|Q98EY4|Q98EY4

3  >smart|TPR-sptrembl|Q98GL0|Q98GL0

4  >smart|TPR-sptrembl|Q98GL1|Q98GL1

6  >smart|TPR-sptrembl|Q98GY3|Q98GY3

6  >smart|TPR-sptrembl|Q98HZ1|Q98HZ1

14 >smart|TPR-sptrembl|Q98IX1|Q98IX1

11 >smart|TPR-sptrembl|Q98IX2|Q98IX2

4  >smart|TPR-sptrembl|Q98J72|Q98J72

3  >smart|TPR-sptrembl|Q98L57|Q98L57

3  >smart|TPR-sptrembl|Q98LH3|Q98LH3

9  >smart|TPR-sptrembl|Q99L66|Q99L66

6  >smart|TPR-sptrembl|Q99L77|Q99L77
```

3. Then pick the ones with "3" in them to get those that, according to SMART, contain three repeats.

```
grep '3' countlist.txt > count_3s.txt
```

The contents of the file count_3s.txt will look like this:

```
3  >smart|TPR-swissprot|P31759|FRZF_MYXXA

3  >smart|TPR-swissprot|P39833|NLPI_ECOLI

3  >smart|TPR-swissprot|P42842|YN53_YEAST

3  >smart|TPR-swissprot|P43988|Y366_HAEIN

3  >smart|TPR-swissprot|P44130|YCIM_HAEIN

3  >smart|TPR-swissprot|P45576|YCIM_ECOLI
```

4. Next change *all* field separators to "|" (in UNIX, do this with, for example, "vi," "emacs," or "jot" text editor, using a macro. You need to do this to the subset, which is *not* from TrEMBL or Swiss-Prot, such that the sequence identifiers are for all sequences in the third field). Then we use "cut" to extract the sequence identifiers for finding the sequences containing three TPR repeats:

```
cut -d'|' -f3 count_3s.txt > count_3s_IDs.txt
```

(this copies the third field (-f3), and fields are separated by "|.")

5. At this point we need to produce the first 3-repeat TPR alignment with PSI-BLAST on the command line:

```
blastpgp -d SMART_TPRs -o smart_psiblasted.out -b 2500 -G 13 -E 3 -e 1
-h 0.1 -P 0 -B 3rep_tpr.aln -i pp5.seq -m 6 -M BLOSUM45 -j 3
```

This actually results in all the sequences being aligned. This alignment is then edited to contain only one alignment per sequence (we simply chose the first alignment for each sequence; this could be changed, to increase the size of the final dataset by some programming, if desired). This is done with command "uniq":

```
uniq -w 18 smart_psiblasted.out > TPRs_uniqs.out
```

We then use command "grep -f" to extract the three-repeat sequences with the sequence identification code list for sequences containing three repeats (file "count_3s_IDs.txt") that we produced in **Note 3**:

```
grep -f count_3s_IDs.txt TPRs_uniqs.out > 3TPR_alignment.txt
```

("grep -f pattern_file" reads one or more patterns from the file named pattern_file. Type "man grep" on your computer, for full manual entry for the command in UNIX or Linux.)
The commands "grep" and "uniq" should keep the line order, so alignment order will not be lost.

6. The alignment is edited by hand with ClustalX *(25)* on a PC to contain only sequences that contain (at least) all 34 positions aligned from the presumed first amino acid of the first TPR motif to the last amino acid of the last TPR motif (position 34 of the third TPR) (*see* **Note 4**).

One could more exhaustively manipulate the data, as needed, to help increase either the size or quality (e.g., in terms of location of insertions) of the alignment by writing small computer programs. However, the procedure described is straightforward and produces good quality results and is generally applicable for generating an alignment for any combination of domains.

3.1.1.2. Using Existing Sequence Alignments or Sorted Sequence Datasets

One may also download the presumably full or complete alignments from available sequence databanks (PFAM or SMART; *see* **Subheading 3.1.1.**). ClustalX could be used for editing alignments, and alignments can be easily viewed on desired computer platform (e.g., PC).

3.1.1.3. Obtaining the Consensus and Amino Acid Distribution Data

For generating the actual statistical data on the alignment, we suggest using a generally available program such as Microsoft Excel on a PC. Although several alignment programs give a consensus (also available for example on the PFAM web site on the alignments available there), it is good to use a software capable of statistical analysis to be able to obtain the frequencies and other data for the different amino acid positions. Here we provide a general example of putting data into Microsoft Excel and extracting the consensus sequence.

1. Read the alignment into Excel data separated by spaces or tabs into columns. Specifically, separate the data on each row (for one sequence) into two columns: first sequence identifier, then the sequence (the whole sequence into one cell).

2. After this, number the following columns (here, from third column onwards) on a row above the first data row, according to the sequence positions: e.g., if there are 46 amino acid positions, number the next 46 columns 1 to 46 (on the first row used).

3. Then using function "MID" in Excel, choose the correct letter to each numbered (1–46) column by specifying, for instance, a function "=MID($B5,$G2,1)." Here, "$B5" might tell the program to read the alignment from the fifth row and column B (cell B5), "$G2" tells it to start from the position in sequence you specify in the cell "G2," and finally "1" tells the program to print one character (more specifically, only the first character from the specified point onwards, i.e., the letter coding the amino acid at this position). Now you have a table with a cell for each amino acid position.

4. Next, obtain the number of occurrences for each amino acid in a certain column (coded by a single a letter) representing the particular position in the alignment. On a separate (or the same) sheet, use the command "COUNTIF (range, criteria)," where "range" is the range of cells from which you want to count cells (e.g., in one particular column). "Criteria" is the criteria in the form of a number, expression, or text that defines which cells will be counted (e.g., cells containing "A"); this will give you the number of occurrences of alanine at this position (e.g., one column representing a sequence position on the previous spreadsheet) in the alignment. An example of such data layout is given in **Table 1**, e.g., second row of second column will in the example give the number of occurrences of alanines (A) in this position (1. position in alignment; *see also* Microsoft Excel help for examples the usage of function "COUNTIF"). From this, it is easy to calculate frequencies of occurrence for each amino acid at a given position and determine the consensus and frequencies of occurrence for all amino acid at each position.

5. The consensus amino acid at each position can be determined in Excel with the function "VLOOKUP", which processes data arrays. Values are chosen from one column and a value in the same row from another column is returned (for example, first the largest number is chosen from a column [using the function MAX], such as those labeled 1–8 in **Table 1**, and then the letter in the first column of **Table 1** representing the amino acid, with that value, could be returned in the cell where the function is coded). *See* Microsoft Excel help on "VLOOKUP" for examples.

3.1.2. Criteria for Design

After the final alignment of the largest dataset is obtained, the positions with a strong preference in the consensus will be identified. One can then choose whether to use (synthesize) the complete consensus sequence or to include only the most conserved residues in a natural template *(11,12)*. It probably makes little difference how the identities of the poorly conserved positions are

Table 1
Example of a Microsoft Excel Table for Tabulating
Occurrences of Amino Acids at Various Positions

Amino acid	Position in motif sequence							
	1	2	3	4	5	6	7	8
A	2232	434	2204	165	307	370	175	2562
C	110	24	150	188	243	53	48	54
D	102	699	72	23	70	185	53	1
E	223	1161	179	105	171	388	150	3
F	171	127	153	347	553	149	42	3
G	402	182	351	19	217	464	16	4000
H	153	91	23	537	208	138	19	1
I	424	236	541	167	340	114	622	32
K	101	772	83	134	431	281	790	1
L	316	353	750	1849	754	577	2769	1
M	52	95	106	131	166	144	313	2
N	48	283	67	109	748	1994	23	0
P	674	173	143	15	4	3	0	0
Q	46	426	80	180	143	421	212	0
R	85	494	63	235	246	582	1202	1
S	537	367	451	115	461	387	50	193
T	266	314	540	26	300	174	48	22
V	620	340	652	102	290	163	164	7
W	86	98	44	865	24	44	31	0
Y	239	218	235	1575	1211	256	160	4

Each cell with a number for an occurrence is coded by the function "COUNTIF."

chosen in this design strategy (but *see* **Note 5**). However, there may be other special considerations on amino acid usage to take into account (*see* **Note 6**).

The size of the protein scaffold is another issue for consideration. For example, in our design, we chose first to make a TPR protein with three repeats *(13)*, because three repeats is the most common size for a TPR domain *(26)* and the smallest verified, natural, functional unit *(27)*. Similarly, Plückthun and coworkers chose to design five and six repeat consensus ankyrin proteins for their functional design *(28)*, because these were the most common sizes of ankyrin domains found in natural proteins.

Additionally, one may also wish to consider whether to use simple frequencies for defining the consensus, or global propensities, which take into account the "rareness" of a given amino acid (for discussion, *see* **refs. *13,26,29***). In the case of consensus TPR domains with three repeats, both kinds of sequence repeats have been constructed and found to be more thermostable than the natural three-repeat TPR domains *(13*; A. L. Cortajarena, T. Kajander, C. Wilson, and L. Regan, unpublished data). Also crystal structures from the global propensity based two- and three-repeat consensus TPR proteins have been solved and shown to fold as do typical TPR structures *(13)*.

Interestingly, all the crystal structures of natural TPRs reveal an additional helix, C-terminal to the last TPR repeat, positioned as a TPR "A" helix, after the last repeat (**Fig. 1**). We have observed that such "capping" helices may have an important role in determining the stability or solubility of the TPR domains (T. Kajander, A. L. Cortajarena, E. Main, and L. Regan, unpublished data). It would seem that the last repeat needs to be capped with and additional hydrophilic "A" helix (**Fig. 1B**). Similarly, for ankyrin repeat proteins, it has been independently observed that solubility is poor without capping structural elements and may be improved by incorporating N- and C-terminal capping repeats from natural proteins *(16,30)*. Thus, this appears to be a general feature of the repeat proteins, and capping elements should be taken into consideration in the design to ensure that the design product is a well-behaved protein.

3.1.3. Gene Synthesis and Cloning

For generating the actual gene, we use synthetic overlapping oligonucleotides of approx 80 to 100 base pairs (bp) in length. For example, a three-repeat TPR gene with a seventh capping helix may be constructed using oligonucleotides, which overlap each other by 18 bp (*see* **Note 7**). Initially, longer dsDNA fragments from oligonucleotide pairs 1-2, 3-4, and 5-6 are generated by Klenow extension and the synthesis is finished with overlap extension polymerase chain reaction (PCR) (**Fig. 3**).

1. To set up the Klenow extension reaction for each oligo-pair, mix 200 pmol of each oligonucleotide (330 Da/bp or calculate exact mass from sequence) with 20 µL 10X EcoPol buffer and add sterile water to 200 µL.
2. Place the sample into boiling water (~250 mL) and let cool down to room temperature (~3–4 h) to anneal the oligonucleotides.
3. Add 10 U of Klenow fragment and 1 mM dNTP (2 µL of a 100 mM dNTP solution). Incubate for 15 min at room temperature (extension reaction of the annealed oligonucleotides), add 4 µL of 0.5 M EDTA and heat to 80°C for 20 min to inactivate the Klenow fragment.
4. Check for the presence of Klenow extension products on a 1.5% agarose gel.
5. Perform overlap extension PCR. The number of cycles depends on the number of oligonucleotides needed to cover the complete gene sequence. For three-repeat TPRs, six oligonucleotides are used to generate three Klenow-extended fragments; therefore, two consecutive PCR reactions are needed to generate the gene. For a 100 µL reaction: templates (Klenow reaction product or overlap PCR reaction product) (1 µL each); primers (1 µM each); 10X ThermoPol buffer (10 µL); dNTP mix (1 µL 100 mM solution); thermostable DNA polymerase (Deep Vent) (2 U); water to 100 µL.

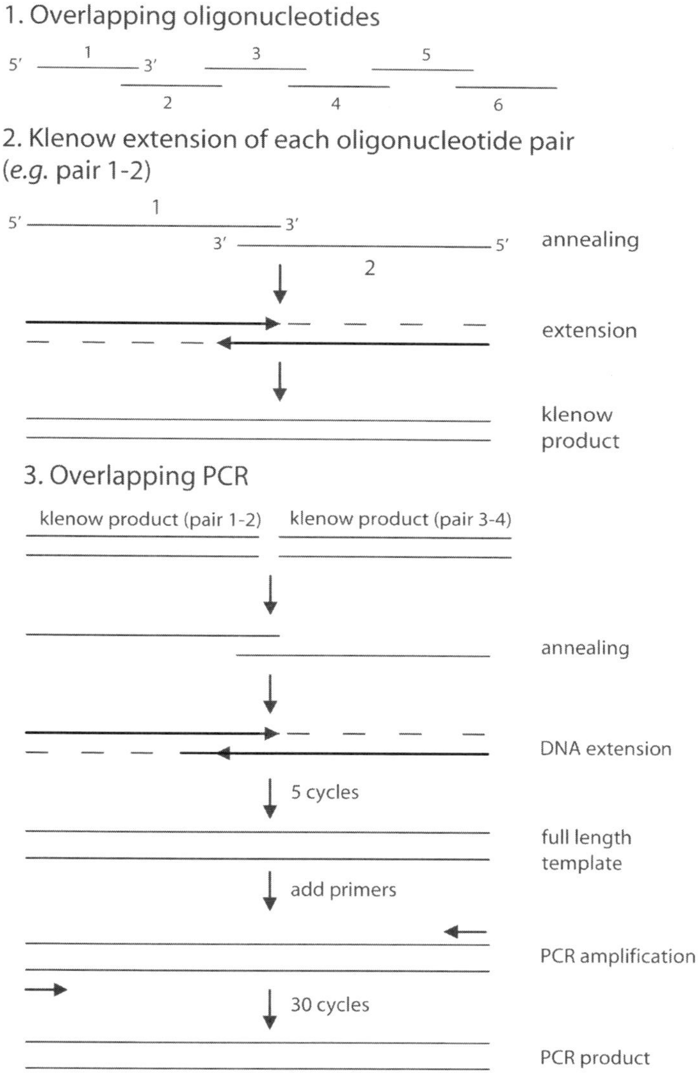

1. Overlapping oligonucleotides

2. Klenow extension of each oligonucleotide pair (*e.g.* pair 1-2)

3. Overlapping PCR

Fig. 3. Schematic representation of gene synthesis using Klenow extension and overlapping PCR. (1) Oligonucleotides that cover the whole gene sequence with an overlap of 18 bp between each other. (2) Klenow extension of each oligonucleotide pair, for example pair 1-2. Denatured oligonucleotides anneal at the region of overlap and are extended to form full-length, double-stranded DNA Klenow product. (3) Overlapping PCR using Klenow reaction products, for example, of pair 1-2 and pair 3-4 as a template. Denatured overlapping Klenow reaction products anneal, and are extended in the first five cycles of the reaction without primers. Double-stranded, full-length DNA is amplified by PCR using primers that anneal at the 5' and 3' end of the fragment.

6. First run 5 cycles without primers to generate a full-length template from the overlapping templates. Add primers required to get the full length fragment (*see* **Note 8**) and run 30 cycles. PCR parameters (times and temperatures) are standard but may need to be optimized for different PCR reactions. Carry out PCR under the following conditions: 5 min at 95°C denaturation at startup; 30 main cycles (or 5 cycles): 1 min at 95°C (denaturation), 45 s at 55°C (annealing), 1 min/kb at 72°C (extension); and last step: 10 min at 72°C (extension).

7. The final product is checked on a 1% agarose gel and further purified from a 1.5% agarose gel by electrophoresis onto Whatman glass fiber paper, backed with glass fiber paper (with a mol wt-cutoff of 1000 Da), and DNA is eluted off the glass fiber paper, for example by centrifugation with an Ultrafree-MC filtration unit (1.5-mL tube, for sample volume <0.5 mL, *see* **Subheading 2.2.**) and ethanol precipitated according to standard procedures. Any commercial methods for purification of DNA from agarose gels can also be used.

8. Digest purified DNA with the appropriate restriction enzymes. Double digestion with *Bam*HI and *Hind*III in *Bam*HI buffer at 37°C for 3 h, is used for TPR genes.

9. To clean the digested inserts, extract with phenol and precipitate with ethanol according to standard procedures, finally dissolve in sterile distilled water and measure concentration by ultraviolet spectroscopy at 260 nm.

10. In the same way, prepare the vector for cloning. For TPR genes we used pProEX-HTa vector (Invitrogen), this vector is no longer commercially available. Any expression vector for *E. coli* (such as pET vectors) can be used, preferably including a tag for ease of protein purification. Digest at 37°C for 3 h, adding 10 U of calf intestinal alkaline phosphatase for the final 30 min. The digested vector is gel purified from 1.5% agarose gel as described in **step 7**.

11. The final products are then ligated with T4 ligase using the following conditions: digested dephosphorylated vector (100 ng or 30 fmol); digested insert (100 ng or 300 fmol); 10X ligase buffer (2 µL); T4 DNA ligase (400 U); water to 20 µL. Incubate overnight at 16°C.

12. Transform 1 µL of each ligation reaction with 30 µL electrocompetent *E. coli* DH10B. Plate the transformations onto LB agar with the appropriate antibiotic.

13. Grow some isolated colonies in LB medium and prepare DNA by standard plasmid Miniprep procedure.

14. Verify the constructs by restriction digestion and agarose gel electrophoresis (1% gel). Confirm clones containing correct size DNA by DNA sequencing.

3.1.4. Protein Expression and Purification

The aim of consensus design is to produce well-behaved, stable, and soluble proteins, which are expressible in bacteria. Below we give an example of typical expression and purification protocol for the consensus TPRs in *E. coli* BL21 (DE3), which could be modified for other proteins. The yields of purified soluble protein we typically get for consensus TPRs are approx 50 mg/L.

1. Transform plasmids encoding desired proteins into *E. coli* BL21 (DE3) cells.

2. Grow cultures in LB medium with appropriate antibiotic, at 37°C to an OD_{600} of 0.6 to 0.8, then induce with 0.6 mM IPTG and grow 3 to 5 h at 30°C.

3. Harvest the cells by centrifugation and resuspend them in lysis buffer (50 m*M* Tris-HCl pH 8.0, 300 m*M* NaCl, 5% glycerol). Add 1 mg/mL lysozyme, ½ tablet of protease inhibitor cocktail, 100 U DNAse I, and incubate on ice for 30 min.

4. Sonicate the suspension and clear the resulting lysate by centrifugation at 25,000*g* for 30 min.

5. Purify His$_6$-tagged proteins from the supernatant by affinity chromatography using Ni-NTA agarose according to manufacturer's (Qiagen) instructions.

6. Cleave His$_6$-tag with TEV protease: incubate at room temperature overnight; add 40 to 50 µL TEV-protease (10 U/µL).

7. Remove uncleaved protein, TEV-protease, and cleaved His$_6$-tag: first dialyze the sample back into the lysis buffer (without inclusion of glycerol, DNAse, lysozyme, or inhibitors) to dilute out excess imidazole, then incubate with 1 mL Ni-NTA agarose in batch at 4°C for 1 h. Apply sample into a column and collect the flowthrough.

8. Finally, to obtain high purity protein, add a size exclusion chromatography step (e.g., HiLoad Superdex HR-75 gel filtration column).

3.2. Considerations for Design of Function

The nonglobular repeat proteins have several features that make them attractive targets for protein engineering studies. First, they are simple in terms of their structure (defined length of the repeat sequence, equivalent key positions in each repeat, and buried interfaces between repeat units conserved similar), and each repeat is related to its nearest neighbor by a simple transformation, creating a structure with a convex and a concave face (**Fig. 1**), typically with ligand binding site on a particular face of the structure. Thus, only certain positions will be available for binding and some others will play structural roles. Second, the structures are "open-ended" and can be extended to desired length, which open the possibility for designing for binding of ligands of different size and shape, or alternatively building either multifunctional or multivalent proteins with several binding sites. Binding functions could be engineered on to the repeat protein scaffolds as on any protein scaffold. The benefit of having a designed consensus scaffold is that it provides a more stable starting point, allowing more (probably destabilizing) mutations to be introduced for engineering a binding function.

For designing of functions on the designed structural scaffolds, most importantly, one must choose which residue positions are important for function; to this end, there have been different methods applied or suggested. Initially, binding residues could simply be grafted from another system with a known binding activity *(31–34)*. We have used statistical approaches (a functional-consensus strategy) to identify the binding residues involved in a particular ligand binding activity (here, Hsp90-binding by TPR domains). We compared the frequency of occurrence of each residue in each of the positions for all the TPR sequences in the PFAM database with the frequency of occurrence of each residue in sequences of a subset of TPR domains with Hsp90-binding specificity. Positions in which there is a significant difference in the amino acid preference for Hsp90-binding TPR domains (compared with all TPR sequences) were assumed to be important for binding. These residues, which are a common feature of only Hsp90-binding TPRs, were introduced into the consensus TPR protein *(13,34)*. Often, results from such grafting experiments may require refinement; for example, by incorporating additional designed changes or employing selection or screening strategies *(33,35)*.

Plückthun and coworkers generated libraries based on consensus designs for both ankyrin and leucine-rich repeat repeats, in which the residues involved in binding were randomized *(14,36)*. In the case of ankyrins, the positions to be randomized in each repeat were chosen by inspection of natural ligand complex structures, giving six positions for randomization in each repeat. The resulting vast libraries were subjected to *in vitro* selection by ribosome display, and nanomolar binders of specific ligands were selected *(28)*. Similarly, in theory, selection could be applied to large libraries via mRNA display *(37)* or smaller libraries could be subjected to selection for function by phage display *(38,39)*.

In this laboratory, approaches based on sequence analysis against a nonbiased reference dataset *(29,40)*, have been applied to identify binding sites as hypervariable regions in datasets of sequences of structurally homologous proteins with differing functions *(41)*. The main requirement for this analysis is a large enough dataset (an alignment of hundreds to thousands of sequences). This is could be a general strategy for a rational and objective way to target the "functional set" of residue positions for engineering binding functions on structural scaffolds.

4. Notes

1. We assume that most institutions have at least access to central (UNIX or Linux) computing facilities. BLAST can be run on any platform, whereas the rest of the work involving UNIX or Linux operating systems deals mainly just with sorting the data and is mostly not computationally expensive, and data could be, for example, moved to a central computer from a PC through an ftp connection for further sorting.

2. The profile, or position specific scoring matrix, is generated from a "seed" alignment with a small number of sequences, which can be obtained by retrieving homologous sequences with PSI-BLAST on-line (http://www.ncbi.nlm.nih.gov/BLAST/) and taking the alignment from PSI-BLAST, or aligning them with, for example, ClustalX or another program. After the position-specific scoring matrix is obtained in this way, the full multiple sequence alignment may be generated, either by using programs that use profile Hidden Markov Models (HMMs, for review, *see* **ref. 24**), such as HMMER (http://hmmer.wustl.edu/) or PSI-BLAST, which uses a related scoring method *(42)* to search the databanks to find as large

a dataset as possible. For PSI-BLAST, one provides the query sequence of choice, and the program will generate the scoring matrix after each round automatically from the sequence homologue set found, without user intervention, and then uses this to search the database again, iteratively, to convergence *(42)*.

3. This is the most computationally expensive step in the procedure, which is why it is a good idea to edit the alignment to contain only one hit per sequence before doing this.

4. We allowed gaps, because otherwise the dataset would have been much smaller. The aligned positions were fairly easily countable from the known nature of the motif positions (e.g., position 32 will be Pro in most cases) and told apart from gap regions by eye. Also in the end-only positions with 50% (this could have been 90% because most chosen positions were there >90% of the time) occupancy were chosen for the consensus analysis; this strategy effectively gets rid off insertions not aligned with rest of the sequences. Gaps could be restricted to only loop regions in cases such as the TPRs, whereas for globular domains it might be more difficult.

5. An important addition to consensus-based design could be covariation analysis to estimate the importance of pairwise, or networks of, interactions present in the alignment. This has been applied to TPRs *(29)*, and it revealed that, although the consensus TPRs are highly negatively charged, the natural sequences are mostly near neutral, often with conserved interaction networks between charged residues, lacking in the consensus sequence. Indeed, engineering one such network to the three repeat consensus TPR protein increased the thermostability significantly, even from the already high value $T_m = 83°C$ to an extrapolated T_m of 103.5°C *(34*; A. L. Cortajarena and L. Regan, unpublished data). Here, certainly, the approach appears powerful. We believe, however, that it is most informative first to construct a straight "consensus protein," and to then include variations to amino acid sequence suggested by covariation analysis as an additional feature.

6. It is worthwhile paying attention to the sequence content at the amino acid level to make sure the sequence is predicted reasonably stable and that, for example, no common protease sites are generated, in particular if protease cleavage of a fusion protein is used as a part of the purification strategy. Also, we typically replace cysteine residues from the consensus with the second most probable residue, to avoid oxidation problems (e.g., nonphysiological oligomerization and aggregation). One may also wish to include amino acids such as methionines to help structure determination efforts (to be able to produce SeMet-labeled protein for structure solution by X-ray crystallography) and tryptophan or tyrosine for ease of quantization and detection at 280 nm.

7. When designing a target gene sequence with several *identical* consensus repeats at the amino acid level, one has to keep the DNA sequence unique for each repeat, taking advantage of the degeneracy of the genetic code. This becomes crucial when a need for incorporation of specific mutations in consensus sequence is anticipated. To design oligonucleotides for gene synthesis, the target gene sequence is divided into the smallest possible number of 80 to 100 bp fragments, which would satisfy the following requirements: (1) The overlapping region between

oligonucleotides should have a minimum length of 18 bp to allow stable anneal-ing and consequently efficient Klenow extension and PCR amplification; (2) the 18-bp overlapping region between different oligonucleotides should be placed in a region of the sequence where the homology with other parts of the sequence is lower. That is especially important in the construction of genes encoding repeat proteins; (3) oligonucleotides should have G or C residue at the 3' end (this helps to ensure correct binding at the 3' from the stronger hydrogen bonding between G and C residues; and (4) stable secondary structures within the oligonucleotide and duplex formation from self-complementarities should be avoided.

8. Considerations for design of the primers for PCR amplification steps: incorpo-rate at the 5' end of both 5' and 3'-flanking primers desired restriction enzyme cleavage sites for consequent cloning into an expression vector. Precede the restriction site by six extra nucleotides to allow efficient cleavage. Both primers should have at least 18 bp homologous to the template DNA and a G or C residue at the 3'-end. In addition, 18-bp internal primers to amplify by PCR gene frag-ments will be needed for intermediate PCR amplification step. In the case of three TPR repeat genes, one additional primer for the first PCR amplification was needed.

Acknowledgments

We wish to thank Dr. Thomas Magliery for providing the Excel protocol for calculating the consensus sequence and for many stimulating discussions. We thank Dr. Chern-Sing Goh for advice on database searches and members of the Regan lab for their input and insight into this work. This work was supported by postdoctoral fellowships from the Helsingin Sanomat Centennial Founda-tion (Finland) (to T. Kajander) and from the Spanish Ministry of Education, Culture and Sports (to A. L. Cortajarena), and by the National Institutes of Health (to L. Regan).

References

1. Regan, L. and DeGrado, W. F. (1988) Characterization of a helical protein designed from first principles. *Science* **241,** 976–978.
2. Hecht, M. H., Richardson, J. S., Richardson, D. C., and Ogden, R. C. (1990) De novo design, expression, and characterization of felix: a four-helix bundle protein of native-like sequence. *Science* **249,** 884–891.
3. Quinn, T. P., Tweedy, N. B., Williams, R. W., Richardson, J. S., and Richardson, D. C. (1994) Betadoublet: de novo design, synthesis, and characterization of a beta-sandwich protein. *Proc. Natl. Acad. Sci. USA* **91,** 8747–8751.
4. Kamtekar, S., Schiffer, J. M., Xiong, H., Babik, J. M., and Hecht, M. H. (1993) Protein design by binary patterning of polar and nonpolar amino acids. *Science* **262,** 1680–1685.
5. Dahiyat, B. I. and Mayo, S. L. (1997) De novo protein design: fully automated sequence selection. *Science* **278,** 82–87.

6. Harbury, P. B., Plecs, J. J., Tidor, B., Alber, T., and Kim, P. S. (1998) High-resolution protein design with backbone freedom. *Science* **282,** 1462–1467.

7. Looger, L. L., Dwyer, M. A., Smith, J. J., and Hellinga, H. W. (2003) Computational design of receptor and sensor proteins with novel functions. *Nature* **423,** 185–190.

8. Kuhlman, B., Dantas, G., Ireton, G. C., Varani, G., Stoddard, B. L., and Baker, D. (2003) Design of a novel globular protein fold with atomic-level accuracy. *Science* **302,** 1364–1368.

9. Krizek, B. A., Amann B. T., Kilfoil, V. J., Merkle, D. L., and Berg, J. M. (1991) A consensus zinc finger peptide: Design, high affinity metal binding, a ph-dependent structure, and a His to Cys sequence variant. *J. Am. Chem. Soc.* **113,** 4518–4523.

10. Steipe, B., Schiller, B., Pluckthun, A., and Steinbacher, S. (1994) Sequence statistics reliably predict stabilizing mutations in a protein domain. *J. Mol. Biol.* **240,** 188–192.

11. Lehmann, M., Kostrewa, D., Wyss, M., et al. (2000) From DNA sequence prove to improved functionality: using protein sequence comparisons to rapidly design a thermostable consensus phytase. *Protein Eng.* **13,** 49–57.

12. Rath, A. and Davidson, A. R. (2000) The design of a hyperstable mutant of the abp1p sh3 domain by sequence alignment analysis. *Protein Sci.* **9,** 2457–2469.

13. Main, E. R., Xiong, Y., Cocco, M. J., D'Andrea, L., and Regan, L. (2003) Design of stable alpha-helical arrays from an idealized TPR motif. *Structure (Camb.)* **11,** 497–508.

14. Stumpp, M. T., Forrer, P., Binz, H. K., and Pluckthun, A. (2003) Designing repeat proteins: modular leucine-rich repeat protein libraries based on the mammalian ribonuclease inhibitor family. *J. Mol. Biol.* **332,** 471–487.

15. Mosavi, L. K., Minor, D. L., Jr., and Peng, Z. Y. (2002) Consensus-derived structural determinants of the ankyrin repeat motif. *Proc. Natl. Acad. Sci. USA* **99,** 16029–16034.

16. Kohl, A., Binz, H. K., Forrer, P., Stumpp, M. T., Pluckthun, A., and Grutter, M. G. (2003) Crystal structure of a consensus ankyrin repeat protein. *Proc. Natl. Acad. Sci. USA* **100,** 1700–1705.

17. Andrade, M. A., Perez-Iratxeta, C., and Ponting, C. P. (2001) Protein repeats: structures, functions, and evolution. *J. Struct. Biol.* **134,** 117–131.

18. Kobe, B. and Kajava, A. V. (2000) When protein folding is simplified to protein coiling: the continuum of solenoid protein structures. *Trends Biochem. Sci.* **25,** 509–515.

19. Sonnhammer, E. L., Eddy, S. R., Birney, E., Bateman, A., and Durbin, R. (1998) *Nucleic Acids Res.* **26,** 320–322.

20. Schultz, J., Milpetz, F., Bork, P., and Ponting, C. P. (1998) *Proc. Natl. Acad. Sci. USA* **95,** 5857–5864.

21. Castiglone Morelli, M. A., Stier, G., Gibson, T., et al. (1995) The KH module has an alpha beta fold. *FEBS Lett.* **358,** 193–198.

22. Musco, G., Stier, G., Joseph, C., et al. (1996) Three-dimensional structure and stability of the KH domain: molecular insights into the fragile X syndrome. *Cell* **85,** 237–245.

23. Gribskov, M., McLachlan, A. D., and Eisenberg, D. (1987) Profile analysis: detection of distantly related proteins. *Proc. Natl. Acad. Sci. USA* **84,** 4355–4358.
24. Eddy, S. R. (1998) Profile hidden markov models. *Bioinformatics* **14,** 755–763.
25. Thompson, J. D., Gibson, T. J., Plewniak, F., Jeanmougin, F., and Higgins, D. G. (1997) The Clustal X windows interface: flexible strategies for multiple sequence alignment aided by quality analysis tools. *Nucleic Acids Res.* **25,** 4876–4882.
26. D'Andrea, L. D. and Regan, L. (2003) TPR proteins: the versatile helix. *Trends Biochem. Sci.* **28,** 655–662.
27. Scheufler, C., Brinker, A., Bourenkov, G., et al. (2000) Structure of TPR domain-peptide complexes: critical elements in the assembly of the Hsp70-Hsp90 multichaperone machine. *Cell* **101,** 199–210.
28. Binz, H. K., Amstutz, P., Kohl, A., et al. (2004) High-affinity binders selected from designed ankyrin repeat protein libraries. *Nat. Biotechnol.* **22,** 575–582.
29. Magliery, T. J. and Regan, L. Beyond consensus: statistical free energies reveal hidden interactions in the design of a TPR motif. (2004) *J. Mol. Biol.* **343,** 731–745.
30. Mosavi, L. K. and Peng, Z. Y. (2003) Structure-based substitutions for increased solubility of a designed protein. *Protein Eng.* **16,** 739–745.
31. Vita, C., Drakopoulou, E., Vizzavona, J., et al. (1999) Rational engineering of a miniprotein that reproduces the core of the CD4 site interacting with HIV-1 envelope glycoprotein. *Proc. Natl. Acad. Sci. USA* **96,** 13091–13096.
32. Domingues, H., Cregut, D., Sebald, W., Oschkinat, H., and Serrano, L. (1999) Rational design of a GCN4-derived mimetic of interleukin-4. *Nat. Struct. Biol.* **6,** 652–656.
33. Chin, J. W. and Schepartz, A. (2001) Concerted evolution of structure and function in a miniature protein. *J. Am. Chem. Soc.* **123,** 2929–2930.
34. Cortajarena, A. L., Kajander, T., Pan, W., Cocco, M. J., and Regan, L. (2004) Protein design to understand peptide ligand recognition by tetratricopeptide repeat proteins. *Protein Eng. Des. Sel.* **17,** 399–409.
35. Chin, J. W. and Schepartz, A. (2001) Design and evolution of a miniature Bcl-2 binding protein. *Angew Chem. Int. Ed. Engl.* **40,** 3806–3809.
36. Binz, H. K., Stumpp, M. T., Forrer, P., Amstutz, P., and Pluckthun, A. (2003) Designing repeat proteins: well-expressed, soluble and stable proteins from combinatorial libraries of consensus ankyrin repeat proteins. *J. Mol. Biol.* **332,** 489–503.
37. Takahashi, T. T., Austin, R. J., and Roberts, R. W. (2003) mRNA display: ligand discovery, interaction analysis and beyond. *Trends Biochem. Sci.* **28,** 159–165.
38. Smith, G. P. (1985) Filamentous fusion phage: novel expression vectors that display cloned antigens on the virion surface. *Science* **228,** 1315–1317.
39. Barbas, C. F. I., Burton, D. R., Scott, J. K., and Silvermann, G. J. (Eds.) (2001) *Phage Display, A Laboratory Manual.* Cold Spring Harbor Laboratory Press, Cold Spring Harbor, NY.
40. Lockless, S. W. and Ranganathan, R. (1999) Evolutionarily conserved pathways of energetic connectivity in protein families. *Science* **286,** 295–299.
41. Magliery, T. J. and Regan, L. (2005) Sequence variation in ligand binding sites in proteins. *BMC Bioinformatics* **6,** 240.
42. Altschul, S. F., Madden, T. L., Schaffer, A. A., et al. (1997) Gapped BLAST and PSI-BLAST: a new generation of protein database search programs *Nucleic Acids Res.* **25,** 3389–3402.

8

Multiple Sequence Alignment as a Guideline for Protein Engineering Strategies

Alan R. Davidson

Summary

Many proteins lack the thermodynamic stability and/or solubility that is required for their use in a desired application. For this reason, it can be advantageous to improve these qualities through rational protein engineering. An effective means for achieving this goal is to use sequence alignment analysis to select amino acid substitutions that are likely to increase the thermodynamic stability or solubility of a protein. Advantages of using this approach are that generally only a small number of substitutions need to be tested, these substitutions are rarely debilitating to protein function, and knowledge of the three-dimensional structure of the protein of interest is not required. This chapter will describe approaches that have been used to exploit the information contained in sequence alignments for the engineering of improved protein properties.

Key Words: Protein solubility; mutagenesis; sequence alignment; protein engineering; protein stability.

1. Introduction

Although every protein present in living organisms has been honed by evolution to fold and function well in its natural environment, many of these same proteins stubbornly refuse to behave as desired by scientists who wish to exploit them for the advancement of their research projects. For this reason, much effort has been expended in developing generalized methods to alter the sequences of unco-operative proteins to make them more amenable to in vitro investigation. A central goal in these studies has been to overcome the two of the most common shortcomings of natural proteins: low thermodynamic stability and low solubility. In this chapter, the use of sequence alignment analysis to design amino acid substitutions that increase the stability or solubility of proteins is discussed. This technique has proven to be both simple and reliable.

From: *Methods in Molecular Biology, vol. 340: Protein Design: Methods and Applications*
Edited by: R. Guerois and M. López de la Paz © Humana Press Inc., Totowa, NJ

Over the years, many different approaches have been used to rationally design stabilizing substitutions in a variety of proteins *(1–3)*. These approaches often involve using the atomic resolution structure of the protein of interest to select substitutions aimed at introducing new favorable interactions or eliminating existing unfavorable ones. Comparison with the structures of thermophilic homologues can aid in such studies *(4)*. However, structure-based approaches have the disadvantage of requiring the three-dimensional structure of the protein of interest to be known. A different approach to protein stabilization is to randomly mutagenize a protein and then perform multiple rounds of in vitro or in vivo selection to evolve the protein towards the desired properties *(5)*. This methodology has also been successful, but requires an easily performed selective assay for the activity or stability of the protein of interest.

The design of stabilizing mutations through sequence alignment analysis provides a very advantageous alternative to other strategies. Sequence alignment-based methods involve comparing the sequence of a protein of interest with an alignment of homologous sequences. Positions in the protein of interest that are occupied by amino acids rarely seen in related sequences are substituted with the amino acid that is observed most often. Variations on this basic approach have been used to stabilize many different proteins including immunoglobulin domains *(6,7)*, an SH3 domain *(8)*, GroEL minichaperone *(9)*, p53 *(10)*, phytase *(11,12)*, WW domains *(13)*, and ankyrin repeats *(14)*. Single mutations have been isolated in these studies that increase the melting temperature of the protein (T_m) by more that 15°C, and increase free energy of unfolding (ΔG_u) by more than 1 kcal/mol. The effects of substitutions identified by this method are generally additive so that multiple mutants have been produce with increases of T_m values of more than 30°C, and increases in ΔG_u of more than 3 kcal/mol *(8)*. The success rate of the method as judged by whether predicted mutants are actually stabilizing ranges between 25 and 50% depending on the study. The advantages of sequence alignment-based methods are that they are simple to carry out, they do not require knowledge of the three-dimensional structure of the protein under investigation, and the resulting mutants generally maintain their biological function. Furthermore, the method has been shown to work with alignments ranging from thirteen to thousands of sequences.

Sequence alignment based methods can also aid in increasing the solubility of some proteins. For example, substitutions selected to stabilize a V_h domain intrabody also greatly increased the solubility of the domain in vivo in *Escherichia coli (7)*. Alignment based methods specifically aimed at identifying substitutions to increase solubility have recently been successfully used *(15–17)*, and these methods will also be discussed here.

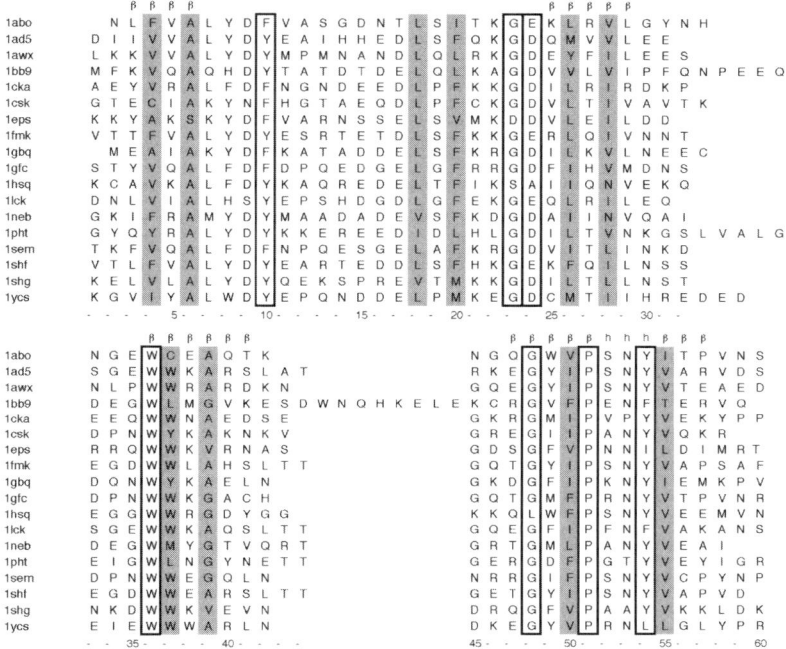

Fig. 1. A portion of an SH3 domain sequence alignment. This small alignment is representative of the complete 266 member SH3 domain alignment used in a protein design project carried out in my laboratory (8). Conserved hydrophobic core positions are shaded; other highly conserved positions are boxed. Examination of the conserved positions is helpful in verifying the correctness of the alignment. It should be noted that gaps are seen between secondary structure elements, but not within in them, and that gap lengths can be quite variable. The PDB accession number is shown for each sequence.

2. Methods

2.1. Construction of a Sequence Alignment

The construction of a high-quality sequence alignment is a key step in any sequence alignment-based protein engineering strategy. Although there are many existing databases containing preassembled sequence alignments for most proteins (*see* **Note 1**), it is often best to construct an alignment from scratch that will be most suited for analysis of a particular protein of interest. The basic steps in constructing an alignment are outlined below. A more detailed description of the construction and analysis of a sequence alignment can be found in a review by Irving et al. (*18*). To illustrate some of the points made, an example of part of a large SH3 domain sequence alignment constructed in my laboratory is shown in **Fig. 1**.

2.1.1. Identification of Sequences Related to the Protein of Interest

1. Go to the BLAST web site (http://www.ncbi.nlm.nih.gov/BLAST/) and select PHI- and PSI-BLAST.
2. Paste the amino acid sequence of your protein of interest into the box and submit using the default settings.
3. Press "Format" to see the results of your search after it has been completed.
4. Run multiple iterations of PSI-BLAST to maximize the chance that a diverse set of sequences related to your sequence of interest will be obtained (*see* **Note 2**).

2.1.2. Select Sequences to Use in the Alignment

1. Examine the pairwise alignments of your sequence with the sequences that have been identified in the PSI-BLAST search.
2. Select sequences of interest for sequence retrieval. This is done by putting a checkmark in the box at the beginning of the sequence entry. In choosing sequences to include in your alignment, avoid including too many obviously redundant sequences and sequences that are so distantly related to your sequence that alignment may be a problem. As a general rule, sequences with greater than 30% identity to your sequence of interest should be included.
3. After sequences have been selected, press "Get Selected Sequences", then select Display "FASTA" and Send To "Text". This should produce a window containing all of your selected sequence in FASTA format. These sequences should be saved as text.

2.1.3. Aligning the Selected Sequences

The most commonly used program for sequence alignment is Clustal *(19)*. This program can be downloaded (ftp://ftp-igbmc.u-strasbg.fr/pub/ClustalX/) or run at a web site (http://www.ebi.ac.uk/clustalw/). Your sequences saved in FASTA format can be loaded directly into this program. There are many parameters that can be adjusted when using Clustal to create an alignment, and discussion of these is beyond the scope of this chapter. Detailed help menus are included with the program. If a three-dimensional structure of your protein or a closely related homologue is known, this structure can be used by the program as an aid in determining gap positions. Phylogenetic trees of your sequences can also be constructed as a means to evaluate the diversity of your alignment.

2.1.4. Refining the Alignment

To obtain reliable results from a sequence alignment, ensure that the sequences are aligned as well as possible. In many cases, sequence alignments must be manually adjusted after the use of an automated alignment program. A very good program to use for the viewing, editing, and analysis of sequence alignments is Jalview (http://www.jalview.org) *(20)*. This program will automatically color conserve positions in the alignment, which is useful for detecting problem areas. Poorly aligned regions can be detected in Clustal by selecting functions under the heading "Quality." In improving an alignment, several factors should be taken into consideration:

1. Sequence gaps (i.e., insertions or deletions) should be minimized as much as possible. Gaps should generally lie between secondary structural elements.
2. The alignment should possess conserved hydrophobic positions at positions corresponding to the hydrophobic core of the protein. Polar residues should rarely be seen at these positions.
3. Sequences possessing unusual residues at highly conserved positions should be examined closely to ensure that they are not misaligned and that they are truly homologous to the other sequences throughout their length.
4. Sequences that are very difficult to align to the rest of the sequences should be eliminated. It is best to err on the side of caution in this respect.
5. Highly redundant sequences should be eliminated. If the alignment contains a large number of sequences that are very closely related, the calculations of the frequency of occurrence of residues at each position in the alignment will be skewed. In Jalview, in the "Calculate" menu, choose "Remove Redundancy." The level of redundancy to be removed can be set at any level. For example, if it is set at 90%, sequences will be eliminated from the alignment so that no two sequences display more the 90% identity. Choosing the level of redundancy to eliminate may depend on the alignment being used. Many groups including my own routinely use 90% as the redundancy cutoff.

2.2. Using the Sequence Alignment to Design Stabilizing Amino Acid Substitutions

The basis of the design methodology is to compare the sequence of your protein of interest to the frequencies of occurrence of amino acids at each position in the sequence alignment. The goal of this comparison is to identify amino acids within the protein of interest that are relatively rare in comparison to what is observed in most homologues. The basic steps are as follows.

1. Calculate the frequency of occurrence of the most commonly observed amino acid at each position of the alignment (*see* **Notes 3** and **4**). Only positions that have been aligned with confidence should be considered in this analysis (e.g., positions near or within gaps may not be suitable).
2. Compare the frequency of occurrence of each residue at each position in your protein of interest to that of the most commonly occurring residue at each position. Positions in your protein that possess relatively rare residues compared with the consensus residue should be targeted for mutagenesis. An example of this type of sequence analysis is shown in **Fig. 2**, in which the sequence of the SH3 domain from the Abp1 protein (Abp1p) of yeast is compared to the consensus sequence derived from a 266 sequence SH3 domain alignment *(8)*. In this case, we chose to mutate all positions at which the WT residue occurred with a relative frequency of 0.15 or less, where the relative frequency is the frequency of occur-

Fig. 2. Alignment of the Abp1p SH3 domain sequence with the SH3 domain consensus sequence. Based on a 266-member sequence alignment, the most frequently observed amino acid residue at each position in the SH3 domain was determined and is given as the consensus sequence. The numbers below each sequence are the occurrence frequency (expressed as %) in the alignment observed for each residue in the Abp1p SH3 domain and in the SH3 domain consensus sequence. The relative frequency of occurrence (Rel. Freq.) was obtained by dividing the Abp1p SH3 domain residue occurrence with that of the consensus residue. The boxed positions had relative frequencies of less than 0.15 were selected for mutagenesis. Three substitutions (E7L, N23G, and V21K) led to significant stabilization of the Abp1p SH3 domain (8).

rence of the residue in Abp1 divided by the frequency of occurrence of the residue that occurred most frequently at the given position, i.e., the consensus residue (*see* **Note 5**).

3. Use a standard polymerase chain reaction-based, site-directed mutagenesis protocol to replace rarely occurring residues in your protein of interest with the residue that occurs most commonly at that position in the alignment.

4. Purify the mutant proteins and assay their thermodynamic stability using spectroscopic or any other amenable techniques.

5. Combine single-site mutants with the largest stabilizing effects into one multiple mutant. Because the effects of mutants designed by sequence alignment are usually additive, the multiple mutant will likely be more stable than the singly substituted mutants (*see* **Note 6**).

2.3. Using Sequence Alignment Analysis to Design Amino Acid Substitutions to Increase Protein Solubility

In at least one case, the use of sequence alignment analysis to design stabilized mutants as described led to the isolation of proteins that were both more stable and more soluble *(7)*. However, a different alignment-based approach has also been successfully used to design mutations specifically aimed at increasing protein solubility *(15–17)*. This approach is carried out as follows.

1. Create an alignment as described in **Subheading 2.1.** containing the most closely related homologues to the protein of interest (a higher redundancy cutoff could be used to create this alignment).

2. Identify positions in the alignment at which your protein of interest possesses a hydrophobic residue and one or more (preferably more than one) closely related homologues possesses a polar residue. These positions are likely to lie on the surface of the protein structure as polar residues are rarely found in buried positions. The presence of a hydrophobic residue on the protein surface could lead to lowered solubility. An example of an alignment that could have been used to pinpoint solubilizing substitutions is shown in **Fig. 3**.

3. Substitute the targeted hydrophobic residue in your protein with a polar residue seen in a closely related sequence. Purify the mutant protein and assess protein solubility.

4. Combine solubilizing substitutions into one multiple mutant. If many potentially solubilizing mutants are identified, it may be more efficient to construct multiple mutants first. These substitutions are unlikely to be detrimental to protein structure or function because the substituted amino acids are seen in very closely related proteins.

3. Conclusions

Sequence alignments undoubtedly provide extremely useful information for the design of mutants that increase both the stability and solubility of proteins. The design of mutants using the methodology described is simple and does not require a known three-dimensional structure for a protein of interest. In addi-

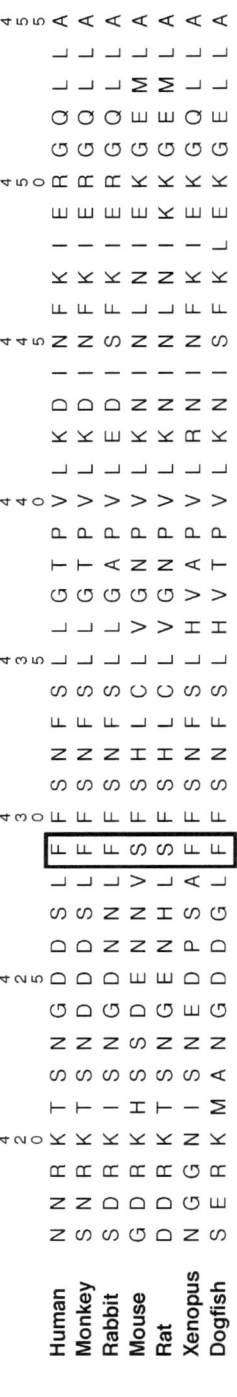

Fig. 3. Identification of a solubility enhancing substitution in the Cystic Fibrosis Transmembrane Conductance Regulator (CFTR). Sequences of a short region of CFTR from the indicated species are shown. This region is within the first nucleotide-binding domain of CFTR (NBD1). A high degree of conservation is seen at most positions. However, position 429 is very hydrophobic in the human sequence (Phe), but Ser, a polar residue, is seen in the mouse and rat sequences. The F429S substitution was found to significantly increase the in vitro purified soluble yield of a human CFTR NBD1 construct (*25*).

tion, mutants selected using these techniques rarely cause an alteration of protein function.

4. Notes

1. Some databases containing alignments for various proteins or domains are Pfam (http://www.sanger.ac.uk/Software/Pfam/) *(21)*, SMART (http://smart.ox.ac.uk/) *(22)*, and HSSP (http://www.cmbi.kun.nl/gv/hssp/) *(23)*. Because alignments in these databases are generated in a fully automated manner, care must be taken to refine them as described in **Subheading 2.1.4.**

2. The size and diversity of alignments that have been used for protein design have varied widely. The number of aligned sequences has ranged from 13 *(11)* to 4000 *(14)*. At this point, the properties of an alignment needed to produce the best results are not known. As long as the sequences used are aligned well and are reasonably diverse, then it should be possible to identify stabilizing substitutions.

3. Unfortunately I do not know of any web-based servers that will automatically calculate the frequency of occurrence of all the amino acids at each position in an alignment. These calculations can be quite tedious for a large alignment. It is fairly straightforward to write a script in Perl (http://www.perl.com/) to perform this task. Another approach is to load the alignment into a Microsoft Excel worksheet. Once loaded, the "Autofilter" procedure (this command is found by selecting "Data," then "Filter") provides a very simple and powerful means to count the occurrences of any residue at any position of the alignment. Although it is rarely used for sequence alignment editing and display, Excel is also useful for these purposes.

4. A complication in calculating residue frequencies from any sequence alignment is in dealing with skew that may result from overrepresentation within an alignment of clusters of highly related sequences. Although using a redundancy threshold as described can help with this problem, further improvement in frequency calculation can be obtained by using a weighting algorithm to downweight the effect of overrepresented groups of sequences. A commonly used weighting algorithm that we have used in our studies is that of Henikoff and Henikoff *(24)*. This procedure can be implemented through Perl scripts. If sequences to be included in an alignment are chosen carefully, weighting should not be essential for the success of the experiments describe in this chapter. Weighting protocols were not used in most of the studies referred to in this chapter.

5. The criteria used for the selection of positions to mutate has varied considerably from study to study. Thus, no strict rule can be stated on this issue. The most appropriate relative frequency cutoff will likely be different depending on the overall diversity of the alignment. The logical approach to this problem is to simply decide how many mutants can be reasonably made and then pick that number of positions displaying the lowest relative frequencies as compared with the consensus residue.

6. Some studies have taken the approach of first creating a "consensus" sequence of the protein of interest by substituting all the most favorable amino acids into it at one time *(11,14)*. However, because approximately half of the

substitutions designed by the sequence alignment based method are generally
found to be destabilizing, it seems better to make single substitutions first to
identify which ones are stabilizing.

Acknowledgments

Work pertaining to this chapter was funded by the Canadian Institutes of
Health Research and by the National Cancer Institute of Canada. I thank Dr.
Karen L. Maxwell for critical reading of the manuscript.

References

1. Eijsink, V. G., Bjork, A., Gaseidnes, S., Sirevag, R., Synstad, B., van den Burg, B.,
   et al. (2004) Rational engineering of enzyme stability. *J. Biotechnol.* **113,** 105–120.
2. Lehmann, M. and Wyss, M. (2001) Engineering proteins for thermostability: the
   use of sequence alignments versus rational design and directed evolution. *Curr.
   Opin. Biotechnol.* **12,** 371–375.
3. van den Burg, B. and Eijsink, V. G. (2002) Selection of mutations for increased
   protein stability. *Curr. Opin. Biotechnol.* **13,** 333–337.
4. Vieille, C. and Zeikus, G. J. (2001) Hyperthermophilic enzymes: sources, uses,
   and molecular mechanisms for thermostability. *Microbiol. Mol. Biol. Rev.* **65,** 1–43.
5. Arnold, F. H., Wintrode, P. L., Miyazaki, K., and Gershenson, A. (2001) How
   enzymes adapt: lessons from directed evolution. *Trends Biochem. Sci.* **26,** 100–106.
6. Steipe, B., Schiller, B., Pluckthun, A., and Steinbacher, S. (1994) Sequence statistics reliably predict stabilizing mutations in a protein domain. *J. Mol. Biol.* **240,**
   188–192.
7. Wirtz, P. and Steipe, B. (1999) Intrabody construction and expression III: engineering hyperstable V(H) domains. *Protein Sci.* **8,** 2245–2250.
8. Rath, A. and Davidson, A. R. (2000) The design of a hyperstable mutant of the
   Abp1p SH3 domain by sequence alignment analysis. *Protein Sci.* **9,** 2457–2469.
9. Wang, Q., Buckle, A. M., Foster, N. W., Johnson, C. M., and Fersht, A. R. (1999)
   Design of highly stable functional GroEL minichaperones. *Protein Sci.* **8,** 2186–2193.
10. Nikolova, P. V., Henckel, J., Lane, D. P., and Fersht, A. R. (1998) Semirational
    design of active tumor suppressor p53 DNA binding domain with enhanced sta-
    bility. *Proc. Natl. Acad. Sci. USA* **95,** 14675–14680.
11. Lehmann, M., Kostrewa, D., Wyss, M., Brugger, R., D'Arcy, A., Pasamontes, L.,
    et al. (2000) From DNA sequence to improved functionality: using protein sequence
    comparisons to rapidly design a thermostable consensus phytase. *Protein Eng.*
    13, 49–57.
12. Lehmann, M., Loch, C., Middendorf, A., Studer, D., Lassen, S. F., Pasamontes,
    L., et al. (2002) The consensus concept for thermostability engineering of pro-
    teins: further proof of concept. *Protein Eng.* **15,** 403–411.
13. Jiang, X., Kowalski, J., and Kelly, J. W. (2001) Increasing protein stability using
    a rational approach combining sequence homology and structural alignment: sta-
    bilizing the WW domain. *Protein Sci.* **10,** 1454–1465.

14. Mosavi, L. K., Minor, D. L., Jr., and Peng, Z. Y. (2002) Consensus-derived structural determinants of the ankyrin repeat motif. *Proc. Natl. Acad. Sci. USA* **99,** 16029–16034.

15. Ito, T. and Wagner, G. (2004) Using codon optimization, chaperone co-expression, and rational mutagenesis for production and NMR assignments of human eIF2 alpha. *J. Biomol. NMR* **28,** 357–367.

16. Malissard, M. and Berger, E. G. (2001) Improving solubility of catalytic domain of human beta-1,4-galactosyltransferase 1 through rationally designed amino acid replacements. *Eur. J. Biochem.* **268,** 4352–4358.

17. Sun, Z. Y., Dotsch, V., Kim, M., Li, J., Reinherz, E. L., and Wagner, G (1999) Functional glycan-free adhesion domain of human cell surface receptor CD58: design, production and NMR studies. *EMBO J.* **18,** 2941–2949.

18. Irving, J. A., Askew, D. J., and Whisstock, J. C. (2004) Computational analysis of evolution and conservation in a protein superfamily. *Methods* **32,** 73–92.

19. Thompson, J. D., Higgins, D. G., and Gibson, T. J. (1994) CLUSTAL W: improving the sensitivity of progressive multiple sequence alignment through sequence weighting, position-specific gap penalties and weight matrix choice. *Nucleic Acids Res.* **22,** 4673–4680.

20. Clamp, M., Cuff, J., Searle, S. M., and Barton, G. J. (2004) The Jalview Java alignment editor. *Bioinformatics* **20,** 426–427.

21. Bateman, A., Birney, E., Durbin, R., Eddy, S. R., Howe, K. L., and Sonnhammer, E. L. (2000) The Pfam protein families database. *Nucleic Acids Res.* **28,** 263–266.

22. Schultz, J., Milpetz, F., Bork, P., and Ponting, C. P. (1998) SMART, a simple modular architecture research tool: identification of signaling domains. *Proc. Natl. Acad. Sci. USA* **95,** 5857–5864.

23. Sander, C. and Schneider, R. (1991) Database of homology-derived protein structures and the structural meaning of sequence alignment. *Proteins* **9,** 56–68.

24. Henikoff, S. and Henikoff, J. G. (1994) Position-based sequence weights *J. Mol. Biol.* **243,** 574–578.

25. Lewis, H. A., Zhao, X., Wang, C., Sauder, J. M., Rooney, I., Noland, B. W., et al. (2005) Impact of the deltaF508 mutation in first nucleotide-binding domain of human cystic fibrosis transmembrane conductance regulator on domain folding and structure. *J. Biol. Chem.* **280,** 1346–1353.

9

Sequence Search Methods and Scoring Functions for the Design of Protein Structures

Hocine Madaoui[§], Emmanuelle Becker[§], and Raphael Guerois

Summary

This chapter focuses on the methods developed for the automatic or semiautomatic design of protein structures. We present several algorithms for the exploration of the sequence space and scoring of the designed models. There are now several successful designs that have been achieved using these approaches such as the stabilization of a protein fold, the stabilization of a protein–protein complex interface, and the optimization of a protein function. A rapid presentation of the methodologies is followed by a detailed analysis of two test case studies. The first one deals with the redesign of a protein hydrophobic core and the second one with the stabilization of a protein structure through mutations at the surface. The different approaches are compared and the consistency of the predictions with the experimental data are discussed. All the programs tested in these protocols are freely available through the internet and may be applied to a wide range of design issues.

Key Words: Protein design; sampling; conformational search; stability; scoring.

1. Introduction

Protein design aims at optimally selecting a protein sequence for a desired structure and function. The selection procedure depends on the ability to efficiently score a large number of protein structures. The scoring issue has been tackled by several research fields ranging from structure prediction to simulation of molecular dynamics. Yet, in contrast to structure prediction algorithms, which can deal with low-resolution models, computational design has to meet stringent precision requirements. At the other extreme, intensive computational approaches, although more precise, are often incompatible with the combinatorial problem of exploring both the sequence and the side-chain conforma-

---

[§]The first two authors contributed equally to this work.

From: *Methods in Molecular Biology, vol. 340: Protein Design: Methods and Applications*
Edited by: R. Guerois and M. López de la Paz © Humana Press Inc., Totowa, NJ

tional spaces. Hence, in the field of protein design, specific scoring methods have been developed to account for the balance between precision and computational efficiency. In this chapter, we present several of these approaches, first by describing the major outlines of the methods and second through a tutorial describing a step-by-step design and scoring of two proteins.

In practice, depending on the complexity of the design project, two strategies can be considered. In the first approach, close inspection of the template structure and of phylogenetic data can help designing the optimized sequence (*see* Chapters 7 and 8). Only a few positions are likely to be mutated and sequence space can be iteratively explored by hand. The structural models of the mutant have to be generated and scored independently using the appropriate energy function. This approach adapted for simple engineering purpose can be carried out with a minimum of computer resources and skills and most programs are available through the internet (**Subheadings 3.3.4.–3.3.7.**)

A more sophisticated approach consists in the use of automated design programs whose academic versions were, for the first time, recently released to the scientific community (**Subheadings 3.2.1.–3.3.3.**). They increase the scope of design projects that can be tackled but usually require the user to be at ease with computer programs running under Linux/Unix environment. Based on a given structural template, these methods explicitly model all the conformations (called "rotamers") of every position mutated during the design process. A specific energy function is used to evaluate the match of the sequence for the template. To cope with the combinatorial complexity of exploring the sequences compatible with a given template, these methods rely on efficient sampling strategies such as Monte Carlo (MC), genetic algorithm (GA), or dead-end elimination (DEE) (for a general description and comparison *see* **ref. *1***). In this chapter, two of these automated design programs are presented: RosettaDesign *(2,3)* and EGAD *(4)*.

Based on these methodologies, there are now many examples of successful designs that illustrate the range of applications that can be tackled. Several studies have shown that the stabilization of a protein based on a high-resolution structural scaffold can be reached using these automated methods *(5–8)*. Far more challenging is the design of new protein folds; only one such example has been recently published *(3)*. Probably a more accessible type of design can be the geometric idealization of an existing structure *(9)* or the local modification of a specific region in a protein *(10)*. The introduction of flexibility in the backbone and between the secondary structure elements during the design process is still a delicate issue but could be addressed for specific folds family *(11,12)*. Because folding and binding process are driven by the same rules, the global design of protein structures has also open the way to the rational design of protein-protein complex interfaces. Increasing the affinity (*see* Chapter 11) and altering the binding specificity between partners can also be undertaken using these design programs *(13–16)*.

In this chapter, we pay particular attention to the scoring functions that can be used to assess the quality of the design and help the selection of the optimal sequence. One of the major differences between the scoring functions is the level of heuristic terms they contain. The methods can be divided into three classes: (1) statistically based methods (SEEF) that rely on the analysis of either sequence or structure databases to transform probabilities into energies; (2) physically based methods (PEEF) that rely on the derivation of energy terms from model compounds and look for the most rigorous treatment of the basic principles of physics to calculate free energy changes; or (3) protein engineering empirically based methods (EEEF), which constitute a hybrid approach that takes advantage of thermodynamic data obtained on protein mutants and of statistical information from the databases. Here we consider each of the three approaches, analyzing their importance to the scope of protein design.

To assess the performance of the various design strategies, two test cases are analyzed in this chapter. The first one deals with the redesign of the hydrophobic core of the GB1 protein and the second with the stabilization of a protein structure (CspB) through mutations at the surface.

2. Materials

2.1. Brief Description of Programs and Downloads or URLs

2.1.1. DFire-Dmutant: EEEF Scoring Method

DMutant *(17)* is part of the DFire package developed in the Zhou laboratory of Biophysics and Bioinformatics at the University of Buffalo. Its goal is to predict stability changes ($\Delta\Delta G$) induced by a single mutation. The mutated position and the wild-type structure need to be given. The potential used by DMutant is DFire; a distance-dependent, residue-specific, all-atom, and knowledge-based potential. The predicted free-energy change caused by a mutation is calculated assuming no structural relaxation after mutation. In the original article *(17)*, the DMutant method was validated with a dataset of 895 large-to-small mutations (the native residue is replaced by a smaller one in the mutant). The DMutant web server is available at: http://phyyz4.med.buffalo.edu/hzhou/dmutation.html. The form is very easy to fill and the server answers very quickly.

2.1.2. PoPMuSiC: SEEF Scoring Method

PoPMuSiC *(18)* stands for Prediction of Protein Mutations Stability Changes. It is developed in the group of M. Rooman at the Université Libre de Bruxelles. As expressed in its name, the aim of PoPMuSiC is to predict stability changes ($\Delta\Delta G$) on mutations. Although it is possible to propose several mutations, the stability changes induced by each mutation are evaluated separately. PoPMuSiC is a database-derived potential that mixes a torsion potential and a C^μ–C^μ potential. The torsion potential describes local interactions along the sequence whereas the C^μ–C^μ potential reflects nonlocal and hydrophobic interactions. The PoPMuSiC program was validated *(18)* with a dataset of 344 experimentally studied mutations in seven different proteins and peptides. One example of successful experimental design was achieved based on PoPMusic predictions *(19)*. PoPMuSiC can be used either to predict the stability changes induced by precise mutations in a structure, or to propose sequence mutations predicted to stabilize the pdb structure. PoPMuSiC web server is available at http://babylone.ulb.ac.be/popmusic. The results are returned by e-mail. The e-mail contains several files; the one that contains the results summarized is the .pdf.

2.1.3. I-Mutant and I-Mutant2.0: SEEF Scoring Method (Neural Networks)

I-Mutant and I-Mutant2.0 *(20,21)* are neural network-based programs developed in the group of R. Casadio at the University of Bologna. The recent development of the ProTherm database *(22)*, collecting thermodynamic data for proteins and mutants, allowed machine-learning methods to emerge. I-Mutant is a neural network that, starting from the protein sequence and structure, classifies single amino acid substitutions into those leading to positive and those leading to negative $\Delta\Delta G$. The learning/training database and the validation one are derived from ProTherm. With a dataset of 1615 mutants (only single mutations are taken into account, *see* **ref. 20**), the accuracy reaches from 80% with I-Mutant alone to more than 90% when I-Mutant is coupled with another energy-based function such as Foldef (*see* **Subheading 2.1.5.**). I-Mutant2.0 is a support vector machine based on the same hypothesis. A web server dedicated to I-Mutant 2.0 is available at http://gpcr2.biocomp.unibo.it/cgi/predictors/I-Mutant2.0/I-Mutant2.0.cgi (click on the "protein structure" option).

2.1.4. Scap: Conformation of Side-Chain Prediction Program

Scap *(23)* is part of the Jackal package developed in B. Honig's laboratory at Columbia University and the Howard Hughes Medical Institute, New York. Its first aim is to predict side-chain conformation over a rigid backbone, but it can also predict the conformation of one or more residues that have been mutated in a particular structure. Different side-chain libraries that do not depend on the backbone and different force fields are available in Scap. Scap can be found at http://trantor.bioc.columbia.edu/programs/sidechain/. The current version only supports the following platforms: SGI 6.5, Intel Linux, and Sun Solaris.

2.1.5. FoldX and Foldef Energy Function: EEEF Scoring Function

FoldX *(24)* is developed in L. Serrano's group at EMBL, Heidelberg. FoldX energy function, Foldef, is an empirical force field for the rapid evaluation of the effect of mutations on the stability of proteins and nucleic acids based on its structure. The mutant's structure needs to be given as an input. This is why, at this stage, FoldX is used downstream as a structure prediction method like Scap to perform design. The predictive power of Foldef has been tested on a very large set of point mutants (1088 mutants) spanning most of the structural environments found in proteins (*24*). Implementation of the full molecular design toolkit capabilities is in development. A web server dedicated to FoldX (*25*) can be found at http://foldx.embl.de/. A former version can also be found at http://fold-x.embl-heidelberg.de:1100/cgi-bin/main.cgi.

2.1.6. RosettaDesign: Sequence Search by MC and EEEF Scoring Function

RosettaDesign, developed in D. Baker's group in the University of Washington, Seattle (*3*), is a program aimed at finding the lowest free energy sequences for a given target protein backbone. The first application of this program is the creation of novel proteins with arbitrarily chosen three-dimensional structures. The procedure was used to design a 93-residue protein called Top7 with a novel sequence and topology (Top7 was found experimentally to be folded. and the X-ray crystal structure of Top7 is very similar to the target model). It can also be used to enhance protein stability and create alternative sequences for naturally occurring proteins. In a recent study (*2*), RosettaDesign was used to predict sequences with low free energy for naturally occurring protein backbones and was able to propose proteins more stable than their wild-type counterparts. A web server dedicated to RosettaDesign can be found at http://rosettadesign.med.unc.edu/. The source codes are available to academics, are located in the directory Rosetta, and can be compiled with the GNU compiler (g77 or gcc) on a computer running Linux. To compile the program, go to the Rosetta directory and type "make."

2.1.7. EGAD: Sequence Search by GA and PEEF Scoring Function

EGAD (a GA for protein Design **ref. 4**) is a protein design application developed in T. Handel's group in the University of California, Berkeley. Its main focus is to perform protein design on rigid backbone scaffolds. EGAD can perform different types of jobs such as protein designing on rigid backbones, predicting mutation effects on protein stability, or minimizing structures of proteins. The EGAD program was validated by predicting the stability of more than 1500 mutants to within 1 Kcal/mol (*4*).

The EGAD program is available at http://egad.berkeley.edu/software.php. The source code for EGAD is located in the directory source_code and can be compiled with the g++ compiler. To compile the program, go to the source_code directory and type "make clean all." This will create the EGAD.exe executable in the EGAD/bin directory.

2.2. Input Formats and Basic Instructions to Run the Programs

2.2.1. DFire-Dmutant

The form is very easy to fill out.

1. The native structure can be either a local pdb file or a code from the pdb (for example 1CSP for CspB or 1PGA for GB1).
2. The mutated position is indicated by its chain identifier and residue number.
3. For the mutated position, the stability change will be predicted for each of the 20 amino acids, so it is not worth indicating precisely the substitution.

DFire-Dmutant predicts stability changes induced by single mutations. As we need to test a mutant with several mutations, the best way (although not rigorous) is to test all the mutations separately and to add the predicted $\Delta\Delta G$ of each single mutation.

2.2.2. PopMusic

The form needs the following fields to be completed.

1. The name, e-mail, and job name: only the e-mail address is necessary for the correct execution of PoPMuSiC (the results will be sent to this address, so it must be checked carefully). The predicted stability changes induced by each single mutation are listed in the result file. To obtain a global score for the mutant, it is possible (although not exact) to add the $\Delta\Delta G$ of each single mutation.
2. The wild-type structure, either a local pdb file or a pdb identifier (1CSP, 1PGA).
3. To score a given mutant, choose the "specific mutations" option. A "mutation file" is then required. It lists all the mutations to score. Note that it is possible to predict the $\Delta\Delta G$ of a mutant with several mutations. The syntax of the "mutation file" is easy and well described. The predicted stability changes induced by each single mutation are listed in the result file. To obtain a global score for the mutant, it is possible (although not exact) to add the $\Delta\Delta G$ of each single mutation.

2.2.3. I-Mutant 2.0

The form needs to be filled out with this information.

1. The pdb code of the native structure (1CSP or 1PGA).
2. The chain and residue number of the mutated position.
3. For a precise substitution—for example, E3R—it is possible to specify that only mutation to arginine should be taken into account by indicating "R" in the "New Residue" field. If no "New Residue" is specified, the position will be screened with all possible amino acids.
4. Select the "Free Energy Change Value ($\Delta\Delta G$)."
5. A valid e-mail address for the results to be returned.

Considering that I-Mutant 2.0 is dedicated to predict the stability change induced by single mutations, the best way to study the stability change of a multiple mutant is to add the $\Delta\Delta G$ of each single mutation.

2.2.4. Scap and Foldx

To predict the three-dimensional structure of the mutants, Scap needs a file that precisely lists all the mutations and all the flexible residues together with these mutations. Here are some tips to write this file.

1. To mutate a position, the syntax is "B,43,G" (residue 43 of chain B is mutated into glycine).
2. To make a position flexible, the syntax is "B,63,V" (the best rotamer for valine 63 of chain B is searched).
3. Only one mutation or one flexible residue per line.
4. The filename must end with "_scap.list."

To run Scap, the command line is:

```
> /usr/local/bin/scap -min 2 -prm 1 -ini 10 -seed 17 1CSP.pdb
                      mutant_scap.list
```

The output structure is written in a file named 1CSP_scap.pdb. To obtain other structures, change the seed number (*see* **Note 1**). To learn more about the other options, type "scap-h" or see the jackal/scap manual (*see* **Note 2**).

To score the structures, we use the FoldX web server. The server requires a login and a password to connect, either as a guest or as a registered user (registration is free). After having registered and connected to the server, do the following.

1. Upload a mutant file produced using Scap (preferentially) or using the whatif server (*see* **Note 3**) (it is possible to upload up to five structures at once).
2. Select the "Stability" option.
3. On the next window, leave unchanged the temperature, van der Waals design, or iron strength.
4. Calculate the stability (*see* **Note 4**).

2.2.5. RosettaDesign

RosettaDesign can be run in different modes that include the following.

1. Repacking side chain on a rigid backbone.
2. Redesigning on a rigid backbone.
3. Redesigning with a flexible backbone.

To run the program in the protein design mode, one needs different files in the running directory: (1) a starting pdb structure; (2) a file that specifies the location of input/output for the program, usually named "path.txt" (a default path.txt file is supplied with the Rosetta source code); (3) a file that specifies the subset of residues to redesign, named "resfile."

Thus, the desired properties for every amino acid to redesign have to be written explicitly in the resfile. All the instructions must be written in a given format (**Table 1**). For example, the following instructions specify to redesign

Table 1
Resfile Format for the RosettaDesign Program

Column	2	4–7	9–12	14–18	21–40
Description	Chain	Sequential residue number	Pdb residue number	Id	Amino acids to be used

the amino acid 3 (columns 9–12) of chain A (column 2), which is the third of the protein (columns 4–7) by using an amino acid between A, V, L, I, F, Y, or W (columns 21–40).

```
A 3 3 PIKA AVLIFYW
```

The identification information (columns 14–18) can take the following values: (1) NATAA (use native amino acid), (2) ALLAA (all amino acids allowed), (3) NATRO (use native amino acid and rotamer), (4) PIKAA (select individual amino acids), (5) POLAR (use polar amino acids), or (6) APOLA (use apolar amino acids).

The program MAKERESFILE (*see* **Note 5**) provided with Rosetta is available from the suite, and can be used to make an initial resfile with the following command line:

```
> makeresfile -p 1pga.pdb -default NATAA -resfile resfile
```

After the initial resfile is made, it can be modified by hand to specify the residues to redesign.

The command line used to redesign a protein considering a rigid backbone is:

```
> pFOLD.gnu —s 1pga.pdb -design -fixbb -resfile resfile -ndruns 1
```

The ndruns option indicates the number of design runs. The energies, structure, and sequence output are written in an output pdb file. A precise description of this file is given in the "README_output" file supplied with the program. To get a score file of the solution built with full atom scoring, the command line is:

```
> pFOLD.gnu -score -s 1pga_0001.pdb -fa_input -scorefile score
```

Considering backbone motion in protein design with RosettaDesign requires fragment files that can be obtained at: http://robetta.bakerlab.org/fragmentsubmit.jsp

Three fields need to be filled out: (1) registered username or registered e-mail address, (2) the target name, and (3) the sequence of the native structure in the fasta format (**Fig. 1**). Two fragment files are then produced.

To move the backbone and design a protein with RosettaDesign, three arguments must be used: (1) a two-letter identification code (e.g., "aa"), (2) a four-character code name for the protein (this must agree with the name of the fragment files and the name of the starting structure xxxx.pdb), and (3) the chain identification.

Fig. 1. Fragment files can be obtained from the Rosetta server to run RosettaDesign in flexible backbone mode.

Moreover, the starting structure must be idealized, with the instruction:

```
> pFOLD.gnu -s 1pga.pdb -idealize -fa_input
```

This command creates an output file 1pga_idl.pdb, which is used as an input to the protein design for a flexible backbone:

```
> pFOLD.gnu aa 1pga A -s 1pga_idl.pdb -design -mvbb -resfile
                                          resfile -nstruct 1
```

2.2.6. EGAD

As with RosettaDesign, the EGAD program takes user-written files as input. An EGAD input file has three sections. The first gives information about the template structure, the energy function, the desired job, and how to run it. The second section lists positions that are allowed to move. The last section, which is optional, lists positions that must be rigid during the design protocol. Here is an example of script used to redesign some positions of a starting structure spga.pdb (**Fig. 2**). The command line used to run the job is:

```
> path_to_EGAD/EGAD.exe 1pga_design
```

All the major rotamer-optimization methods (GA, MC-simulated annealing, self-consistent mean-field optimization, DEE, fast and accurate side-chain topology, and energy refinement) have been implemented in the EGAD program.

```
START
TEMPLATE_PDB ./1pga.pdb
FORCEFIELD_FILE ./EGAD/examples/energy_function/forcefield
JOBTYPE mc_ga
OTHER_RESIDUES none
OUTPUT_PREFIX gb1.mc_ga
END
VARIABLE_POSITIONS
    3A   AVLIFYW
    5A   AVLIFYW
    7A   AVLIFYW
   20A   AVLIFYW
   26A   AVLIFYW
   30A   AVLIFYW
   34A   AVLIFYW
   39A   AVLIFYW
   52A   AVLIFYW
   54A   AVLIFYW
END
```

Fig. 2. Example of script 1pga-design used by EGAD to redesign specific positions of a starting structure template.pdb.

For our study, we have used the MC_GA method (in which solutions from MC runs are used to seed a population for genetic algorithm) (*see* **Note 6**).

2.3. pdb Used in the Example

1. Protein G (B1 immunoglobulin G-binding domain) from *Streptococcus* sp. (PDB code: 1pga.pdb).
2. Major cold shock protein (CspB) from *Bacillus subtilis* (PDB code: 1csp.pdb).

3. Methods

3.1. A Tutorial-Based Analysis of the Methods: Presentation of Two Case Studies

The β1 immunoglobulin-binding domain of streptococcal protein G (GB1) is a relevant model to study the core packing mechanisms in proteins. It has been used several times to validate design algorithms since the first studies of automatic protein design *(8,26)*. GB1 consists of 56 residues arranged in one α-helix and two β-hairpins (**Fig. 3A**). The contiguous core of the GB1 protein is formed by 11 residues whose accessible surface area is smaller than 10% (**Table 2**). Experimentally, mutants of GB1 were found to be more stable than the wild-type GB1 *(12)*. With three mutations in the core of the protein GB1 (**Table 3**), the midpoint of the thermal unfolding temperature can increase by up to 6°C. The stability of the second example analyzed in the chapter, the cold shock proteins (Csp), has been widely studied during the past 10 yr. They are

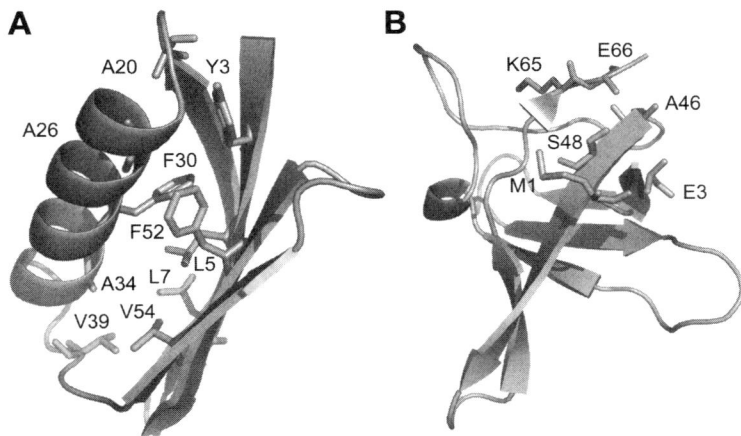

Fig. 3. Ribbon representation of the structural templates with residues mutated in the design tutorial as sticks (**A**) the GB1 protein (**B**) the CspB protein.

Table 2
Buried Residues of β1 Immunoglobulin-Binding Domain of Streptococcal Protein G (GB1)

Secondary structures	Core positions
β-sheet	3 – 5 – 7 – 20 – 43 – 52 – 54
α-helix	26 – 30 – 34
Coil	39

Their accessible surface area is <10%.

found in mesophilic, thermophilic, and hyperthermophilic bacteria, and consist of 65 to 70 residues arranged in a five-stranded antiparallel β-sheet (**Fig. 3B**). Studies proved that very few mutations at the surface of Csp proteins can lead to large differences in stability *(27)*, making them relevant examples for studying the design of protein surface. A recent article *(28)* identified stabilized variants of CspB of *Bacillus subtilis* involving the surface positions 1, 3, 46, 48, 65, and 66 using three different methods: site-specific randomization, site-directed mutations, and spontaneous mutations (error-prone polymerase chain reaction). The midpoint of the thermal unfolding transition was measured for each mutant (**Table 4**).

3.2. Sequence Search: Automatic Design Using RosettaDesign and EGAD

Our first goal was to redesign with EGAD and RosettaDesign both the core of GB1 and the surface of CspB, and to compare the sequences proposed by these programs with those known to stabilize the wild-type proteins.

Table 3
Tm is the Midpoint of Thermal Unfolding Transition (in °C)

Structures	Mutation(s)	Tm in °C
Wild-type GB1		**85.0**
Mutant GB1_1	Y3F – L7V – V39I	89.0
Mutant GB1_2	Y3F – L7I – V39I	91.0

Measures are given for the wild type β1 immunoglobulin-binding domain of streptococcal protein G and for two stabilizing mutants experimentally found (data taken from **ref. *12***).

Table 4
Tm is the Midpoint of Thermal Unfolding Transition (in °C)

Structures	Mutation(s)	Tm in °C
Wild-type CspB		**53.8**
Mutant CspB _1	N55S – Q59R	51.8
Mutant CspB _2	E43S	54.7
Mutant CspB _3	M1R – E3K – K65I	83.7
Mutant CspB _4	M1R – E3K – K65I – E66L	85.0

Measures are given for the wild type protein CspB of *Bacillus subtilis* and the four mutants studied (data taken from **ref. *28***).

In the case of GB1, positions that constitute the core, including residues 3, 5, 7, 20, 26, 30, 34, 39, 52, and 54 were redesigned (hydrophobic amino acids A, V, L, I, F, Y, and W were considered at these positions) according to the study of Su and Mayo *(12)*. All the other positions were kept in their native conformation.

For CspB, we tested if the protein design programs were able to propose protein variants involving the surface positions 1, 3, 46, 48, 65, and 66. Therefore, these positions were selected to be redesigned (all amino acids were considered at these positions, whereas native amino acids and rotamers were kept for other positions). To check the convergence of the programs, we did 30 different runs considering rigid or flexible backbone for every case study (*see* **Note 1**).

3.2.1. Sequence Search: Considering Rigid Backbone Design

3.2.1.1. REDESIGNING THE CORE OF GB1 USING ROSETTADESIGN OR EGAD IN RIGID BACKBONE MODE

From the different runs, only one solution is proposed by the program RosettaDesign with three mutations compared to the native structure (Y3F, L7V, F52A). These mutations were not reported in previous studies of GB1, and no experimental data can check the stabilizing effects of these mutations. We can note however that the RosettaDesign score of the produced solution (−133.42), is higher than that of the native structure (−135.27), which means that the output solution produced considering rigid backbone is less stable than the native structure.

Using the same protocol to design the core of protein GB1 with EGAD, one sequence is proposed by the program: Y3F/V39I/V54I. As with the RosettaDesign program, the output solution (1966.96) proposed has a higher energy than the native structure (1878.56) (*see* **Note 7**).

In the study by Su and collaborators, the best protein variant found, mutant GB1_2, is more stable than the native protein. This mutant, resulting from Y3F, L7I, and V39I mutations (Tm = 91°C), has neither been proposed by the EGAD nor by the RosettaDesign program. Yet, one can note partial agreement on the Y3 and V39 positions from the EGAD results. In **Subheadings 3.3.1.** and **3.3.2.**, we analyze the scoring of the experimentally determined most stable mutants to decipher if the lack of consistency with experimental data can be explained by either a problem of conformational search or a problem in the scoring function of the programs.

3.2.1.2. Redesigning the Surface of CspB Using RosettaDesign or EGAD in Rigid Backbone Mode

With RosettaDesign, 30 different runs considering rigid backbone produce four sequences compatible with the backbone fold of the protein CspB. The EGAD program suggests one sequence (the resulting structure is less stable [2191.34] than the native structure according to the EGAD energy function [2165.92]). Unfortunately, none of the proposed sequence has been analyzed experimentally in the study of Wunderlich and collaborators *(28)*.

3.2.2. Sequence Search: Considering Flexible Backbone Design

3.2.2.1. Redesigning the Core of GB1 Using RosettaDesign in Flexible Backbone Mode

From the different runs, four sequences compatible with the backbone fold of the protein GB1 are proposed by RosettaDesign when considering backbone motion:

1. mutant GB1_rD_1: Y3F, L7I
2. mutant GB1_rD_2: Y3F, L7V
3. mutant GB1_rD_3: L7V
4. mutant GB1_rD_4: Y3F, L7V, V39I

Several solutions are proposed emphasizing the need of launching different runs in the flexible backbone mode. For sequences 1 to 3, no experimental data are available from previous studies. However, the structure resulting from sequence 4, which was proposed by Su and Mayo in 1997, is actually more stable that the native structure of protein GB1. Indeed, the melting temperature (Tm) measured for the sequence 4 (89°C) is higher than the Tm of the native sequence

(85°C). This example highlights the interest of running RosettaDesign with the flexible backbone mode rather than the rigid one.

3.2.2.2. REDESIGNING THE SURFACE OF CspB USING ROSETTADESIGN IN FLEXIBLE BACKBONE MODE

A total of 30 runs using flexible backbone have been done with RosettaDesign, each of them producing an output solution. From these 30 runs, 30 sequences compatible with the backbone fold of the protein CspB were proposed. However, none of the solutions was tested experimentally *(28)*. The Shannon entropy (Hx) for every redesigned position over the 30 runs ranges from 2.31 to 3.37, which emphasizes a high variability for each position (which probably results from a problem of convergence when considering surface positions with flexible backbone).

This example highlights the difficulties of selecting a sequence when designing solvent exposed positions and stress the importance of generating multiple runs by varying the seed number (*see* **Note 1**).

3.3. Evaluating the Stabilized Protein Variants

In the previous section, the selection of sequences proposed by the RosettaDesign and EGAD were found difficult to evaluate because of the lack of experimental data validating the proposed sequences. To further assess the predictive power of the design programs, we evaluate now the scoring functions of several programs. The tests are based on both the wild-type proteins GB1 and CspB, and their respective proteins variants shown experimentally to be more stable (the mutants GB1_1, GB1_2 for GB1 and the mutants CspB_2 to CspB_4 for CspB). The CspB_1 is less stable than the wild type and has been chosen as a negative control of the prediction.

3.3.1. Scoring Design Models: RosettaDesign With Rigid Backbone

To estimate the scoring function of RosettaDesign considering rigid backbone, we have compared the score of the best mutants found for GB1 (mutants GB1_1 and GB1_2) with the wild-type protein GB1 and the score of the mutants CspB_1 to CspB_4 with the wild-type CspB.

To do so, we have set up a protocol for evaluating the mutants using the scoring function of the program, which consists of the following.

1. The wild-type protein is mutated using the WHATIF server to replace the correct atoms in the pdb (for example, we have done the Y3F, L7V, V39I mutations for the mutant GB1_1) (*see* **Note 8**).
2. Thirty different runs are launched with the mutated structure as an input.
3. A distribution of the scores from every output solution is produced in each of the run.
4. Last, every suggested mutant is ranked according to the median of its score distribution.

Table 5
Scoring of the GB1 and the CspB Optimized Protein Variants
With the RosettaDesign Score Considering the Rigid Backbone Mode

Structures	RosettaDesign score	Tm (°C)
Wild-type CspB	**−111.80**	**53.8**
Mutant CspB_4	−113.86	85.0
Mutant CspB_3	−120.03	83.7
Mutant CspB_2	−124.11	54.7
Mutant CspB_1	−129.92	51.8
Mutant GB1_2	−113.60	91.0
Mutant GB1_1	−114.59	89.0
Wild-type G B1	**−135.27**	**85.0**

The lowest score refers to the most stable model. Tm is the midpoint of thermal unfolding transition (in °C).

For the evaluation of the GB1 mutants, sequence positions of the core including residues 3, 5, 7, 20, 26, 30, 34, 39, 52, and 54 were kept identical but were allowed to move. For the evaluation of the CspB mutants, mutated residues and amino acids within 8Å of them were kept identical, but were allowed to move. In the case of GB1, the mutant GB1_2 gives a structure which is less stable than the wild-type protein when considering this protocol and the RosettaDesign rigid backbone mode. In the case of CspB, we have also ranked the mutant CspB_1 to CspB_4 and the wild-type Bs-CspB. We can conclude that the ranking of these protein variants does not agree with experimental data because all the mutants are predicted to be more stable than the wild-type protein (**Table 5**).

3.3.2. Scoring Design Models: EGAD

The same protocol as in **Subheading 3.3.1.** for evaluating the mutants (considering this time the energy function of EGAD) has been used.

We found that the mutant GB1_1 is less stable than the wild-type protein. Furthermore, the ranking of the CspB protein variants doesn't agree with experimental data because all the mutants are predicted to be more stable than the wild-type protein (**Table 6**).

3.3.3. Scoring Design Models: RosettaDesign With Flexible Backbone

The same protocol as in **Subheading 3.3.1.** for evaluating the mutants (this time running RosettaDesign in flexible backbone mode) has been used. This procedure (except from s**tep 1**) has been used also to score the wild-type proteins (*see* **Note 9**).

Launching 30 different runs in the second step is a way of exploring different conformations for a single sequence (**Fig. 4A**) (*see* **Note 1**). For example,

Table 6
**Scoring of the GB1 and the CspB Optimized Protein
Variants With the EGAD Energy Function**

Structures	EGAD energy Kcal/mol	Tm (°C)
Wild-type CspB	**2165.92**	**53.8**
Mutant CspB_1	2107.22	51.8
Mutant CspB_2	2105.02	54.7
Mutant CspB_4	2065.78	85.0
Mutant CspB_3	2063.31	83.7
Mutant GB1_1	1886.11	89.0
Wild-type G B1	**1878.56**	**85.0**
Mutant GB1_2	1878.41	91.0

The lowest score refers to the most stable model. Tm is the
midpoint of thermal unfolding transition (in °C).

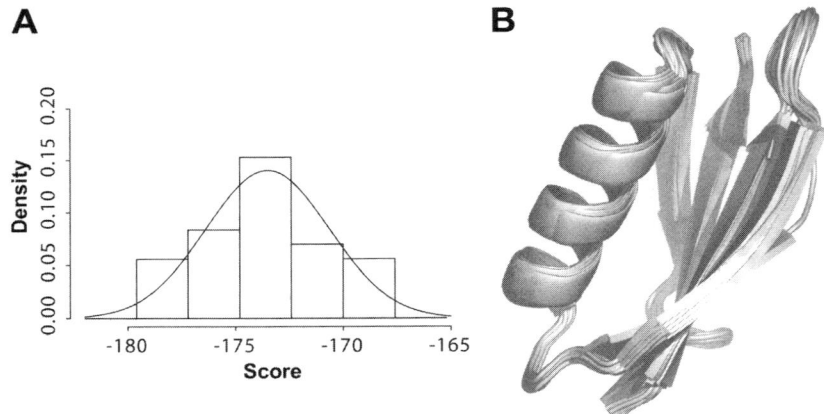

Fig. 4. Analysis of the mutant GB1_2 using RosettaDesign in flexible backbone
mode. (**A**) Distribution of the RosettaDesign scores obtained with 30 different runs
(varying the random seed number). (**B**) Superimposition of the ribbon representation
of the 30 models, which illustrates the backbone moves explored during the design.

all the conformations generated for the mutant GB1_2 are shown on the **Fig.
4B** when considering flexible backbone with RosettaDesign (the root mean
square deviation [RMSD] between all the output solutions is equal to 0.41).

This protocol has been followed for the mutants GB1_1, GB1_2, and for the
wild-type protein GB1. The results are available in the **Table 7**.

According to the RosettaDesign score, the mutant GB1_1 and mutant
GB1_2, which are known to be more stable than the wild-type protein, are

Table 7
**Scoring of the Four Mutants of the Protein CspB and the Two
Mutants of Streptococcal Protein G With the RosettaDesign
Score Considering the Flexible Backbone Mode**

Structures	RosettaDesign score	Tm (°C)
Mutant CspB_1	−154.12	51.8
Wild-type CspB	**−154.66**	**53.8**
Mutant CspB_2	−157.58	54.7
Mutant CspB_4	−157.74	85.0
Mutant CspB_3	−159.74	83.7
Wild-type G B1	**−172.05**	**85.0**
Mutant GB1_2	−173.67	91.0
Mutant GB1_1	−175.10	89.0

The lowest score refers to the most stable model. Tm is the midpoint of
thermal unfolding transition (in °C).

predicted to stabilize the core of protein GB1. However, the mutant GB1_2
which is known to be more stable than mutant GB1_1 *(12)* has a lower score.

When considering flexible backbone with the RosettaDesign program, the
scoring function is able to predict the stability of the protein variants with respect
to the wild-type CspB. The mutant CspB_3 and the mutant CspB_4 are indeed
more stable than the native protein, whereas the mutant CspB_1 is less stable
(**Table 7**).

These results show a good ability of RosettaDesign to predict the stability of
protein variants compared to the wild type proteins. Interestingly, good results
are reported not only for the prediction of core mutations but also for mutations
involving solvent exposed amino acids. These results highlight (1) the importance of considering backbone flexibility in protein design to sample a larger
search space and (2) the importance of considering alternative conformations
compatible with one sequence to predict the possible effect of one mutation.

3.3.4. DFire-Dmutant Scoring Function

At first, DFire-DMutant seems to have problems to deal with protein surface
design of CspB (*see* **Note 10**). The mutations predicted to be the most stabilizing do not seem to be relevant.

The four mutants of CspB and the two mutants of GB1 presented previously
can be scored by DFire-DMutant (**Table 8**). For the experimentally highly
stabilizing mutants (*see* mutant CspB_3 and mutant CspB_4), DFire-DMutant
correctly assigns a negative ΔΔG. For the other mutants, the predictions are
less reliable (as an example, *see* mutant GB1_1, experimentally found more
stable than wild-type GB1, but predicted to be less stable). The quality of the

Table 8
Results Observed for the Four Mutants of the Protein CspB
Previously Described and the Mutants of Streptococcal Protein G
With the DFire-DMutant Program

Structures	DFire-DMutant predicted ΔΔG (kcal/mol)	Tm (°C)
Mutant CspB_2	+0.24	54.7
Wild-type CspB	**0.0**	**53.8**
Mutant CspB_1	−0.63	51.8
Mutant CspB_3	−2.38	83.7
Mutant CspB_4	−3.64	85.0
Mutant GB1_1	+0.25	89.0
Wild-type GB1	**0.0**	**85.0**
Mutant GB1_2	−0.21	91.0

The predicted ΔΔG is given in kcal/mol. Negative ΔΔG correspond to stabilizing mutations. Tm is the midpoint of thermal unfolding transition (in °C).

Table 9
Results Observed for the Four Mutants of the Protein CspB
and for the Two Mutants of Streptococcal Protein G Using PoPMuSiC

Structures	PoPMuSiC predicted ΔΔG (kcal/mol)	Tm (°C)
Mutant CspB_1	+1.03	51.8
Mutant CspB_2	+0.23	54.7
Mutant CspB_3	+0.04	83.7
Wild-type CspB	0.0	53.8
Mutant CspB_4	−0.70	85.0
Mutant GB1_1	+2.56	89.0
Mutant GB1_2	+2.40	91.0
Wild-type GB1	0.0	85.0

The predicted ΔΔG is given in kcal/mol. Negative ΔΔG correspond to stabilizing mutations. Tm is the midpoint of thermal unfolding transition (in °C).

prediction seems to be related to the amplitude of the ΔTm between the wild-type protein and the mutants.

3.3.5. PoPMuSiC Scoring Function

The four mutants of the *B. subtilis* protein CspB and the two mutants of GB1 were submitted to PoPMuSiC (**Table 9**). For GB1, the two mutants are predicted to be significantly less stable than the wild-type CspB, although these mutants were experimentally found more stable. For CspB mutants, all mutants are predicted to be destabilizing except from one of the highly stabilizing mutant, mutant CspB_4. The quality of the predictions do not seem to correlate with the ampli-

Table 10
Results Observed for the Four Mutants of the Protein CspB and the Two
Mutants of Streptococcal Protein G Using the I-Mutant2.0 Server

Structures	I-Mutant 2.0 predicted $\Delta\Delta G$ (kcal/mol)	Tm (°C)
Mutant CspB_1	−1.03	51.8
Wild-type CspB	**0.0**	**53.8**
Mutant CspB_3	0.19	83.7
Mutant CspB_2	0.46	54.7
Mutant CspB_4	1.49	85.0
Wild-type GB1	**0.0**	**85.0**
Mutant GB1_2	0.29	91.0
Mutant GB1_1	0.64	89.0

The predicted $\Delta\Delta G$ is given in kcal/mol[-1].By convention in I-Mutant2.0, negative $\Delta\Delta G$ correspond to nonstabilizing mutants, whereas positive $\Delta\Delta G$ correspond to stabilizing ones. Tm is the midpoint of thermal unfolding transition (in °C).

tude of the ΔTm between the wild-type protein and the mutants (for example, mutant CspB_3, the most stabilizing one, is predicted as slightly destabilizing).

3.3.6. I-Mutant2.0 Scoring Function

Table 10 summarizes I-Mutant2.0 results obtained for our two case studies (*see* **Note 11**). The sign of the $\Delta\Delta G$ is correct for all CspB and GB1 mutants. I-Mutant2.0 correctly identifies stabilizing and nonstabilizing mutants for both surface and core design.

3.3.7. Scap and Foldef

3.3.7.1. SCAP AND FOLDX TO SCORE GB1 MUTANTS

The two mutants of GB1 experimentally found to be more stable than the wild-type protein are Y3F-L7V-V39I and Y3F-L7I-V39I. The three mutated positions are the same and are buried (their accessible surface area is smaller than 10%).

The strategy we used to construct these mutants with Scap was the following: (1) we mutated the positions 3, 7, and 39 with amino acids F,V,I for mutant GB1_1 and F,I,I for mutant GB1_2; (2) to allow repacking in the core of the protein, we left the side chains flexible of the residues around—in a sphere of 8 Å—each mutated position; and (3) we minimized the structures with 200 steps of steepest descent using Gromos96 vacuum force field (http://www.igc.ethz.ch/gromos/) in Gromacs (http://www.gromacs.org/) to release van der Waals clashes.

Because Scap is a program that searches for a local optimum and not a global one, we recommend to do several runs of Scap with different seeds (*see* **Note 1**). We built 20 different structures for each mutant.

Table 11
Results Obtained for the Four Mutants of the Protein CspB and the Two Mutants of Streptococcal Protein G With the Strategy Combining Scap for the Prediction of Sidechain Conformations and FoldX for the Scoring Function

Structures	FoldX predicted energy—minimalist strategy	FoldX predicted energy—thorough strategy	Tm (°C)
Mutant CspB_2	18.68	19.63	54.7
Wild-type CspB	**18.11**	**17.04**	**53.8**
Mutant CspB_1	18.37	16.33	51.8
Mutant CspB_3	16.31	14.79	83.7
Mutant CspB_4	15.20	15.43	85.0
Wild-type GB1	1.86		**85.0**
Mutant GB1_1	1.84		89.0
Mutant GB1_2	1.14		91.0

The smaller the energy, the more stable the structure. For a mutant to stabilize the structure, its energy must be smaller than the energy of the wild type protein. Tm is the midpoint of thermal unfolding transition (in °C).

These 40 structures were scored by the FoldX server, and we selected the structure with the lowest FoldX score for each mutant (corresponding to the lowest energy). **Table 11** presents the results.

3.3.7.2. Scap and FoldX to Score CspB Mutants

The case of the CspB mutants is different from GB1 mutants because we aimed at comparing structures mutated at different positions located at the surface. To deal with the flexibility around the mutated positions (*see* **Note 12**), we propose two different strategies, a minimalist one and a thorough one: (1) mutate the positions needed, for example 55 and 59 for mutant CspB_1; (2) in the minimalist strategy, residue side chains are kept rigid around the mutated position, in the thorough strategy, all the side chains of the protein are defined as flexible; and (3) minimize the structures with 200 steps of steepest descent using Gromos96 vacuum force field (http://www.igc.ethz.ch/gromos/) in Gromacs (http://www.gromacs.org/) to release van der Waals clashes.

For each mutant, 20 structures were built and scored using the FoldX server. The lowest score for each mutant can be found in **Table 11**. The minimalist strategy and the thorough one gave quite good results. In the minimalist strategy, the highly stabilizing mutants CspB_3 and CspB_4 are predicted to be more stable than the wild-type protein with a significant energy gap (>1.5), whereas the mutants CspB_2 and CspB_1, whose Tm are close to that of the wild type are predicted to be slightly destabilizing. In the thorough strategy, the full side-chain flexibility may account for the more disperse values. Yet, when the experimental energy gap is large, the stability of the mutants is well predicted and the hierarchy between the most stabilizing mutations is well respected.

3.4. Concluding Remarks

The results of the blind design protocols presented in this chapter show that, on average, the current algorithms offer a reliable predictive power. Design of densely packed hydrophobic regions is generally quite reliable. Stabilizing effects at the surface are more difficult to predict. Yet, when the free energy difference is large enough between the mutant and the wild type, good correlations between theory and experiments can be observed. From the global output results, the RosettaDesign software only when used with the flexible backbone option performs better. Although sequence search at the surface cannot be easily evaluated, the scoring function coupled to the flexible backbone mode gives reliable results. The support vector machine I-mutant-2.0 coupled to the Foldx program also provides interesting results in its ability to discriminate stabilizing from destabilizing mutations. It should be particularly useful for simple and fast design applications.

4. Notes

1. When running algorithms based on stochastic search methods (such as MC, GA), it is critical to run the programs several times with different random seed numbers to best explore the solutions generated by the conformational search algorithm (**Fig. 4A**). We could find dramatic variations in the interpretation of the results whether or not this repetition procedure was included in the protocols.

2. In scap "-prm 1" indicates to use Charmm22 (visit http://www.charmm.org/) with an all-atom model for the force parameters (torsion energy, van der Waals radius, or charge parameters). Other force parameters are available as "-prm 2" for Amber (visit http://amber.scripps.edu/) for all-atom model.

3. As an alternative to using Scap, the WHATIF server can be used to build simple mutations (http://swift.cmbi.kun.nl/WIWWWI/) with the menu "mutate a residue." In future version, the mutant generation facility will directly be included in the Foldx program.

4. It is possible that a pdb structure is not accepted by the FoldX server if its format is unusual. In this case, you can try to standardize the pdb file by simply opening and saving it with either Swiss PDB Viewer (*29*) or WHATIF (*30*).

5. To run these programs, make sure the most recent version of the gcc compilers is installed; otherwise, the executable file does not run properly.

6. With the version of EGAD we tested, we could not run the DEE algorithm.

7. When scoring the native structure, EGAD extracts the side-chain dihedrals and rebuilds them with ideal geometry (which causes slight deviations between the idealized and the native structure).

8. This step can be done more simply by using the PIKAA flag for every position to mutate in the resfile used by RosettaDesign.

9. The flexible backbone mode is an important feature of the RosettaDesign program that allows slight structural variations in the backbone and optimization of the solutions energy (**Fig. 4**). The studies presented in this chapter highlight the interest of activating this option. The differences between the energies of the rigid and the flexible backbone mode are huge and comparison with the wild type should be done with the same calculation conditions.

10. We noticed that the most stabilizing mutation predicted for all the positions screened is always the mutation to tryptophan. This is a frequent bias of design programs and mutations to tryptophan or other large hydrophobic residues, such as tyrosine or phenylalanine, should always be considered with caution.

11. Contrary to the methods presented in this chapter, I-Mutant and I-Mutant2.0 both predict a positive $\Delta\Delta G$ for stabilizing mutations. This singularity comes from the use as a training dataset of the ProTherm database that lists the free energy of unfolding instead of the free energy of folding.

12. In the case of CspB, it is not advised to restrict the flexible side chains to the residues neighboring the mutated positions. Indeed, flexible residues would be different for each mutant. Some mutants, such as CspB_3 and CspB_4, would have nearly all residues flexible, whereas mutant CspB_2 would have only a small part of its residues flexible. Because this can introduce significant noise, we propose either to keep all the neighboring residues rigid or to make all the side chains of the protein flexible.

References

1. Voigt, C. A., Gordon, D. B., and Mayo, S. L. (2000) Trading accuracy for speed: a quantitative comparison of search algorithms in protein sequence design. *J. Mol. Biol.* **299,** 789–803.

2. Dantas, G., Kuhlman, B., Callender, D., Wong, M., and Baker, D. (2003) A large scale test of computational protein design: folding and stability of nine completely redesigned globular proteins. *J. Mol. Biol.* **332,** 449–460.

3. Kuhlman, B., Dantas, G., Ireton, G. C., Varani, G., Stoddard, B. L., and Baker, D. (2003) Design of a novel globular protein fold with atomic-level accuracy. *Science* **302,** 1364–1368.

4. Pokala, N. and Handel, T. M. (2005) Energy functions for protein design: adjustment with protein-protein complex affinities, models for the unfolded state, and negative design of solubility and specificity. *J. Mol. Biol.* **347,** 203–227.

5. Filikov, A. V., Hayes, R. J., Luo, P., Stark, D. M., Chan, C., Kundu, A., and Dahiyat, B. I. (2002) Computational stabilization of human growth hormone. *Protein Sci.* **11,** 1452–1461.

6. Korkegian, A., Black, M. E., Baker, D., and Stoddard, B. L. (2005) Computational thermostabilization of an enzyme. *Science* **308,** 857–860.

7. Ventura, S., Vega, M. C., Lacroix, E., Angrand, I., Spagnolo, L., and Serrano, L. (2002) Conformational strain in the hydrophobic core and its implications for protein folding and design. *Nat. Struct. Biol.* **9,** 485–493.

8. Dahiyat, B. I. and Mayo, S. L. (1997) De novo protein design: fully automated sequence selection. *Science* **278,** 82–87.

9. Offredi, F., Dubail, F., Kischel, P., Sarinski, K., Stern, A. S., Van de Weerdt, C., et al. (2003) De novo backbone and sequence design of an idealized alpha/beta-barrel protein: evidence of stable tertiary structure. *J. Mol. Biol.* **325,** 153–174.

10. Nauli, S., Kuhlman, B., and Baker, D. (2001) Computer-based redesign of a protein folding pathway. *Nat. Struct. Biol.* **8,** 602–605.

11. Harbury, P. B., Plecs, J. J., Tidor, B., Alber, T., and Kim, P. S. (1998) High-resolution protein design with backbone freedom. *Science.* **282,** 1462–1467.

12. Su, A. and Mayo, S. L. (1997) Coupling backbone flexibility and amino acid sequence selection in protein design. *Protein Sci.* **6,** 1701–1707.

13. Shifman, J. M. and Mayo, S. L. (2002) Modulating calmodulin binding specificity through computational protein design. *J. Mol. Biol.* **323,** 417–423.

14. Reina, J., Lacroix, E., Hobson, S. D., Fernandez-Ballester, G., Rybin, V., Schwab, M. S., et al. (2002) Computer-aided design of a PDZ domain to recognize new target sequences. *Nat. Struct. Biol.* **9,** 621–627.

15. Kortemme, T., Joachimiak, L. A., Bullock, A. N., Schuler, A. D., Stoddard, B. L., and Baker, D. (2004) Computational redesign of protein-protein interaction specificity. *Nat. Struct. Mol. Biol.* **11,** 371–379.

16. Havranek, J. J. and Harbury, P. B. (2003) Automated design of specificity in molecular recognition. *Nat. Struct. Biol.* **10,** 45–52.

17. Zhou, H. and Zhou, Y. (2002) Distance-scaled, finite ideal-gas reference state improves structure-derived potentials of mean force for structure selection and stability prediction. *Protein Sci.* **11,** 2714–2726.

18. Gilis, D. and Rooman, M. (2000) PoPMuSiC, an algorithm for predicting protein mutant stability changes: application to prion proteins. *Protein Eng.* **13,** 849–856.

19. Gilis, D., McLennan, H. R., Dehouck, Y., Cabrita, L. D., Rooman, M., and Bottomley, S. P. (2003) In vitro and in silico design of alpha1-antitrypsin mutants with different conformational stabilities. *J. Mol. Biol.* **325,** 581–589.

20. Capriotti, E., Fariselli, P., and Casadio, R. (2004) A neural-network-based method for predicting protein stability changes upon single point mutations. *Bioinformatics* **20(Suppl 1),** I63-I68.

21. Capriotti, E., Fariselli, P., and Casadio, R. (2005) I-Mutant2.0: predicting stability changes upon mutation from the protein sequence or structure. *Nucleic Acids Res.* **33,** W306–W310.

22. Gromiha, M. M., Uedaira, H., An, J., Selvaraj, S., Prabakaran, P., and Sarai, A. (2002) ProTherm, thermodynamic database for proteins and mutants developments in version 3.0. *Nucleic Acids Res.* **30,** 301–302.

23. Xiang, Z. and Honig, B. (2001) Extending the accuracy limits of prediction for side-chain conformations. *J. Mol. Biol.* **311,** 421–430.

24. Guerois, R., Nielsen, J. E., and Serrano, L. (2002) Predicting changes in the stability of proteins and protein complexes: a study of more than 1000 mutations. *J. Mol. Biol.* **320,** 369–387.

25. Schymkowitz, J., Borg, J., Stricher, F., Nys, R., Rousseau, F., and Serrano, L. (2005) The FoldX web server: an online force field. *Nucleic Acids Res.* **33,** W332–W388.

26. Dahiyat, B. I. and Mayo, S. L. (1997) Probing the role of packing specificity in protein design. *Proc. Natl. Acad. Sci. USA* **94,** 10172–10177.

27. Perl, D., Mueller, U., Heinemann, U., and Schmid, F. X. (2000) Two exposed amino acid residues confer thermostability on a cold shock protein. *Nat. Struct. Biol.* **7,** 380–383.

28. Wunderlich, M., Martin, A., and Schmid, F. X. (2005) Stabilization of the cold shock protein CspB from *Bacillus subtilis* by evolutionary optimization of Coulombic interactions. *J. Mol. Biol.* **347,** 1063–1076.

29. Guex, N. and Peitsch, M. C. (1997) SWISS-MODEL and the Swiss-PdbViewer: an environment for comparative protein modeling. *Electrophoresis* **18,** 2714–2723.

30. Vriend, G. (1990) WHAT IF: a molecular modeling and drug design program. *J. Mol. Graph.* **8,** 52–56.

10

Prediction of Protein–Protein Interaction Based on Structure

Gregorio Fernandez-Ballester and Luis Serrano

Summary

A great challenge in the proteomics and structural genomics era is to predict protein structure and function from sequence, including the identification of biological partners. The development of a procedure to construct position-specific scoring matrices for the prediction and identification of sequences with putative significant affinity faces this challenge. The local and web applications used for sequence and structure search, sequence alignment, protein modeling, molecule edition and modification, and scoring matrices construction are described in detail. The methodology is based on the information contained in structural databases and takes into account the subtle conformational and sequence details that characterize different structures within a family. Using the matrices, the protein sequence databases can be easily scanned to locate putative partners of biological significance. The success of this methodology opens the way for the prediction of protein–protein interaction at genome scale.

Key Words: Bioinformatics; protein–protein interaction; protein modeling; protein prediction; positional scoring matrix; pattern search.

1. Introduction

A major fraction of the genomes has now been sequenced, and this vast amount of data opens a way for novel methods of analysis of all genes and their products. One of these methods, particularly important, is the prediction or characterization of functional interactions between proteins on a genome-wide level. Currently of interest are those parts of proteins ("domains") involved in protein interactions that fold independently of the rest of the molecule. They are remarkable because they have a fully functional interaction activity at high levels and are commonly crystallized alone or in complex with a polypeptide ("ligands"). However, for most of these domains, the structural and functional features are completely unknown, and their putative roles are only suggested by homology. The information regarding protein–protein interactions and multicomponent systems formation is limited and interaction network maps are incomplete.

From: *Methods in Molecular Biology, vol. 340: Protein Design: Methods and Applications*
Edited by: R. Guerois and M. López de la Paz © Humana Press Inc., Totowa, NJ

Protein–protein interactions are ubiquitous in biology: transient associations between proteins support a broad range of biological processes, including hormone–receptor binding, protease inhibition, antibody–antigen interaction, signal transduction, correction of misfolding by chaperones, and even enzyme allostery. On the contrary, permanent associations are essential for proteins whose stability or function is defined by multimeric states, such as viral capsides, oligomeric enzymes, and channel proteins (*1*).

Protein–protein interactions occur at the surface of a protein and are biophysical phenomena, governed by shape, chemical complementarity, and flexibility of the molecules involved. Assemblies involving proteins that must be independently stable before association, referred as "transient," have interfaces that differ from those in oligomer complexes, referred as "permanent." The permanent association of an oligomer interface tends to be planar, roughly circular in shape, with a high abundance of hydrophobic groups and depletion in charged groups (*2*). On the contrary, transient interfaces more closely resemble the protein exterior, containing a higher proportion of polar and charged groups, with salt bridges and hydrogen bonding networks playing a more important role in stabilizing these complexes (*3*).

Nevertheless, both types of interfaces are closely packed and exhibit a high degree of geometric and electrostatic complementarity. The observations of these interfaces and the apparent consistencies found have led some groups to suggest simple rules for the prediction and location of putative interfaces (*3–7*). However, the properties that make a good interface depend on the type of complex and should be ranked by different criteria; also, the predictions were more powerful when applied to homodimer than to transient dimers (*2*). The geometric and electrostatic complementarity observed within interfaces has been also the basis of docking studies using proteins of known structure, and putative complexes are refined with electrostatic or chemical criteria to predict the "best" complex.

There are several sequence and structure methods for predicting protein–protein interactions: as examples, (1) SPOT, an algorithm to predict ligands (*8*), does not need explicit three-dimensional (3D) structures because the interaction database is a multidimensional array containing frequencies at position-specific contacts that explore the probability for a given ligand to bind to a domain; (2) VIP creates virtual interactions profiles (position-specific scoring matrices) from a 3D structure in complex with ligand, and scans sequence databases to seek binding partners of biological relevance (*9*); and (3) DOCK is an algorithm that accounts for flexible ligand docking, either with small ligand or protein, followed by virtual screening (*10*).

Analysis of conservation patterns in binding sites benefits from the fact that the residues involved are on the molecular surface and surface conservation is generally low. This potentially high signal-to-noise ratio arises because

changes in surface residues do not generally influence folding and overall stability as much as changes in residues at the structural core, so any mutational
intolerance that does exist can be detected more easily. Several groups have
explored patterns of conservation at binding sites in a systematic way using
multiple alignments, and sometimes phylogenetic trees of homologous sequences
to map evolutionary information onto datasets of protein structures. If a
protein's function is common within a homologous family and essential or
advantageous for the survival of the host organism, the maintenance of that
function describes the limits to which mutational variation in the sequence may
be tolerated. So, if a protein–protein interaction plays an important functional
role, it is interesting to study how patterns of evolutionary conservation in the
protomer sequences relate to the maintenance of this interaction *(2,11)*.

Thermodynamic studies in which the interface is systematically mutated reveal
that the distribution of energetically important residues can be uneven across
interfaces and concentrated in "hot spots" of binding energy *(12)*. There is also
evidence that residues distant from the interface can play a critical role in stabilizing protein-protein interactions. Such residues are believed to be energetically coupled with those directly involved in binding and allow binding energy
to propagate through tertiary structure.

In addition to theoretical analyses of crystal structures, a large quantity of
experimental studies has provided insights into protein–protein interactions
(3,13–17). Recently, however, there has been a large increase in the number of
known 3D structures that contain protein–protein recognition sites, covering a
much broader range of activities than ever, and allowing us to determine the
extent of generalization of these rules based on a few structures. The aspects of
structure that must be taken into consideration are those related to the stabilization of protein association: the size and chemical character of the protein surface that is buried at interfaces; the packing density of atoms that make contacts
across the interface, which expresses complementarity; and polar interactions
through hydrogen bonds and interface water molecules. Each of these aspects
can be described at the level of the individual atom that forms protein–protein
recognition sites, where three classes can be distinguished: atoms that lose
accessibility but do not make direct contacts across the interface; atoms that
make direct contacts but remain partly accessible; and atoms that become buried.

For all this, in this chapter we focus on the detailed use and development of
bioinformatic applications for the prediction of protein–protein interaction on
a structural basis to guess the potential partners of known proteins. The accurate computational measurements of the stability of protein–protein complexes
and the improvements of the software for *in silico* protein engineering, drug
design, mutagenesis, and dynamics allow us to tackle the exploration of these
surfaces for protein engineering and the prediction of partners, that should help
in interpreting experimental data. The methodology focuses on the application

of molecular modeling to calculate, manipulate, and predict protein structures and functions. Concepts of structure similarity/overlap, sequence alignment, structure superposition, homology modeling, and molecular docking, which are special concerns of protein biochemists, are considered. Approaches to protein modeling by the use of programs such as Swiss PDB Viewer, and online servers (Swiss Pdb Servers; FoldX) are described. The study is centred in transient domain-ligand interaction, making special emphasis on problems associated with sequence and structure alignments. The success of these methods provides an invaluable tool to select accurate pools of putative partners for further biochemical/biophysical characterization of proteins. The ultimate goal of these studies is the prediction of protein function at genome scale.

2. Methods

The prediction of protein interactions is a multiple-step methodology that requires the comprehension of theoretical concepts and the use of several software applications, either local or from the internet *(18)*. Because all these methodologies could be extended to entire genomes, it is strongly recommended to use scripting languages to automatize the usually tedious and monotonous jobs on each molecule. To launch multiple local jobs we recommend Python (http://www.python.org/) as an effective tool for the automatization of the work. For internet protein databases search and other purposes, we recommend Perl (http://www.perl.com/) because of its powerfulness and simplicity.

The flow diagram in **Fig. 1** comprises all steps and methodologies to accomplish our objectives.

2.1. Isolation of Domain Sequences and Domain Assignment

The SMART database at EMBL (http://SMART.embl-heidelberg.de) is a very powerful tool when working with protein–protein interaction domains. The database SMART *(19)* allows the identification and annotation of genetically mobile domains and the analysis of domain architectures, containing more than 400 domain families, extensively annotated with respect to functional class, tertiary structures, and functionally important residues. Each domain from nonredundant protein databases is stored in a relational database system with search parameters and taxonomic information. In addition, the web user interface allows searches for proteins containing specific combinations of domains in defined taxa. Useful examples are:

1. Isolation of a given domain from the whole protein sequence:
   - Go to SMART web page.
   - Paste the protein ID or the sequence of your protein (one letter code) under "Sequence Analysis".
   - Press Sequence SMART.

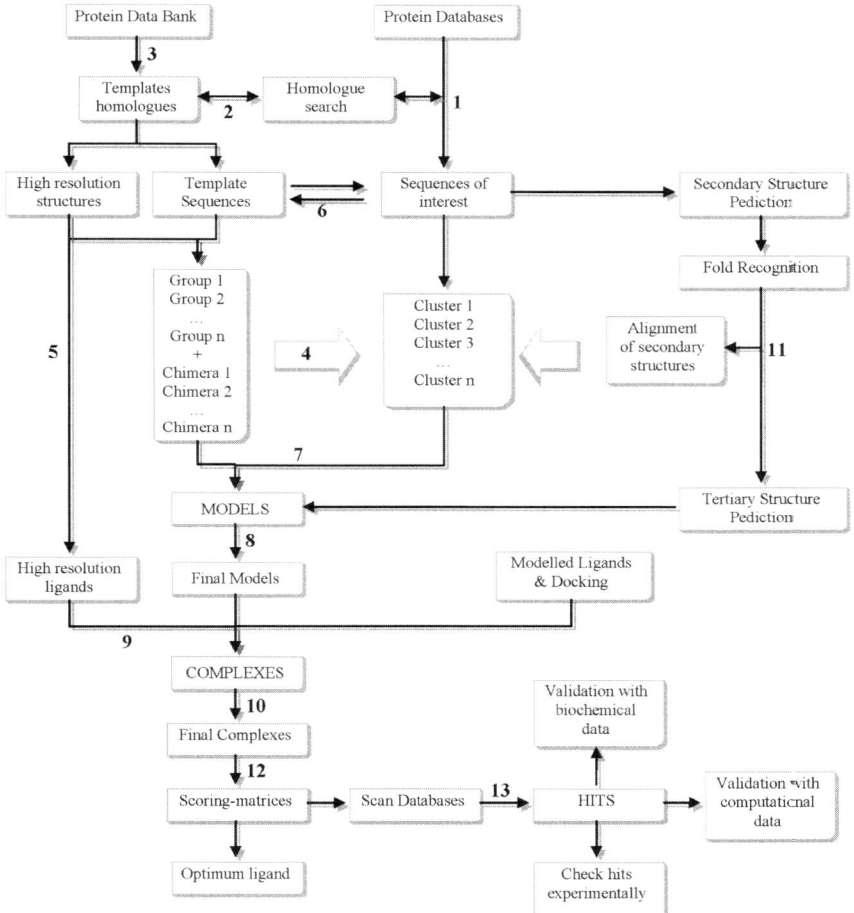

Fig. 1. Flow chart of methodologies for protein prediction based on structure. (1) Isolation of domain sequences and domain assignment. (2) Homology search. (3) Edition of the molecules and template selection. (4) Clustering of templates. (5) Selection of ligands. (6) Sequence alignment. (7) Homology modeling. (8) Evaluation of the models in terms of energy. (9) Ligand superposition. (10) Selection of complexes. (11) Modeling from secondary and tertiary structure predictions. (12) Scoring matrices construction. (13) Database search and hits filtering.

The output gives a picture and a table with the list of functional domains found, the scores and the boundaries. The isolated sequence domains can be accessed individually and stored in text files.

2. Isolation of all sequences of a particular domain found in the whole proteome of a taxon:

- Write the domain ID of your interest in Domain selection under "Architecture analysis."
- Select the taxonomic range in the selection box.
- Press Domain selection.

The output is a list of proteins from a selected organism that contains the domain of interest. Now you can:

- Get all the sequences of the whole protein or get all the sequences of the isolated domains.
- Save as text file.

The ADAN database (http://adan-embl.ibmc.umh.es/) contains biochemical and structural information on different modular protein domains implied in protein-protein interactions (SH3, SH2, WW, PDZ, PH, methyl transferase, acetyl transferase, WD40, VHS, protein tyrosine phosphatase, PTB, FHA, BRCT, and 14-3-3). The records in the database contain a useful collection of the most common domain features, as well as links to other databases such as PDBsum (http://www.ebi.ac.uk/thornton-srv/databases/pdbsum/), Protein Data Bank (http://www.rcsb.org/pdb/), and Swiss-Prot (http://au.expasy.org/). The database also offers the scoring matrices for ligand predictions and putative partners from protein domains when available (*see* **Subheadings 2.12.** and **2.13.**).

2.2. Homology Search

One of the most common analyses done on protein sequences is the similarity/homology search. It allows a mapping of information from known sequences to novel ones, especially the functional sites of the homologous proteins. Dynamic programming has been applied to sequence alignment and related computational problems: A dynamic algorithm finds the best solution by breaking the original problem into smaller sequentially dependent subproblems that are solved first. Each intermediate solution is stored in a table along with a score and the sequence of solutions that yields the highest score is chosen. The search for similarity/homology is well supported by internet resource tools. A query sequence can be entered to conduct the homology search using BLAST servers at Expasy (http://au.expasy.org/tools/blast/), NCBI (http://www.ncbi.nlm.nih.gov/BLAST/), or EBI (http://www.ebi.ac.uk/blastall/), or WU-BLAST servers at EMBL (http://dove.embl-heidelberg.de/Blast2/), or EBI (http://www.ebi.ac.uk/blast2/).

1. Go to web tool.
2. Paste or upload the query sequence, normally in FASTA format (sequence of amino acids in one letter code preceded by ">sequence_name").
3. Select the program "blastn" for nucleotides or "blastp" for amino acids.
4. Select database "nr" for nonredundant.
5. Check other fields/parameters of interest.
6. Press Search.

Table 1
Some Protein Modeling Commercial and Noncommercial Packages

Modeling package	URL
Insight II (Accelrys, Inc.)	http://www.accelrys.com/sim/
Chem3D (CambridgeSoft Corp)	http://www.camsoft.com
HyperChem (Hypercube, Inc.)	http://www.hyper.com
SYBYL (Tripos, Inc.)	http://www.tripos.com
SPDBV (GSK)	http://www.expasy.org/spdbv/
Wavefunction, Inc.	http://www.wavefun.com
MOE (CCG, Inc.)	http://www.chemcomp.com
Modeller (UCSF)	http://salilab.org/modeller/modeller.html
WHAT IF (CMBI)	http://swift.cmbi.ru.nl/whatif/

You will receive a submit confirmation, and in a few minutes an output results. The output includes alignment scores (sometimes as a plot), the list of sequences with significant E-values, and the corresponding information (e.g., sequences ID, name, score), and pairwise alignments (if chosen).

The homologue search can be done in most of the BLAST web servers selecting the pdb (Protein Data Bank) as the target database, thus directly obtaining the homologues whose structures are already known. The structures can be downloaded and edited with public software to model the domains of interest.

2.3. Edition of the Molecules and Template Selection

Most comprehensive software programs suitable for protein modelling are commercial packages, some of which are listed in **Table 1**. The protein modeling is illustrated with freeware programs and online servers, such as Swiss PDB Viewer (http://www.expasy.org/spdbv/), which is an application program that provides a user interface for visualization and analysis of biomolecules in particular proteins *(20)*. The program (Swiss PDB viewer or Spdbv) can be downloaded for free, and a user guide is also available. Spdbv implements GROMOS96 force field *(21)* to compute energy and to execute energy minimization by steep descent and conjugate gradient methods.

Typically, the workspace consists of a menu bar, tool icons, and four windows: Main (Display) windows, Control panel, Layers Info, and Align window. The Control panel, Layers Info, and Align window can be turned on and off from the Window menu. The Control panel provides a convenient way to select and manipulate the attributes of individual residues. The first column shows groups included in the structure, including chain name in uppercase letters (e.g., A, B, C), secondary structure recognized by Spdbv, represented by lowercase letters (s, strand; h, alfa-helix), and the names and numbers of all amino acid residues/nucleotides/heteroatoms. The second column and subsequent have check marks that toggle between display/hide the whole group, the side chains, residue labels, dot surfaces (VDW or accessibility), and ribbon (if the check marks are activated). The last column with small square boxes is used to highlight the residue(s) with colors.

The Layers Info Window allows the management of entire layers by turning on and off layer visibility, movement, displayed carbonyl groups, hydrogens, hydrogen bonds, and so on. The Align window permits an easy means of manage protein sequences for the homology modeling.

Briefly, the common tools are located at the top of the display window under the menu bar. These include move molecule (translate, zoom, and rotate) tools and general usage (bond distance, bond angle, torsion angle, label, center, fit, mutation, and torsion) tools. The menus provide a full set of utilities to load, display, select, edit, measure, superpose, or modify the molecules. Additionally, the scripting language provides an invaluable tool to simplify and automatize the repetitive edition or modification of molecules (*see* http://www.usm.maine.edu/~rhodes/SPVTut/text/DiscuSPV.html to post or exchange scripts). This is especially important for studies at genome-wide level.

The selection of templates starts in the Protein Data Bank (http://www.rcsb.org/pdb/), and this is the limiting step for the consecution of a successful prediction on a protein–protein interaction domain. If the sequence homology is low, it is mandatory the existence of several X-rays, high-resolution structures of the domain of interest, either alone or (better) in complex with natural or designed partners, the reason being that small or large conformational rearrangements could take place in the structure of the target sequence that will not be present in the model structure. Thus, the existence of several domain-ligand complexes is of capital importance in the reliability and accuracy of the predictions. Those complexes involving proteins with low sequence homology are more useful because they provide a glimpse of the conformational variability of the protein family. In the absence of this kind of complexes, docking techniques should be used (*see* **Subheading 2.9.**).

The structures of interest can be easily grouped in SMART:

1. Go to web tool.
2. Under "Domains detected by SMART," in "Display domain annotation" fill Domain Name (e.g., SH2, SH3, PDZ, etc.).
3. Press Display.
4. Select "Structure" and get the list of all structures available for the domain of interest.
5. Download the pdb files. Each pdb name in the previous list is directly linked to PDBsum database (http://www.ebi.ac.uk/thornton-srv/databases/pdbsum/) which, in turn, connects with Protein Data Bank.

The structural coordinate files contain sometimes much more structural information than needed for prediction. As an example, SH2 and SH3 domains are usually forming part of tyrosine kinases crystals, including the catalytic domain. Many other times, the coordinates of the domains appear several times in the same pdb file as a consequence of crystal symmetry properties. Other accessory molecules such as heteroatoms can be removed from the structure file if they are not involved in the protein interaction. These "cleaned" structures can be directly downloaded from the ADAN database, or can be easily isolated with Spdbv following these steps.

1. Load the molecule in Swiss PDB viewer: *File / Open PDB file*.
2. Select the residues comprising the domain of interest and the natural ligand, if present.
3. Use Control Panel or Alignment window. Activate first these windows in Window/Control Panel or Window/Alignment.
4. Save the selected part of the molecule. *File / Save / Save selected residues*.

It is necessary to stress the importance of keeping the structures of the natural or designed ligands in complex with the domain of interest, when present in the crystal.

2.4. Clustering of Templates

At this point, we have a set of structures representative of the domain of interest that can be compared, grouped, and classified according to different structural or sequence features. This important step allows the connection between proteins with known and unknown structures through homology modeling. A failed clustering gives unrealistic models and unsuccessful predictions even when the target sequences and templates are homologous. Unfortunately, there is no automatic way to carry out these steps, and it should be made by hand after careful observation of structural motives or key positions repetition. Thus, a previous study of the templates is required, including either multiple superposition or multiple sequence alignment of the structures and their sequences.

Superposition can be easily made with Spdbv following these steps.

1. Open the template1 molecule and color the whole molecule if desired.
2. Open the template2 molecule.
3. Make active the template2 layer by clicking its name in Layer Info, or Alignment, or Control Panel windows (three ways to do the same).
4. Invoke the *Iterative Magic Fit* tool from the *Fit* menu.
5. Choose the Auto Fit Options (α-carbon or backbone), and press OK. The template2 molecule is superimposed on template2 molecule.
6. Check the root mean square distance (RMSD <1.5Å is better), and press *Fit/ Improve Fit* to improve the superimposition (lowering rms distance).
7. Get the correct alignment by clicking *Fit / Generate Structural Alignment*. The movement of the cursor on the residues in Alignment window causes the residue in the Display window to blink to orange color, allowing the easy visualization of the overlapping molecules.

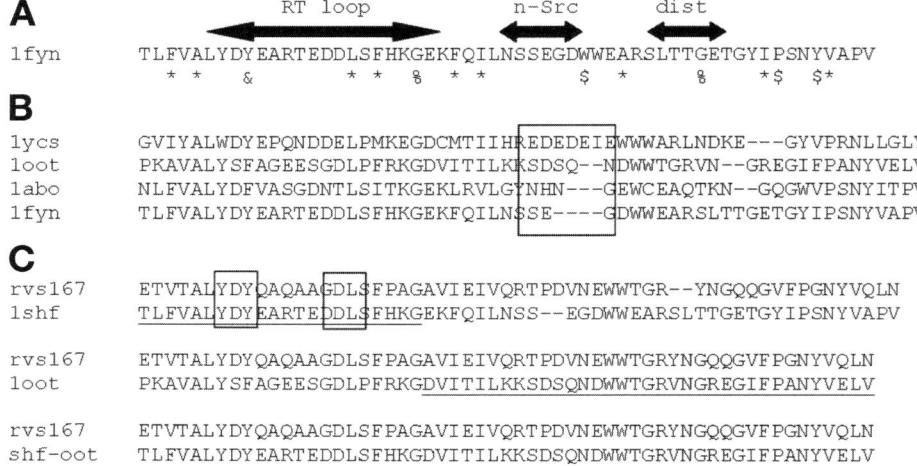

Fig. 2. Sequence characteristics tested for clustering of SH3 templates. (**A**) The key positions in SH3 core (*), Gly (%), binding ($), and motive YXY (&). (**B**) Different *n-Src* loop lengths in different SH3 templates. (**C**) The criteria used for chimera construction from two SH3 templates, 1SHF.pdb, and 1OOT.pdb. The resulting chimera was optimal to model the yeast SH3 rvs167 protein.

8. Save each of the overlapped molecules separately (*File / Save Layer*) as individual structure or save the superimposed molecules together (*File / Save Project*) as overlapped structures.

Interesting examples of previous studies are provided by the SH3 domain. There are two characteristics of SH3 sequences taken into account. Key positions that determine the protein folding, and later on the function, thus making the sequence belong to a given domain. In SH3, nine core positions, two very well conserved Gly and three binding positions *(22,23)* can be distinguished. These positions must be conserved (hydrophobic residues in the core) or identical (Glys and Trp, Pro and Phe in the binding pocket), and should be checked manually after sequence alignment (**Fig. 2A**). Second is the length of the loops *RT* and *n-Src* involved in binding. Because this feature can be involved in ligand recognition, it is important the clustering of templates by loop length. **Figure 2B** illustrates the groups obtained for SH3 structures classified according to *n-Src* length, since the *RT* loop is quite constant. Other insertions and deletions in the SH3 sequence not involved in binding (i.e., *distal* loop) are not taken into consideration.

Relevant structural characteristics of SH3 domains are: the Trp switch, where a very well-conserved Trp in the binding pocket of SH3 can tilt a few degrees depending of the kind of residues in the immediate vicinity and the

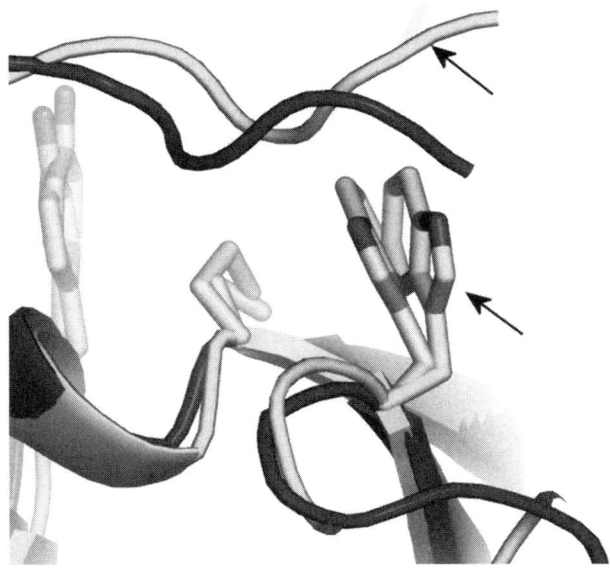

Fig. 3. Structural characteristics tested for clustering of SH3 templates. The figure shows the different conserved Trp orientation observed upon binding of ligand type I to Abl SH3 domain (template 1ABO.PDB, black arrows) or ligand type II to C-Crk N-terminal SH3 domain (template 1CKA.PDB).

nature of the ligand *(24)*. The consequence of this small Trp movement governs in part the binding specificity of SH3 to poly-Pro ligands type I or type II. So, depending on the orientation of this Trp in our template, we will be able to predict binding to type I or type II, but not both with the same structure (**Fig. 3**). Second is motive YXY or YXF in the *RT*-loop of SH3 domain. The second aromatic in this motive is part of the binding pocket of SH3 and is pointing to other important residues in the pocket. Some templates having Phe in the motive cannot accommodate Tyr (after homology modeling, *see* **Subheading 2.7.**) because either the extra hydroxyl group could clash with the conserved Trp, or the hydroxyl becomes uncompensated into the pocket, or both. On the contrary, the templates having Tyr in the motive do not have steric clashes after substitution by Phe, but residues previously forming hydrogen bonds with Tyr could become uncompensated into the binding pocket.

As a consequence of both sequence and structural features, the construction of chimeras from structural templates would be necessary to fulfill all the requirements needed for modeling a sequence target. A good example is shown in **Fig. 2C**, where the yeast protein rvs167 SH3 domain is modeled with a chimera formed with 1SHF.PDB and 1OOT.PDB templates. The former (1SHF.PDB) fits very well with protein yeast in the *RT* loop (motives YDY and DL), but presented a deletion in *n-Src* and a insertion in *distal* loops. The later (1OOT.PDB) fits well in *n-Src* and *distal* loop, but fails in the motive YDY to YSF (*see* boxes in **Fig. 2C**).

The construction of chimeras can be easily accomplished with Spdbv as follows.

1. Load the two templates selected to generate the chimera. Load first the template contributing with the N-terminal fragment.
2. Superimpose following the fully automatic method (*see* earlier in this subheading) or, if necessary, the manual one, to get a good overlapping of the connection point between the two templates to build the chimera.
3. For manual superimposition select suitable parts of the molecules, and take care of selecting the same number of residues in both layers (the number of residues selected can be followed in the last column of Layers Info window).
4. Select *Fit Molecules (from selection)* in the *Fit* menu. Select *Auto Fit Options* (backbone) and press OK.
5. Align sequences with *Fit / Generate structural alignment*.
6. In the Align window select the residues in the first molecule contributing with the N-terminal.
7. Select the residues of the second molecule contributing with the C-terminal. They should superimpose well to avoid peptidic bond distortion.
8. Assemble the two selected fragments with *Create Merged Layer from Selection* in the *Edit* menu. An extra layer ("_merge_") appears with the chimera.
9. Activate the merged layer and select all residues in *Select / All*.
10. Renumber and rename the chain with *Edit / Rename Current Layer*.
11. Save the merged layer with *File / Save Layer*.

2.5. Selection of Ligands

The importance of the availability of high-resolution domain-ligand complexes for prediction of protein interaction is already mentioned. These structures help to understand in detail the hydrophobic and hydrophilic interaction map and confirm the key residues important for binding *(6)*. At the same time, these structures validates the surface prediction studies *(5,7)*, and allows the accurate prediction of potential partners based on structure.

The templates containing ligands are filtered according to the resolution quality, trying to recruit ligands with 2.5Å resolution or better. These ligands are "cleaned" to remove parts of the protein not interacting with the domain, and comprising no more than 7 to 10 residues long for extended ligands. The isolated ligands are usually grouped into categories, depending of the nature and family features, and stored for later use (*see* **Subheading 2.9.**).

As an example, ligands binding to SH3 are grouped into type I, type II, type I' *(24)*, or other types *(25)* up to 24 different ligands.

Table 2
Clustalw Servers for Sequence Alignment
(Some Servers for Structure Alignments Also Included)

CLUSTALW servers	URL
CLUSTALW (EBI)	http://www.ebi.ac.uk/clustalw/
CLUSTALW (PBIL)	http://npsa-pbil.ibcp.fr/cgi-bin/npsa_automat.pl?page=npsa_clustalw.html
MUSCA	http://cbcsrv.watson.ibm.com/Tmsa.html
T-Coffee	http://www.ch.embnet.org/software/TCoffee.html
Dialign	http://bibiserv.techfak.uni-bielefeld.de/dialign/

Structure alignments	URL
CE / CL	http://cl.sdsc.edu
Dali	http://www.ebi.ac.uk/dali/
TMAP	http://www.mbb.ki.se/tmap/
MAMMOTH	http://ub.cbm.uam.es/mammoth/mult/index.php

2.6. Sequence Alignment

This step allows the assignments of the target sequences to the clusters previously obtained for the structures, so that each sequence is only linked to the templates included in the cluster, but not to all templates. The alignment of the target sequence inside its group permits a more accurate and realistic alignment, which is complemented with the chimera construction when there are no good candidates as templates.

The pairwise or multiple sequence alignment can be accomplished with ClustalW, the most commonly used program (**Table 2**).

1. Go to web tool to get the query form.
2. Paste or upload the sequences in FASTA format.
3. Press Send.

The output includes pairwise alignment scores, multiple sequence alignment and tree file. Check carefully if the new alignment fits all the requirements of your domain.

2.7. Homology Modeling

Protein modeling aims to predict the 3D structures of proteins from their amino acid sequences, using related sequences for which structures are available. The prediction of protein structures is based on two complementary approaches that can be used in conjunction: (1) Knowledge-based model combining sequence data to structure information, such as homology modeling *(26,27)*. The methodology modifies closely related functionally analogous sequence molecules (orthologous) whose 3D structure has been previously elucidated, and a putative 3D structure (model) of a protein from a known 3D structure is obtained. Thus, functionally analogous proteins with homologous sequences will have closely related structures with identical tertiary folding patterns. (2) Energy-based calculations through theoretical models and energy minimization, such as *ab initio* prediction *(28)*. Energy-based structure prediction relies on energy minimization and molecular dynamics. The method is faced with the problem of a large number of possible multiple minima, making the traversal of the conformational space difficult, and making the detection of the real energy minimum or native conformation uncertain. Residues are changed in the sequence with minimal disturbance to the geometry, and energy minimization optimizes the altered structure.

Although the success of homology modeling is satisfactory when sequence homology is greater than 50%, the structure homology may remain significant even if sequence homology is low: 3D structures are better conserved than the residue sequence. The region between 20% and 30% sequence identities is less certain, because part of the problem of homology modeling is the correct alignment of unknown and target proteins. Care should be taken with the important key residues within a structurally conserved common fold.

A methodology to explore sequence identities below 25% (remote homologues) is threading techniques *(29)*, in which a sequence of unknown structure is threaded into a sequence of known structure, and the fitness of the sequences for that structure is assessed.

Basically, the homology modeling consists of four steps:

1. Start from the known sequences.
2. Assemble the new sequence onto the template backbone from different, known homologous structures.
3. Optimize the structures.
4. Select the structures with better stability energy.

The modeling of a protein structure from its sequence against the known 3D structure of homologous protein(s) in the homology modelling can be attempted with Spdbv as follows:

1. Prepare your previously obtained alignment between the target sequence and the homologue pdb structure. You will use it later.
2. Save your target sequence in FASTA format in a text file (i.e., "target.txt"). Add extra sequence if you want to model the ligand in complex. In this case your template should have the ligand coordinates.
3. Choose the *Load Raw Sequence to Model* tool from the *SwissModel* menu and open "target.txt." The target sequence appears on the Display window as a long helix.
4. Open the template structure of the reference molecule (template.pdb).

5. Select the *Fit Raw Sequence* tool from the *Fit* menu. The α-helix changes to the structure overlapping the reference structure. Center the molecule in the Display window.

6. Click the little arrow beside the question mark of the Alignment window to view a plot of threading energies. Click *smooth* to set smooth to 1 and check *SwissModel / Update threading display now* tool. Select *Color / By Threading Energy* to display threading energy profile of the structure (the mean force potential energy of the polypeptide chain increases with color varying from blue to red).

7. The alignment of the target sequence onto the template can be manually refined on the Alignment window by translating residues or inserting and removing gaps.

8. Select a residue in the Alignment window and use the space bar to insert a gap or the backspace key to remove a gap. Select a group of residues and use left/right arrow keys to displace a gap. Do the same for the template if necessary. A long bond appears in the Display window.

9. Displace the ligand sequence in the target to align with the ligand sequence in the template. The ligand sequence can be identical or different to the template since it will be explored later.
   Break the connection between domain and ligand in the target. Use menu *Build*, then *Break backbone* and pick one of the backbone atoms connecting domain and ligand target.

10. Activate the target structure and change chain name (i.e., A for domain and B for ligand) and numbering, if desired.

11. Perform two operations: *Select / Select_aa Making Clashes*, and *Tools / Fix Selected Sidechains / Quick and Dirty*. Repeat the process to decrease the number of amino acid residues making clashes.

12. Make sure that Swiss Model settings, under *Preferences* menu has the correct server information: Modeling server (http://swissmodel.expasy.org/cgi-bin/sm-submit-request.cgi); template server (http://swissmodel.expasy.org/cgi-bin/blastexpdb.cgi), your name and your e-mail correctly written (very important).

13. Submit your model to SwissPdbServer using Optimise (project) mode for optimization. Go to *Submit Modeling Request* under *SwissModel* menu. You are asked to give a name for your model (i.e., my_project.htm in html format). An additional project file is automatically saved as proj_my_project.pdb as coordinates.

14. Your Internet browser opens and shows the web page to be submitted. Fill the information regarding your Swiss Model Project and check the appropriate Result options (usually Swiss PDB viewer Mode and What-if check) and press Send Request. You will get a notification of successful submission.

15. You will receive several messages in your e-mail box. One of them has your model in complex with the desired ligand. The two structures (domain and ligand) are already merged if you have decided to model both together. Open the received project and save the first layer to isolate your final model.

The models at this step require visual inspection and evaluation in terms of energy (*see* **Subheading 2.8.**) to check the quality of the models. We should look for inconsistencies in the core (i.e., polar groups, strong clashes between core residues), in the binding pocket (i.e., incompatible residues, bad rotamers), or in the loop movements among different templates (i.e., clashes, obstacle for ligand binding).

2.8. Evaluation of the Models in Terms of Energy

The use of fast and reliable protein force field is an efficient tool to evaluate the delicate balance between the different energy terms that contribute to protein stability *(30,31)*. Many different force fields are constructed for predicting protein stability changes, ranging from force fields based on pure statistical analysis of structural sequence preferences *(32,33)*, or force fields based on multiple sequence alignments *(23,34)*, to detailed molecular dynamics force fields *(35,36)*. These force fields can be divided into three major categories: (1) those using a physical effective energy function (PEEF), (2) those based on statistical potentials for which energies are derived from the frequency of residue or atom contacts in the protein database (SEEF), and (3) those using empirical data obtained from experiments on proteins (EEEF).

There are several molecular modelling packages that implement force fields that can be used for the evaluation of the models in terms of energy: Spdbv implements Gromos96 force field; InsightII (Accelrys, Inc.) implements CVFF, CFF91, and AMBER force fields; Sybyl (Tripos, Inc.) implements MM2, amberall40, and amberuni40 force fields.

In addition, FoldX *(37,38)*, a computer algorithm, provides a fast and quantitative estimation of the interactions contributing to the stability of proteins and protein complexes. The different energy terms taken into account in FoldX have been weighted using empirical data from protein engineering experiments, and the predictive power has been tested on a very large set of protein mutants covering most of the structural environments found in proteins.

The FoldX energy function includes terms that have been found to be important for protein stability. The free energy of unfolding (ΔG) of a target protein is calculated using **Eq. 1**:

$$\Delta G = \Delta G_{vdw} + \Delta G_{solvH} + \Delta G_{solvP} + \Delta G_{hbond} + \Delta G_{el} + \Delta G_{kon}$$
$$+ \ T\Delta S_{mc} + T\Delta S_{sc} + T\Delta S_{tr}$$

where ΔG_{vdw} is the sum of the van der Waals contributions of all atoms with respect to the same interactions with the solvent. ΔG_{solvH} and ΔG_{solvP} are the difference in solvation energy for apolar and polar groups respectively when going from the unfolded to the folded state. ΔG_{hbond} is the free energy difference between the formation of an intramolecular hydrogen-bond compared with intermolecular hydrogen-bond formation (with solvent). ΔG_{wb} is the extra stabilizing free energy provided by a water molecule making more than one hydrogen-bond to the protein (water bridges) that cannot be taken into account with non-explicit solvent approximations. ΔG_{el} is the electrostatic contribution of charged groups, including the helix dipole. ΔG_{kon} reflects the effect of electrostatic interactions on kon *(39)* term for protein complexes. ΔS_{mc} is the entropy cost for fixing the backbone in the folded state. This term is dependent on the intrinsic tendency of a particular amino acid to adopt certain dihedral angles. ΔS_{sc} is the entropic cost of fixing a side chain in a particular conformation. Finally, ΔS_{tr} is the loss of translational and rotational entropy upon making the complex.

The FoldX web server (http://foldx.embl.de) gives the user the stability energy, the interaction network of the different energy components per interaction type and per residue, or the energy contribution per chain in a protein–protein complex.

The careful examination of these energy values permits the selection of high-quality molecular models and opens the way for successful prediction.

2.9. Ligand Superposition

When a large amount of structures in complex with ligands are available, it is quite easy to add these ligands to models to perform predictions: just follow the protocol used for generating chimeras to create the complexes (*see* **Subheading 2.4.**). It should be noted that the superposition has to be accurate and realistic, and has to be adapted to the actual target. For SH3 domains as an example, manual superposition is used by selecting nine residues in the core and four residues in the binding pocket, but this can be quite different for other domains, such as PDZ or SH2.

When complexes are not available, it is necessary to use docking techniques, thus assuming a great problem inherent to the methodology: The peptidic ligand conformation. Molecular docking aims to fit two interacting molecules by exploring the binding modes of their topographic or energy-based features consideration that lead to favorable interactions. Ligand-receptor interaction is an important initial step in protein prediction and function. The structure of ligand-receptor complex profoundly affects the specificity and efficiency of protein action. The molecular docking performs the computational prediction of the ligand-receptor interaction and the structures of ligand-receptor complexes, usually computing the van der Waals and the Coulombic energy contributions between all atoms of the two molecules.

There are two classes of strategies for docking a ligand to a receptor. The first class uses a whole ligand molecule as a starting point and employs a search algorithm to explore the energy profile of the ligand at the binding site, looking for optimal solutions for a specific scoring function. The search algorithms include geometric complementary match, simulated annealing, molecular dynamics, and genetic algorithms. Representative examples are DOCK3.5 *(40)*,

Table 3
Available Resources for Docking Purposes

Docking applications	URL
GRAMM	http://www.bioinformatics.ku.edu/vakser/gramm/
HEX	http://www.csd.abdn.ac.uk/hex/
FlexX	http://www.biosolveit.de/software/
DOCK	http://dock.compbio.ucsf.edu/
AUTODOCK	http://www.scripps.edu/mb/olson/doc/autodock/
LIGIN	http://swift.cmbi.kun.nl/swift/ligin/
ICM-Docking	http://www.molsoft.com/docking.html
3D-Dock	http://www.bmm.icnet.uk/docking/
ZDOCK, RDOCK	http://zlab.bu.edu/zdock/index.shtml
Bielefeld Software	http://www.techfak.uni-bielefeld.de/~sneumann/agaiprot/
Molfit	http://www.weizmann.ac.il/Chemical_Research_Support//molfit/
3D-JIGSAW	http://www.bmm.icnet.uk/~3djigsaw/

AutoDock *(41)*, and GOLD *(42)*. The second class starts by placing one or several fragments (substructures) of a ligand into a binding pocket, and then it constructs the rest of the molecule in the site. Representative examples are DOCK4.0 *(43)*, FlexX *(44)*, LUDI *(45)*, and GROWMOL *(46)*. **Table 3** shows some Internet resources for molecule docking.

The use of docking methods from isolated structures is, however, very difficult and risky because the conformations of the isolated ligand and the receptor can be different (or very different) after complex formation. For this reason, it is strongly recommended the use of ligand-receptor structure complexes of high resolution. The methodological use of the docking methods is out of the scope of this chapter.

2.10. Selection of Complexes

Finally, we have a set of models in complex with ligands that should be evaluated. First of all, the positions in the ligand should be mutated to Ala to minimize the clashes within the complex, but also to normalize all the positions in the same ligand, and among different ligands.

Mutations can be easily accomplished with Swiss PDB viewer following these steps in the active molecule:

1. Press the mutation icon.
2. Pick the amino acid to be mutated. The list of amino acids is displayed.
3. Select the desired amino acid. A rotamer is selected automatically.
4. Change the rotamer by clicking the small black arrows under the mutation icon.
5. Press the mutation icon again and select OK.
6. Refine the structure by removing clashes:

- Press *Select / aa making clashes.*
- Fix clashing residues with *Tools / Fix selected Sidechains* and *Quick and Dirty* or *Exhaustive Search* (no more than 10 residues).
- Save your mutated structure (*File / Save / Layer*).

The complexes must be explored, looking for incompatibilities between domains and ligands. Mainly, strong clashes between ligand and domain backbones should be avoided, and in this case, the complexes should be discarded. Regarding side chains, small clashes can be tolerated for external residues that are able to adopt different rotamers, but cannot for residues important for binding.

Again, the energy calculation with FoldX provides a great tool to explore the complexes in terms of energy. The stability energy of the complex (*Es*) and the binding energy (*Eb*) between domain and ligand can be calculated in FoldX web server. The *Eb* can be calculated under AnalyseComplex mode as:

$$Eb = Es - \sum (EsA + EsB)$$

where *EsA* and *EsB* refer to the stability energy of isolated chains A (domain) and B (ligand), respectively.

Finally, we have a set of high-quality complexes, selected for ligand and domain compatibility, and prepared to predict the optimal ligand, construct the scoring matrices, and search in the databases.

2.11. Modeling From Secondary and Tertiary Structure Predictions

The modeling of proteins that have homologues in the Protein Data Bank is very easy. However, not all proteins have homologues of known structure. This case requires an additional secondary or tertiary prediction to build models and guess the putative partners. The flow diagram in **Fig. 1** shows an alternative route when the sequences of interest have no homologue structures. First of all, a prediction of secondary structure is required, an old technique not exempt of many problems (the early methods suffered from a lack of data, predictions were performed on single sequences rather than families of homologous sequences, and there were relatively few known 3D structures to derive parameters). Now, this technique is more accurate and aims to determine the probable placement of secondary structural elements along the sequence. The prediction is made at several levels: (1) secondary structure prediction, expecting three states (helix, strand, rest) with an accuracy ranging 72 to 76% for water-soluble globular proteins; (2) solvent accessibility prediction; (3) transmembrane helix prediction, expecting overall two states (transmembrane, nontransmembrane) with an accuracy higher than 95%; and (4) globularity, which identifies interdomain segments containing linear motifs and apparently ordered regions that do not contain any recognized domain. Most common servers for protein secondary structure are depicted in **Table 4**.

Table 4
Secondary Structure Prediction Servers and Links

Secondary structure prediction	URL
GOR	http://npsa-pbil.ibcp.fr/cgi-bin/npsa_automat.pl?page=npsa_gor4.html
HNN	http://npsa-pbil.ibcp.fr/cgi-bin/npsa_automat.pl?page=npsa_nn.html
DSC	http://www.aber.ac.uk/~phiwww/prof/
PredictProtein	http://www.embl-heidelberg.de/predictprotein/predictprotein.html
nnPredict	http://www.cmpharm.ucsf.edu/~nomi/nnpredict.html
PSA	http://bmerc-www.bu.edu/psa/
BCM-PSSP	http://dot.imgen.bcm.tmc.edu/pssprediction/pssp.html
JPred	http://www.compbio.dundee.ac.uk/~www-jpred/submit.html
Predator	http://bioweb.pasteur.fr/seqanal/interfaces/predator-simple.html

Table 5
Folding Recognition Servers and Links

Folding recognition	URL
3D-PSSM	http://www.sbg.bio.ic.ac.uk/~3dpssm/
UCLA-DOE	http://fold.doe-mbi.ucla.edu/
LOOPP	http://cbsu.tc.cornell.edu/software/loopp/
PROSPECT Pro!	http://www.bioinformaticssolutions.com/products/prospect.php
123D	http://123d.ncifcrf.gov/123D+.html
UCSC HMM	http://www.cse.ucsc.edu/research/compbio/HMM-apps/
FFAS03	http://ffas.ljcrf.edu/ffas-cgi/cgi/ffas.pl
PSPC	http://globin.bio.warwick.ac.uk/%7Ejones/threader.html

Even with no homologue of known 3D structure, it may be possible to find a suitable fold for you protein among known 3D structures by folding recognition methods (**Table 5**). There are many approaches, but the unifying theme is to try and find folds that are compatible with a particular sequence. These methods combine 1D (or even 3D) sequence profiles coupled with secondary structure and contact capacity potential information to thread a protein sequence through the set of structures and predict the fold (*47*).

The alignments of sequence onto tertiary structure from fold recognition methods may be inaccurate. After the identification of a remote homologue, it is convenient to edit the alignment around variable regions and consider the alignment of secondary structures. Check that: (1) the residues predicted to be buried or exposed are aligned with those known to be buried/exposed in the template structure; (2) hydrogen bonds networks are not disrupted in β-sheet structures; and (3) the residues properties (i.e., size, polarity, hydrophobicity) are conserved as much as possible in the alignment.

Table 6
***Ab initio* Structure Prediction Sites**

Ab initio prediction	URL
Rosetta/HMMSTR	http://www.bioinfo.rpi.edu/%7Ebystrc/hmmstr/server.php
I-Sites	http://www.brunel.ac.uk/depts/bio/project/biocomp/mak_fan/isites.htm

If the fold can not be yet recognized, there are methods to try *ab initio* structure predictions or at least predictions that do not rely on a template. *Ab initio* prediction of protein 3D structures is not "possible" at present, and a general solution to the protein folding problem is not likely to be found in the near future. However, some methods have been developed to try the prediction of the structure of proteins starting from the sequence by calculating: (1) secondary structure in the form of three states (helix, extended, loop); (2) local conformation (backbone torsion angles phi and psi); (3) supersecondary structure for strands an β-turns; and 4) tertiary structure, in the form of coordinates. These methods (**Table 6**) are based on hidden Markov model for local and secondary structure prediction, based on the I-sites Library.

Finally, we have models that can be evaluated in terms of energy, then following the route in the prediction diagram (**Fig. 1**).

2.12. Scoring Matrices Construction

The models in complex with poly-Ala ligands are now ready to use for scoring matrices construction (**Fig. 4**). Basically, each position in the ligand is explored by systematic mutation of the Ala to the 20 natural amino acids and further energy evaluation of the resulting structures (stabilization and binding energies). The energies obtained are correlated with the ability of a residue (in a ligand, in a position) to improve the ligand-domain interaction. Several assumptions have to be made.

1. Every position in the ligand is treated and computed separately within the ligand. This simplification is of great importance to save hard disk space and computational time and is based on the fact that most ligands binds to the domain in an "extended" conformation (e.g., SH2, PDZ, poly-Pro helix in SH3), that makes contiguous residues point to different directions and to interact with different set of residues in the domain. In the case in which there is energy coupling between two positions in the ligand, other more sophisticated approaches like mean field, dead-end elimination, branch and bound, or Monte Carlo techniques could be used to explore many-fold sequence combinations.

2. The domain residues in the immediate vicinity of the mutating ligand position should be allowed to change their rotamer to accommodate the new environment after mutation. This is also important to avoid strong van der Waals clashes and to optimize hydrophobic/hydrophilic interactions. The most important positions

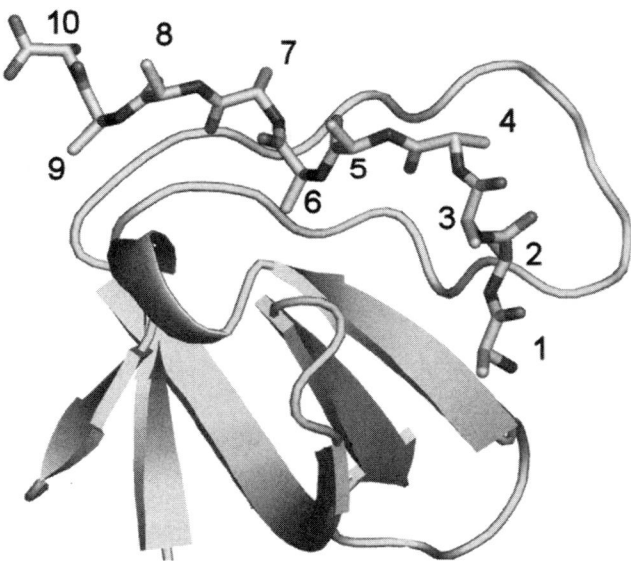

Fig. 4. A yeast SH3 domain in complex with a type I ligand 10 residues long. All 20 natural amino acids are placed in each position, and the surrounded residues are allowed to relax. The positions in the ligand are processed independently, and the coordinate files are generated and evaluated in terms of energy.

in the binding pocket probably will not change their rotamers (i.e., W, P, and Y in the SH3 pocket), but other positions involved in binding could be adapted to improve the interaction.

Methodologically, there is not a direct way to build these structures easy an automatically with commercial applications. The construction of 20 structures per position, with a mean of 9 positions per ligand, and with the overall use of 10 ligands, gives a total of 1800 structure files. This number is big enough to dissuade anybody to build these structures by hand. Most modeling packages include scripting capabilities that can be adapted to:

1. Mutate a position in a ligand to the first of 20 amino acids.
2. Relax the surrounded domain positions.
3. Mutate the same position to the first amino acid again. This is made to avoid conformational traps.
4. Relax again the surrounded domain positions.
5. Save the coordinates of the new structure.
6. Repeat 1–5 for every position in a ligand.
7. Repeat 1–6 for all ligands.
8. Repeat 1–7 for all complexes.
9. Evaluate the stabilization and binding energies for all structures with FoldX.

At this point, we have a set of structures that represents a complete screening of the ligand-domain interactions. The evaluation of the binding energy of these structures provides the link between ligand position, type of residue and binding improvement. This quantification results in scoring matrices (**Fig. 5**).

The matrices are corrected by adding internal van der Waals clashes of the interface residues with their own chains to the binding energy and normalized to the lower value (becoming 0). The lower the energy value, the better the ligand-domain interaction. After having the matrices, we model the best ligand by taking the most favorable amino acid at each position (now all positions of the ligand at the same time) and evaluating its binding energy. This will be used as a reference for a particular matrix. Note in the example (**Fig. 5**) that some positions are more tolerant than others to accommodate different residues. This is a reflection of the role of the positions in the binding interaction, being more permissive the positions that point to the solvent than those pointing to the hydrophobic pocket.

2.13. Database Search and Hits Filtering

The scoring matrices provide the link between computational predictions of partners and the localization of these resulting motives in the genome databases, so that we can use these data to scan de genome, looking for the sequences that better fit in the modeled SH3 domains, thus guessing function. The simplest way to do this is the generation of Prosite patterns and the use of the web application in Prosite web page (http://au.expasy.org/tools/scanprosite/). Prosite patterns should follow a specific format (*see* Prosite *format* link in the web page), and can be easily derived from the scoring matrices. As an example, this pattern is obtained from the scoring matrix of **Fig. 5**:

Pattern: [PIL]-X-[HKRM]-X-[PHQ]-P-[HDPWFRS]-[MPLKR]-[PW]

The ScanProsite tool allows to look protein sequence(s) for the occurrence of patterns, profiles and rules stored in the Prosite database, but also to search protein database(s) for hits by specific motif(s) *(48)*. The last feature is the one used for our purposes to scan genomes. To search a genome with a pattern:

1. Go to web application.
2. Enter one pattern under *PROSITE pattern(s)/profile(s) to scan for.*
3. Select the database to search: Swiss-Prot and TrEMBL.
4. Write your taxon in *Taxonomic lineage (OC) / species (OS).*
5. Press *START THE SCAN.*

After genome scanning, the results are presented in the screen. The information included contains protein name, SwissProt code numbers, protein length, position and sequence of the hits, as well as the link to get the whole protein. Depending on the patterns entered, the results obtained can vary from a few of proteins to several hundreds. The results should be filtered because probably

#	ALA	ARG	ASN	ASP	CYS	GLN	GLU	GLY	HIS	ILE	LEU	LYS	MET	PHE	PRO	SER	THR	TRP	TYR	VAL	optimum
1	1.60	2.18	1.22	2.59	1.16	1.92	2.51	1.58	0.00	2.16	3.24	0.60	1.51	1.23	1.69	2.26	2.26	2.60	1.34	1.80	**H**
2	1.21	0.79	1.31	2.62	0.70	1.47	1.10	1.48	0.52	0.22	0.27	1.44	1.21	2.45	0.00	1.97	2.22	0.87	2.75	0.80	**P** IL
3	0.52	0.18	0.56	1.00	0.71	0.47	0.64	0.66	0.43	0.48	0.45	0.13	0.49	0.46	0.00	0.52	0.59	0.58	0.51	0.57	**P** KRHLFQIM
4	1.15	0.38	0.87	0.70	1.06	0.81	1.09	0.77	0.00	0.54	0.78	0.29	0.37	0.91	3.25	0.71	0.99	0.98	0.97	0.95	**H** KMR
5	0.66	0.23	0.30	0.34	0.83	0.23	0.36	0.19	0.32	0.41	0.34	0.10	0.18	0.27	0.38	0.58	0.73	0.00	0.36	0.65	**W** KMGQRFNHDLYEPI
6	1.46	4.0E+005	1.59	3.33	0.99	0.38	1.41	2.83	0.18	17.34	5.0E+003	4.61	114.13	1.12	0.00	1.90	11.09	4.0E+005	2.79	13.03	**P** HQ
7	0.91	0.76	0.67	1.15	1.16	1.03	1.19	1.22	0.63	0.92	0.85	0.70	0.84	0.63	0.00	0.84	1.46	0.76	0.62	1.18	**P**
8	0.61	0.47	0.77	0.20	0.78	0.60	0.78	0.65	0.00	0.68	0.65	0.57	0.59	0.47	0.29	0.48	1.04	0.46	0.66	0.94	**H** DPWFRS
9	1.20	0.41	1.48	1.55	1.02	0.54	1.07	1.45	85.87	0.51	0.31	0.39	0.00	5.0E+004	0.24	1.71	1.75	9.8E+004	2.0E+007	0.60	**M** PLKR
10	1.53	0.58	1.23	1.85	1.19	1.58	1.64	1.59	0.98	1.30	1.10	0.86	0.87	0.73	0.00	1.63	1.82	0.40	1.20	1.75	**P** W

Fig. 5. Scoring matrix from a yeast SH3 domain. Rows represent position in the ligand, and columns represent the 20 natural amino acids. The matrices are built by generating a set of structures containing systematic mutations in the ligand and further evaluation of the binding energy. The values are usually corrected and normalized to the lower one (the lower, the better).

most of the hits obtained are not real. There is not a fixed rule to do this filtering step. What follows is just an example that can be modified depending of the domain, the taxon, or the available biochemical data.

The simplest way to filter the hits is follows:

1. Get the sequences of the proteins found by ScanProsite.
2. Send the sequences to a protein secondary structure prediction server.
3. Send the sequences to predict globularity/disorder (http://globplot.embl.de/). GlobPlot is a web service that allows the user to plot the tendency within the query protein for order/globularity and disorder.
4. Match secondary structure and order/disorder predictions (per protein).

The hit sequence is localized in the match and checked for coincidences. The hit must be located in a region corresponding to both a disordered secondary structure prediction and to an unstructured part of the protein *(50)*. Whether a hit missing one residue in these regions is selected or not depends on each case. Again, there are no fixed rules and the limits can be established and modified according to the necessities.

The remaining hits can be filtered again with more criteria: the most important is probably the experimental data, when available. All panning experiments to search putative specific ligands of the domains are suitable to validate the prediction. As an example, peptide libraries displayed in filamentous phages or large-scale two-hybrid interaction tests are used for these purposes *(49)*. There are also available, for some domains (i.e., SH3), different computational prediction data that serve for comparisons and mutual validation *(8,9)*. A further filtering criterion can be the data regarding the subcellular localization of the proteins in a taxon. It is possible that a predicted hit that fulfils all the conditions to interact with the domain, is confined to a different compartment, so that they never meet in the cell. In this case, the hit should be discarded.

The scoring matrices can be also used to evaluate the probability of interaction between a domain and a given peptide, by summing the positional free energy of each amino acid in the sequence and comparing with the optimum ligand.

3. Conclusion

The prediction of protein-protein interactions based on structure should be viewed as a new tool to guess protein function, to improve database annotation, and to design rational experiments to understand protein network interactions.

Acknowledgments

The authors thanks Dr. Pilar Aguado Giménez for editing the manuscript.

References

1. Lo, C. L., Chothia, C., and Janin, J. (1999) The atomic structure of protein-protein recognition sites. *J. Mol. Biol.* **285,** 2177–2198.

2. Valdar, W. S. and Thornton, J. M. (2001) Protein-protein interfaces: analysis of amino acid conservation in homodimers. *Proteins* **42,** 108–124.
3. Jones, S. and Thornton, J. M. (1997) Analysis of protein-protein interaction sites using surface patches. *J. Mol. Biol.* **272,** 121–132.
4. Jones, S. and Thornton, J. M. (1997) Prediction of protein-protein interaction sites using patch analysis. *J. Mol. Biol.* **272,** 133–143.
5. Aloy, P., Querol, E., Aviles, F. X., and Sternberg, M. J. (2001) Automated structure-based prediction of functional sites in proteins: applications to assessing the validity of inheriting protein function from homology in genome annotation and to protein docking. *J. Mol. Biol.* **311,** 395–408.
6. Stein, A., Russell, R. B., and Aloy, P. (2005) 3did: interacting protein domains of known three-dimensional structure. *Nucleic Acids Res.* **33,** D413–D417.
7. Aloy, P., Bottcher, B., Ceulemans, H., Leutwein, C., Mellwig, C., Fischer, S., et al. (2004) Structure-based assembly of protein complexes in yeast. *Science* **303,** 2026–2029.
8. Brannetti, B., Via, A., Cestra, G., Cesareni, G., and Helmer-Citterich, M. (2000) SH3-SPOT: an algorithm to predict preferred ligands to different members of the SH3 gene family. *J. Mol. Biol.* **298,** 313–328.
9. Wollacott, A. M. and Desjarlais, J. R. (2001) Virtual interaction profiles of proteins. *J. Mol. Biol.* **313,** 317–342.
10. Meng, E. C., Gschwend, D. A., Blaney, J. M., and Kuntz, I. D. (1993) Orientational sampling and rigid-body minimization in molecular docking. *Proteins* **17,** 266–278.
11. Lichtarge, O. and Sowa, M. E. (2002) Evolutionary predictions of binding surfaces and interactions. *Curr. Opin. Struct. Biol.* **12,** 21–27.
12. Bogan, A. A. and Thorn, K. S. (1998) Anatomy of hot spots in protein interfaces. *J. Mol. Biol.* **280,** 1–9.
13. Schreiber, G. and Fersht, A. R. (1995) Energetics of protein-protein interactions: analysis of the barnase-barstar interface by single mutations and double mutant cycles. *J. Mol. Biol.* **248,** 478–486.
14. Janin, J. and Chothia, C. (1990) The structure of protein-protein recognition sites. *J. Biol. Chem.* **265,** 16027–16030.
15. Janin, J. (1995) Principles of protein-protein recognition from structure to thermodynamics. *Biochimie* **77,** 497–505.
16. Jones, S. and Thornton, J. M. (1996) Principles of protein-protein interactions. *Proc. Natl. Acad. Sci. USA* **93,** 13–20.
17. Tsai, C. J., Lin, S. L., Wolfson, H. J., and Nussinov, R. (1996) A dataset of protein-protein interfaces generated with a sequence-order-independent comparison technique. *J. Mol. Biol.* **260,** 604–620.
18. Tsai, C. S. (2002) Molecular modelling: protein modelling, in *An Introduction to Computational Biochemistry* (Han, L., ed.), John Wiley & Sons, Inc., New York, pp. 315–342.
19. Letunic, I., Copley, R. R., Schmidt, S., Ciccarelli, F. D., Doerks, T., Schultz, J., et al. (2004) SMART 4.0: towards genomic data integration. *Nucleic Acids Res.* **32,** D142–D144.

20. Guex, N. and Peitsch, M. C. (1997) SWISS-MODEL and the Swiss-PdbViewer: an environment for comparative protein modeling. *Electrophoresis* **18,** 2714–2723.
21. van Gunsteren, W. F. and Mark, A. E. (1992) Prediction of the activity and stability effects of site-directed mutagenesis on a protein core. *J. Mol. Biol.* **227,** 389–395.
22. Northey, J. G., Di Nardo, A. A., and Davidson, A. R. (2002) Hydrophobic core packing in the SH3 domain folding transition state. *Nat. Struct. Biol.* **9,** 126–130.
23. Larson, S. M. and Davidson, A. R. (2000) The identification of conserved interactions within the SH3 domain by alignment of sequences and structures. *Protein Sci.* **9,** 2170–2180.
24. Fernandez-Ballester, G., Blanes-Mira, C., and Serrano, L. (2004) The tryptophan switch: changing ligand-binding specificity from type I to type II in SH3 domains. *J. Mol. Biol.* **335,** 619–629.
25. Cesareni, G., Panni, S., Nardelli, G., and Castagnoli, L. (2002) Can we infer peptide recognition specificity mediated by SH3 domains? *FEBS Lett.* **513,** 38–44.
26. Hilbert, M., Bohm, G., and Jaenicke, R. (1993) Structural relationships of homologous proteins as a fundamental principle in homology modeling. *Proteins* **17,** 138–151.
27. Chinea, G., Padron, G., Hooft, R. W., Sander, C., and Vriend, G. (1995) The use of position-specific rotamers in model building by homology. *Proteins* **23,** 415–421.
28. Bonneau, R. and Baker, D. (2001) Ab initio protein structure prediction: progress and prospects. *Annu. Rev. Biophys. Biomol. Struct.* **30,** 173–189.
29. Bryant, S. H. and Altschul, S. F. (1995) Statistics of sequence-structure threading. *Curr. Opin. Struct. Biol.* **5,** 236–244.
30. Murphy, K. P. and Freire, E. (1992) Thermodynamics of structural stability and cooperative folding behavior in proteins. *Adv. Protein Chem.* **43,** 313–361.
31. Pace, C. N., Shirley, B. A., McNutt, M., and Gajiwala, K. (1996) Forces contributing to the conformational stability of proteins. *FASEB J.* **10,** 75–83.
32. Sippl, M. J. (1995) Knowledge-based potentials for proteins. *Curr. Opin. Struct. Biol.* **5,** 229–235.
33. Topham, C. M., Srinivasan, N., and Blundell, T. L. (1997) Prediction of the stability of protein mutants based on structural environment-dependent amino acid substitution and propensity tables. *Protein Eng.* **10,** 7–21.
34. Bordo, D. and Argos, P. (1991) Suggestions for "safe" residue substitutions in site-directed mutagenesis. *J. Mol. Biol.* **217,** 721–729.
35. Prevost, M., Wodak, S. J., Tidor, B., and Karplus, M. (1991) Contribution of the hydrophobic effect to protein stability: analysis based on simulations of the Ile-96–Ala mutation in barnase. *Proc. Natl. Acad. Sci. USA* **88,** 10880–10884.
36. Pitera, J. W. and Kollman, P. A. (2000) Exhaustive mutagenesis in silico: multicoordinate free energy calculations on proteins and peptides. *Proteins* **41,** 385–397.
37. Guerois, R., Nielsen, J. E., and Serrano, L. (2002) Predicting changes in the stability of proteins and protein complexes: a study of more than 1000 mutations. *J. Mol. Biol.* **320,** 369–387.
38. Kiel, C., Serrano, L., and Herrmann, C. (2004) A detailed thermodynamic analysis of ras/effector complex interfaces. *J. Mol. Biol.* **340,** 1039–1058.

39. Vijayakumar, M., Wong, K. Y., Schreiber, G., Fersht, A. R., Szabo, A., and Zhou, H. X. (1998) Electrostatic enhancement of diffusion-controlled protein-protein association: comparison of theory and experiment on barnase and barstar. *J. Mol. Biol.* **278,** 1015–1024.
40. Kuntz, I. D., Blaney, J. M., Oatley, S. J., Langridge, R., and Ferrin, T. E. (1982) A geometric approach to macromolecule-ligand interactions. *J. Mol. Biol.* **161,** 269–288.
41. Morris, G. M., Goodsell, D. S., Huey, R., and Olson, A. J. (1996) Distributed automated docking of flexible ligands to proteins: parallel applications of AutoDock 2.4. *J. Comput. Aided Mol. Des.* **10,** 293–304.
42. Jones, G., Willett, P., Glen, R. C., Leach, A. R., and Taylor, R. (1997) Development and validation of a genetic algorithm for flexible docking. *J. Mol. Biol.* **267,** 727–748.
43. Ewing, T. J., Makino, S., Skillman, A. G., and Kuntz, I. D. (2001) DOCK 4.0: search strategies for automated molecular docking of flexible molecule databases. *J. Comput. Aided Mol. Des.* **15,** 411–428.
44. Rarey, M., Kramer, B., Lengauer, T., and Klebe, G. (1996) A fast flexible docking method using an incremental construction algorithm. *J. Mol. Biol.* **261,** 470–489.
45. Bohm, H. J. (1992) The computer program LUDI: a new method for the de novo design of enzyme inhibitors. *J. Comput. Aided Mol. Des.* **6,** 61–78.
46. Bohacek, R. S. and McMartin, C. (1997) Modern computational chemistry and drug discovery: structure generating programs. *Curr. Opin. Chem. Biol.* **1,** 157–161.
47. Jones, D. and Thornton, J. (1993) Protein fold recognition. *J. Comput. Aided Mol. Des.* **7,** 439–456.
48. Gattiker, A., Gasteiger, E., and Bairoch, A. (2002) ScanProsite: a reference implementation of a PROSITE scanning tool. *Appl. Bioinformatics* **1,** 107–108.
49. Tong, A. H., Drees, B., Nardelli, G., Bader, G. D., Brannetti, B., Castagnoli, L., et al. (2002) A combined experimental and computational strategy to define protein interaction networks for peptide recognition modules. *Science* **295,** 321–324.
50. Beltrao, P. and Serrano, L. (2005) Comparative genomics and disorder prediction identity biologically relevant SH3 protein interactions. *PLoS Comput. Biol.* **1,** e26.

11

Electrostatic Design of Protein–Protein Association Rates

Gideon Schreiber, Yossi Shaul, and Kay E. Gottschalk

Summary

De novo design and redesign of proteins and protein complexes have made promising progress in recent years. Here, we give an overview of how to use available computer-based tools to design proteins to bind faster and tighter to their protein-complex partner by electrostatic optimization between the two proteins. Electrostatic optimization is possible because of the simple relation between the Debye-Huckel energy of interaction between a pair of proteins and their rate of association. This can be used for rapid, structure-based calculations of the electrostatic attraction between the two proteins in the complex. Using these principles, we developed two computer programs that predict the change in k_{on}, and as such the affinity, on introducing charged mutations. The two programs have a web interface that is available at www.weizmann.ac.il/home/bcges/PARE.html and http://bip.weizmann.ac.il/hypare. When mutations leading to charge optimization are introduced outside the physical binding site, the rate of dissociation is unchanged and therefore the change in k_{on} parallels that of the affinity. This design method was evaluated on a number of different protein complexes resulting in binding rates and affinities of hundreds of fold faster and tighter compared to wild type. In this chapter, we demonstrate the procedure and go step by step over the methodology of using these programs for protein-association design. Finally, the way to easily implement the principle of electrostatic design for any protein complex of choice is shown.

Key Words: Electrostatic rate enhancement; protein complexes; protein–protein interactions; protein association; protein engineering; simulation; proteomics.

1. Introduction

De novo design of proteins and protein complexes has advanced tremendously in recent years. Maybe the most exciting development was the *de novo* design of a new α/β protein from scratch *(1)*, and the redesign of protein-interaction sites *(2)*. These and many other studies have shown that computationally based methods for protein design are maturing and can be used with a better degree of confidence to modulate, reengineer and design individual proteins as well as proteins–protein complexes.

From: *Methods in Molecular Biology, vol. 340: Protein Design: Methods and Applications*
Edited by: R. Guerois and M. López de la Paz © Humana Press Inc., Totowa, NJ

One of the main functions of proteins is to "communicate" with each other. The communication between proteins is based on their ability to interact fast, efficiently, and specifically. The formation of protein–protein interactions is one of the most fundamental requirements for life. Moreover, the interaction network grows with increased complexity of the living organism. The better understanding in how to redesign protein–protein interactions will allow us to better study and eventually to redesign complete protein–complex networks.

1.1. How Proteins Interact

Protein complexes are stabilized by the same interactions that also determine protein folding, which are noncovalent interactions such as van der Waals, H-bonds, electrostatic Coulombic interactions, and the hydrophobic effect. The relative contribution of each of those to specific complexes varies *(3–7)*. Proteins form complexes with other macromolecules on a fine-tuned time scale without sacrificing the specificity of interaction. This is true even in crowded solutions, in which other macromolecules are added at large concentrations (mimicking the cellular environment) *(8,9)*. The association reaction begins by a random search of the two molecules for each other within the vast space of solution, followed by a precise docking of the two interfaces. Brownian motion dictates the rate of collision for a pair of globular proteins.

According to the Smolochowski/Einstein equation, a collision rate of 10^9 to 10^{10} M^{-1}/s^{-1} is calculated independent on the exact size of the two molecules *(10)*. Because only certain relative orientations lead to a productive collision, the rate of association is decreased by up to five orders of magnitude because of angular constraints. Prealigning the proteins in the potential field of the complex partner can partially compensate for the reduced phase space of productive collisions. This prealignment is caused by favorable long-range electrostatic forces *(10,11)* and has been described as electrostatic steering. The association of a protein complex (AB) from the unbound components (A+B) can be best described using a four-state model (**Fig. 1**). Here A and B represent two diffusing proteins in solution, with the rate of translational diffusion following the predicted rate according to the Stokes-Einstein equation. Translational diffusion results in random collisions between the proteins (A:B is the collision complex). Only a small proportion of collisions will lead to the formation of a complex. This is due to very stringent rotational constraints, which require an exact match between the binding sites. The collision complex (A:B) will mostly dissociate back to the unbound proteins, so that $k_{-1} >> k_2$. From time to time, A:B evolves into an encounter complex (A::B), which is characterized by the two binding sites being prealigned against each other.

To obtain the final complex, the binding site may have to undergo structural rearrangement (which can be small or larger, depending on the case) and the interface has to be desolvated. This can cause an energy barrier, the transition

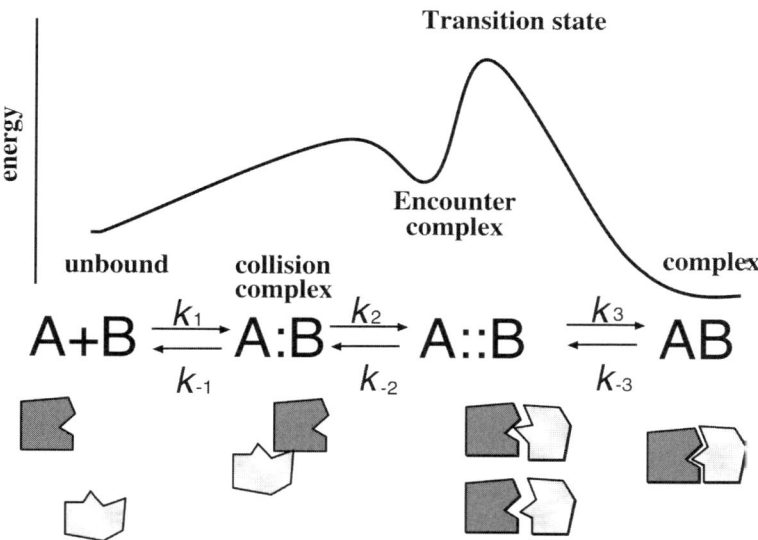

Fig. 1. Energy diagram of a protein–protein interaction. Collisions between the unbound proteins (A and B) are driven by translational diffusion. After collision, some of the pairs proceed to form an encounter complex (A::B). In the encounter complex, the binding sites are preoriented, but the interface is solvated and has to undergo final rearrangement before complexation (AB).

state. Additional intermediates may well exist following the transition state, but these are mostly not observed in the binding kinetics and therefore can be ignored. The encounter complex A::B and the transition state have been experimentally observed for protein–protein interactions. The nature of the encounter complex and following transition state are still under debate, but they can be described as a partially solvated state with the two proteins being roughly aligned one toward the other as in the complex *(12–18)*. Our data indicate that a transition state exist between the encounter complex and the final complex (**Fig. 1**) *(17,19,20)*; however, this point is still under debate *(21–23)*. Direct structural information on the structure of the transition state was obtained by measuring activation interaction energies, using double mutant cycles. These studies have shown that only charged residues, which are in close proximity in the final complex, interact already at the activated complex *(20,24)*. No interaction was measured between uncharged residues at this stage, again pointing at the importance of electrostatic interaction for long-range effects. Masking these specific long-range electrostatic interactions by increasing the concentration of salt causes the loss of some, but not all, pairwise charge–charge interactions at the activated complex, suggesting that structural specificity of the activated complex is maintained even at high salt *(20,25)*.

A somewhat different approach to probe association trajectories experimentally is using Φ value analysis ($\Phi = \Delta\Delta G^{\ddagger}_{on}/\Delta\Delta G_D$) *(16)*. A Φ value close to one indicates that a specific interaction is formed at the transition state, whereas a Φ value close to zero indicates that the interaction is formed after the transition state. In a study of the HyHEL-10 FAB complex, multiple replacements were made in two positions, with most of the replacements having Φ values close to zero *(26)*. This was interpreted as the transition state being early along the reaction trajectory, before short-range interactions (which have the largest contribution on $\Delta\Delta G_D$) being formed. The notion that short-range interactions affect k_{off}, whereas long-range electrostatic interactions affect k_{on}, was directly tested by introducing charged mutations at the vicinity, but outside the binding site of TEM1-BLIP. These mutations did increase specifically k_{on} by 250-fold, but did not affect k_{off} (thus, the increase in k_{on} equals the increase in K_D and $\Phi = 1$) *(19)*. These data suggest that the contribution of the long-range electrostatic interactions to the rate of association is the lowering the free energy of the transition-state by the same magnitude as the equilibrium constant. While mutations of noncharged residues do not significantly affect the transition state for association, they change the energy level of the bound complex and can thus significantly alter k_{off} and K_D.

Protein–protein interactions are often termed as being diffusion-limited. A hallmark of a diffusion limited association reaction is that A::B is not accumulated, and thus, $k_1[C] << k_{-1}+k_2$, in other words (assuming that k_{-2} is small), the rate of formation of the encounter complex A::B is significantly smaller then the rate of its dissipation (either by complex formation or by its dissociation). The linear dependence of k_{obs} on the protein concentration is not always observed. A nonlinear dependence between k_{obs} and the protein concentrations was observed at high protein concentrations for the interaction between Ral or Raf with Ras, making this a partially reaction limited reaction *(16)*.

1.2. Rate of Association of a Protein Complex Directly Related to Charge Complementarity

Mutagenesis studies performed for many different protein–protein complexes have shown that the rate of association is most readily changed by changing the charge of one or both proteins *(11,18,21,27–29)*. This can be done either by introducing a mutation or by changing the ionic strength of the solution. Indeed, it was shown that the rate of association of a protein-complex can be deduced from the following simple equation *(19)*:

$$\ln k_{on} = \ln k^0_{on} - \frac{U}{RT}(\frac{1}{1+\kappa a})$$

where k_{on} and k^0_{on} are the rates of association in the presence and absence of electrostatic forces respectively, U is the electrostatic energy of interaction, κ is the inverse Debye length, and a is the minimal distance of approach. Hence, k_{on} is the sum of two components: (i) the basal rate of association in the absence

of electrostatic forces (k^0_{on}) and (ii) the contribution of the electrostatic forces between the proteins. The latter can be attended by mutation (changing U) or changing solution conditions. According to **Eq. 1**, plotting the salt-dependence $(1/(1+\kappa a))$ of an association reaction vs the experimentally measured $\ln k_{on}$ should produce a linear plot with the slop equal to U/RT and the

intercept at k^0_{on} *(30)*. **Equation 1** also shows that if we can calculate U, we can

actually calculate $\ln k_{on}$ for any mutant complex assuming that $\ln k^0_{on}$ can either be measured for the wild type, or calculated or guessed as having an average value of 10^5 M^{-1}s^{-1}. In the following equations we summarize the method developed by us to calculate the change in k_{on} on introducing a mutation.

The Coulombic energy, U, is calculated using:

$$U = \frac{1}{2} \sum_{i,j} \frac{q_i q_j}{4\pi\varepsilon_0 \varepsilon r_{ij}} \frac{e^{-\kappa(r_{ij} - a)}}{1+\kappa a}$$

where i and j are the charged atoms in the proteins, e is the dielectric constant set to 80, a is set to 6 Å, and κ is the Debye-Huckel screening parameter that relates to the ionic strength of the solution (for further explanations, *see* **ref. *19*** and http://bip.weizmann.ac.il/hypare). The electrostatic interaction energy is calculated from the difference between the electrostatic energy of the complex and the electrostatic energy of the two individual proteins:

$$\Delta U = U_{complex} - U_{protein1} - U_{protein2}$$

The relation between k_{on} and ΔU was calibrated using experimental data obtained from a number of different systems *(30)*. The best fit was obtained when the calculations were done in a fixed ionic strength of 0.022 *M*, and a correction factor of 0.9 (which was determined from the best fit of the calculations to the experimental data for a number of different experimental systems) was used *(19)*. The association rate of a mutant protein at 0.022 *M* is calculated from **Eq. 3**:

$$\ln k_{on(mut)(I=0.022)} = \left(\Delta U_{wt} - \Delta U_{mut} \right)/0.9 + \ln k_{on(wt)}$$

where $\ln k_{on}(wt)$ is determined experimentally at $I = 0.022\ M$. In the case that the association rate is given at a different ionic strength is can be transformed using **Eq. 4**:

$$\ln k_{on(I=0.022)} = \ln k_{on} - \Delta U \left[(\frac{1}{1+\kappa a})_{I=0.022} - (\frac{1}{1+\kappa a})_{lc} \right] / 0.9$$

where Ic is the ionic strength at which the experimental k_{on} was measured. The basal rate for association (in the absence of any electrostatic forces) is found using the relation:

$$\ln k_{on}^{0} = \Delta U_{I=0.022} / 0.9 + \ln k_{on(I=0.022)}$$

The rate of association at any ionic strength Ic can be then calculated using:

$$\ln k_{on(lc)} = \ln k_{on}^{0} - \Delta U \left[1 + (\frac{1}{1+\kappa a})_{lc} - (\frac{1}{1+\kappa a})_{I=0.022} \right] / 0.9$$

Although it is possible to calculate the basal rate and $k_{on(Ic)}$ directly from the experimental k_{on}, it is convenient to do the extra step of calculation at 0.022 M because it fits the experimental procedure. **Equations 1–7** were incorporated into two computer algorithm (*PARE* and *HyPare*), which calculate the electrostatic component of the rate of association of a protein complex from the Coulombic energy of interaction between two proteins (including the effect of screening of charges by salt) *(18,19)*. The method presented here for calculating k_{on} of any mutant protein complex at any salt concentration has proven to be accurate, and is at the basis of our protein design method.

Additional methods to calculate the rate of association of protein complexes, which are not discussed in this chapter, include: Brownian dynamic simulation, which is a method to calculate the rate of diffusive bimolecular encounters *(28,31)*; calculation of the average Boltzmann factor at the area of the binding site *(21)*; and estimating protein–protein association kinetics based on diffusion association on free energy landscapes obtained by sampling configurations within and surrounding the native complex binding funnels *(32)*.

1.3. Electrostatic Design of Protein–Protein Association Rates

The method presented in **Subheading 1.2.** to calculate the effect of a charge mutation on the rate of association works independent of the location of the residue in question, whether it is within or outside the protein–protein binding site *(19)*. For most protein complexes, it is possible to identify residues that, on mutation, will increase (or decrease) association significantly. In a proteomic study conducted on 68 different protein–protein interactions, 866 hot spots were identified (hotspots being residues that, on mutation, change binding more than fourfold [at $I = 0.15$]). A total of 481 of those potential mutations were shown to decrease association and 385 increased k_{on}. In 12 of 68 complexes, no residue is anticipated to increase k_{on} more than fourfold. This study shows that for most protein–protein complexes, residues can be identified that will significantly enhance association *(33)*.

Ideally, we would like to change association specifically, without changing the rate of dissociation. This was found to be feasible using electrostatic design by mutating only residues located outside the physical binding site of the two proteins. Using this design strategy the rate of association and affinity between TEM1-β-lactamase and its protein inhibitor BLIP was enhanced 250-fold, whereas the dissociation rate constant was kept unchanged, resulting in an increase in the binding affinity between those two proteins *(19)* similar to the increased association. Similarly, electrostatic optimization was used to design a set of mutants of the Ras effector protein RalGDS with optimized electrostatic steering. The fastest binding RalGDS mutant, M26K,D47K,E54K, binds Ras 14-fold faster and 25-fold tighter compared with wild type *(16)*. The increased rate of association could be directly mapped to an increase in the stability of the encounter complex, whereas the rate of formation of the final complex from the encounter complex (k_3 in **Fig. 1**) was unchanged. As can be seen, the method results not only in faster, but also in tighter binding. The results emphasize that long-range electrostatic forces specifically alter k_{on}, but do not affect k_{off}. The design strategy presented here is applicable for increasing rates of association and affinities of protein complexes in general.

2. Methods
2.1. Design Methodology

Here we are presenting a step-by-step guide of how to use *PARE* and *HyPare* for the electrostatic design of faster and tighter binding protein complexes.

The program *PARE* (http://www.weizmann.ac.il/home/bcges/PARE.html) calculates the rate of association (k_{on}) of mutant protein–protein complexes. In other words, the user is choosing which mutations he or she wants to calculate, produces the relevant mutant protein coordinate file in pdb format (for example using swiss-pdb-viewer: http://www.expasy.org/spdbv/) and uploads the mutant files and the wild-type file into the program. In the case that the mutation is the addition of a charged residue (and not a charge deletion mutation), it is desired to model multiple rotamers of the introduced charged side chain and to calculate the expected rates of association for the different rotamers *(18)*. In some cases, significant differences will be obtained in the rate of association between the different rotamers.

HyPare (http://bip.weizmann.ac.il/*hypare*) automatically estimates the impact of mutations on a per-residue basis for all residues of a protein–protein interaction. In other words, *HyPare* scans the protein sequence of the pair of proteins in question automatically and calculates the change in the rate of association on mutation for all positions. The major difference between these two programs is that *PARE* is a somewhat more accurate for mutations of addition of charged residue (the precision of calculations for deletions of charges is the same for both programs), but *HyPare* is much faster and easier to use for scanning entire proteins for potential sites for mutations that potentially will alter rates of association. The output file of *HyPare* will also provide information of the location of the mutation relative to the binding site and present all the results in a tabular format and a convenient graphical output. After a *HyPare* output file is generated, one should decide whether the pair of proteins in question fits for electrostatic design.

Figure 2 shows an example of an output file as obtained by *HyPare*. Here, the interaction between TEM1 and BLIP was probed for mutants that will confer faster binding. On the left side of the figure is a residue-by-residue analysis of the expected change in k_{on} on mutation to a charge (positive or negative); on the right side, the results are drawn in an interactive Java-based viewer (Jmol). At the top of the view, general information concerning the complex and links to all results are provided in a downloadable, tabular format. The graphical results show that two hot spot residues are found on BLIP, D73, and D163. For both of those residues, a change in k_{on} of > 10-fold is predicted for the charge-reverse mutant (for example, to Lys). Both of these residues are located outside the physical binding site for TEM1. On TEM1, on the other hand, no mutation is expected increase k_{on} by more than fourfold, and also those are located within the binding site. Therefore, BLIP is in this case a much better candidate for mutagenesis. The calculated results shown here were verified experimentally and found to be accurate *(33)*.

2.2. Generating an Input File

To run either *PARE* or *HyPare*, an input file in pdb format of the protein–protein complex has to be generated. The source of the file can be either from the pdb database (using the pdb accession code) or uploaded from a computer. In generating an input file, it is crucial to introduce the correct charges for the calculation.

Therefore:

1. The user must be sure that all residues are present and if not, they should be added using any program that is suitable for this task (for example, SwissPdbViewer). Missing residues, or missing side chains, will not be taken into account in the calculation.
2. Each protein in the complex should be designated by a different chain identifier.

HyPare© results

The following calculation of energy and hotspots using residues and heteroatoms, is for the experimentally measured k_{on} 2.5e5 $M^{-1}s^{-1}$, determined at 0.022 M ionic strength:

	Ionic Strength (M)				Basal rate (M⁻¹s⁻¹)	Charge		
File name	**Experimental**		**Target**					
	0.022		0.022					
export 🖺 files	DU (kcal/mol)	k_{on} (M⁻¹s⁻¹)	DU (kcal/mol)	k_{on} (M⁻¹s⁻¹)		1st chain	2nd chain	Complex
1jtg.pdb	-1.62e+00	2.5e5	-1.62e+00	2.50e+05	4.14e+04	A -6.00e+00	B -2.00e+00	-8.00e+00

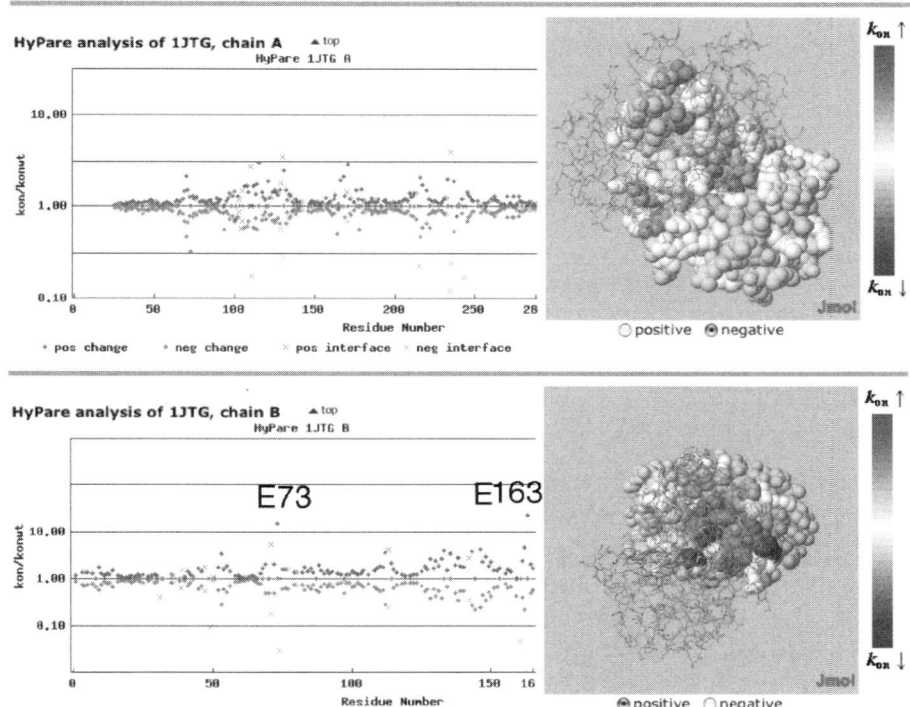

Fig. 2. *HyPare* result sheet as obtained by using the interactive web-based program found in http://bip.weizmann.ac.il/*hypare*. The table at the top gives the value of k_{on} as given by the user (at the appropriate ionic strength) and the calculated electrostatic energy of interaction (ΔU) for this complex, and both values as calculated at any desired ionic strength. To the right there is the calculated basal rate of association (which is the rate in the absence of any electrostatics) and the protein charges as used by the program. The per-residue results of the calculations are below the table for both proteins. The figure shows the expected change in k_{on} on mutation to a positive or negatively charged residue. Amino acids located in the binding site are marked by an x. On the right of the figure is an interactive Jmol picture of the results, with the proteins colored according to the expected change. All results can be downloaded in tabular form.

3. The second backbone oxygen at the C termini of both proteins in the protein complex should be given the name OXT (or else no charge will be assigned to them). The charge on the N-ter is assigned automatically.
4. Many proteins have cofactors bound to them (e.g., ATP, GTP, GDP, SO₄, Mg). *PARE* and *HyPare* include a standard charge-file. In *PARE*, the file includes only the charge amino acids, whereas in *HyPare*, it includes many additional heavy atoms and cofactors (for a complete list, *see* http://bip.weizmann.ac.il/*hypare/*input_opt.html#charges_list). However, the user can upload his preferred charge-file, as long as the following format is used:

ATOM_NAME RESIDUE CHARGE,

where each charge rule entry is separated by a new line and each column is separated by space or tab.

For example:

NH1 ARG 0.5
NH2 ARG 0.5
CA HIS 1.0

To be sure that the charges are inserted correctly, it is advisable to always check the total sum of the charges, and see whether the *PARE* or *HyPare* outcome fits the real number. If the total charge does not fit the expectations, review the file and correct it.

2.3. Using HyPare for a First Approximation

As a first step in the calculations, we would always suggest using *HyPare*, because it gives a complete picture of the expected change in k_{on} at all protein positions. *HyPare* applies a simplistic simulation approach by which no "real" mutations are simulated, but instead the program introduces charges automatically. The location of the pseudo-charge on the amino acid are tabled in http://bip.weizmann.ac.il/*hypare*/input_opt.html#charges_list. We tried to mimic the position of the pseudo-charge to be as close as possible to that of a charge in a real charged residue. For some amino acids (e.g., Gln, Asn, Leu, Val), the task is relatively easy. For others (such as Ala) the location of the pseudo-charge is farther away from the location of a charge in a charged-residue (Asp, Glu, Lys, or Arg). Still, we feel confident that this is a good approximation. It has been shown that the exact location of the charge is not critical in many cases *(29)*.

Second, the exact location of the charge after mutation cannot be known *a priori*, because the precision of modeling the correct rotamer of surface side chains beyond Cβ is problematic *(34)*. The approach used in *HyPare* for charge allocation, and its precision in estimating the k_{on} values of mutant complexes, were evaluated for five different protein–protein interactions, including 60 different surface mutations (**Fig. 3**). The experimental and *PARE* data shown in the figure were previously published. **Figure 3** clearly shows that *HyPare* performs almost as good as *PARE*, without the need to manually mutate the resi-

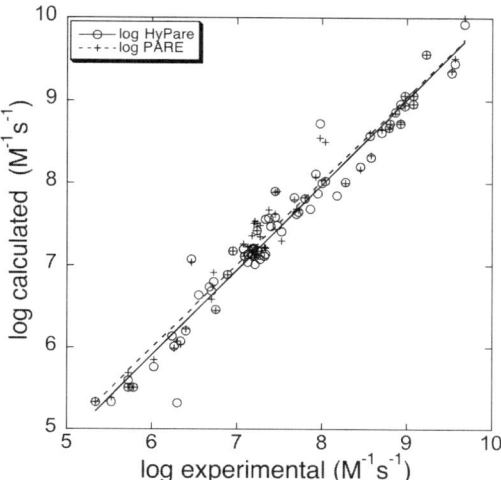

Fig. 3. Validating the calculated rates of association using *HyPare* against experimental data for five different protein complexes. Experimental wild-type and mutant data were plotted against calculated *HyPare* rates (+) using the charge rules given in Table 1 of **ref. 33** and at http://bip.weizmann.ac.il/hypareb/main. For comparison, *PARE* prediction is also shown (o), using side-chain mutations followed by a step of minimization. The data shown in the figure are from the following complexes: barnase/barstar, TEM1/BLIP, acetylcholinesterase/fasciculin, hirudin/thrombin, and Ras/Ral. A correlation coefficient of 0.98 and a slope of 1 were determined between both calculations and the experimentally determined rates.

dues, and feed in new pdb files with modified coordinates into the calculations. Therefore, *HyPare* is a good high-throughput tool to investigate the role of electrostatics in protein complex association and as a first-estimate for protein design. Electrostatic optimization will in most cases give the correct trend for the calculation, but the estimate in the magnitude of change may not be always precise. An example for such case is the interaction between Ras and Ral, where we got the correct trend but the exact data differed. In that case, we explain the difference between calculation and experimental results by the assumption that this association reaction is partially reaction limited, and not fully diffusion limited *(16)*.

2.4. HyPare*Algorithm*

Here, a brief pseudo-code description of the *HyPare*Algorithm is given. Input:

1. Separate pdb file into chains.
2. Combine desired chains (pairwise if complex has more than two chains).

3. Assign charges to atoms (according to Table 1 in **ref. 33** and at http://bip.weizmann.ac.il/hypareb/main).
4. Find missing atoms (incomplete residues in the pdb file). Print to log file.
5. Find interface residues (at least one atom ≤4 Å from second chain).
6. Reduce calculation set to charged atoms only.
7. Calculate internal Coulombic energy of first chain (**Eq. 1**).
8. Calculate internal Coulombic energy of second chain (**Eq. 1**).
9. Calculate Coulombic energy of the complex (**Eq. 1**).
10. Calculate energy difference (**Eq. 2**).
11. Calculate association rate at 0.022 *M* (**Eq. 4**).
12. Extract basal rate (**Eq. 5**).
13. Calculate k_{on} at target salt (**Eq. 6**).
14. **If mode of work is finding hot spots [**

> **For each residue position [**
> > **For each charge type (positive and negative) [**
> > > 1. Assign charge to selected atoms.
> > > 2. Calculate Coulombic energy of first chain (**Eq. 1**).
> > > 3. Calculate Coulombic energy of second chain (**Eq. 1**).
> > > 4. Calculate Coulombic energy of the complex (**Eq. 1**).
> > > 5. Calculate energy difference (**Eq. 2**).
> > > 6. Calculate k_{on} at target salt (**Eq. 6**).
> > > 7. Print rate increase (the new rate/experimental rate).]]

Else do next pdb (do not find hot spots, just calculate ΔU).]

2.5. Using PARE to Obtain Further Precision

The method of using multiple rotamer entries per charge residue is described. Both *PARE* and *HyPare* calculate the Debye-Huckel energy of interaction of each charge with all surrounding charges, correcting the outcome by a factor that takes into account the nonlinearity of the distance-dependence of the charges because of the globular shape of protein surfaces. However, using *PARE*, it is possible to generate multiple rotamer input files, something not feasible using *HyPare*. Clearly the "best" calculated results using multiple rotamer entries for mutations where charges are added are better than *HyPare* results. However, the worst results, using multiple rotamers for calculation are worse than those obtain using *HyPare*. This means that only if we have a way to choose the correct rotamer of the mutation, the use of *PARE* and the correct rotamer is justified. If we do not know which rotamer is better, we can use *HyPare* alone. One should bear in mind that in the case of charge-deletion mutations, *HyPare* will in any case perform as good as *PARE*.

2.6. How to Choose Mutations for Mutagenesis

The *HyPare* output file provides information about the closeness of each residue to the interface, but not whether it is buried within a protein. One should not mutate buried residues, but only well-solvated surface residues. This is important to avoid structure perturbations of the protein. Second, ask what is the purpose of the mutagenesis study. If the purpose is to change the rate of association independent of the rate of dissociation (and thus change the binding affinity accordingly), try and mutate hotspot residues (as identified by *HyPare*) located outside the physical binding site. It is not recommended to alter charged residues into opposite charges or to introduce new charges within the physical binding site, because these mutations are of an extreme nature and are expected to cause major changes in k_{off}.

3. Summary

This chapter explains in detail how to design faster and tighter binding of protein–protein interactions using electrostatic optimization. The method was fine tuned for increasing the rate of association of a protein complex through increasing specific long-range electrostatic interactions of charge residues. This method was proven to work well for a number of different systems. It does not optimize the short-range electrostatic interactions within a protein–protein interface (also termed salt bridges), but only the long-range electrostatic attraction that leads to faster binding. The task of interface improvements to achieve slower dissociation is much more demanding.

References

1. Kuhlman, B., Dantas, G., Ireton, G. C., Varani, G., Stoddard, B. L., and Baker, D. (2003) Design of a novel globular protein fold with atomic-level accuracy. *Science* **302**, 1364–1368.
2. Kortemme, T., Joachimiak, L. A., Bullock, A. N., Schuler, A. D., Stoddard, B. L., and Baker, D. (2004) Computational redesign of protein-protein interaction specificity. *Nat. Struct. Mol. Biol.* **11**, 371–379.
3. Sheinerman, F. B. and Honig, B. (2002) On the role of electrostatic interactions in the design of protein-protein interfaces. *J. Mol. Biol.* **318**, 161–177.
4. Lee, L. P. and Tidor, B. (2001) Barstar is electrostatically optimized for tight binding to barnase. *Nat. Struct. Biol.* **8**, 73–76.
5. Chakrabarti, P. and Janin, J. (2002) Dissecting protein-protein recognition sites. *Proteins* **47**, 334–343.
6. Nooren, I. M. and Thornton, J. M. (2003) Structural characterisation and functional significance of transient protein-protein interactions. *J. Mol. Biol.* **325**, 991–1018.
7. Reichmann, D., Rahat, O., Albeck, S., Meged, R., Dym, O., and Schreiber, G. (2005) The modular architecture of protein-protein binding interfaces. *Proc. Natl. Acad. Sci. USA* **102**, 57–62.
8. Minton, A. P. (2000) Implications of macromolecular crowding for protein assembly. *Curr. Opin. Struct. Biol.* **10**, 34–39.
9. Kozer, N. and Schreiber, G. (2004) Effect of crowding on protein-protein association rates: fundamental differences between low and high mass crowding agents. *J. Mol. Biol.* **336**, 763–774.

10. Berg, O. G. and von Hippel, P. H. (1985) Diffusion-controlled macromolecular interactions. *Ann. Rev. Biophys. Biophys. Chem.* **14,** 131–160.

11. Sheinerman, F. B., Norel, R., and Honig, B. (2000) Electrostatic aspects of protein-protein interactions. *Curr. Opin. Struct. Biol.* **10,** 153–159.

12. Xavier, K. A. and Willson, R. C. (1998) Association and dissociation kinetics of anti-hen egg lysozyme monoclonal antibodies HyHEL-5 and HyHEL-10. *Biophys. J.* **74,** 2036–2045.

13. Rajpal, A., Taylor, M. G., and Kirsch, J. F. (1998) Quantitative evaluation of the chicken lysozyme epitope in the HyHEL-10 Fab complex: free energies and kinetics. *Protein Sci.* **7,** 1868–1874.

14. Sydor, J. R., Engelhard, M., Wittinghofer, A., Goody, R. S., and Herrmann, C. (1998) Transient kinetic studies on the interaction of Ras and the Ras-binding domain of c-Raf-1 reveal rapid equilibration of the complex. *Biochemistry* **37,** 14292–14299.

15. Li, Y., Lipschultz, C. A., Mohan, S., and Smith-Gill, S. J. (2001) Mutations of an epitope hot-spot residue alter rate limiting steps of antigen-antibody protein-protein associations. *Biochemistry* **40,** 2011–2022.

16. Kiel, C., Selzer, T., Shaul, Y., Schreiber, G., and Herrmann, C. (2004) Electrostatically optimized Ras-binding Ral guanine dissociation stimulator mutants increase the rate of association by stabilizing the encounter complex. *Proc. Natl. Acad. Sci. USA* **101,** 9223–9228.

17. Schreiber, G. (2002) Kinetic studies of protein-protein interactions. *Curr. Opin. Struct. Biol.* **12,** 41–47.

18. Selzer, T. and Schreiber, G. (2001) New insight into the mechanism of protein-protein association. *Proteins* **45,** 190–198.

19. Selzer, T., Albeck, S., and Schreiber, G. (2000) Rational design of faster associating and tighter binding protein complexes. *Nat. Struct. Biol.* **7,** 537–541.

20. Frisch, C., Fersht, A. R., and Schreiber, G. (2001) Experimental assignment of the structure of the transition state for the association of barnase and barstar. *J. Mol. Biol.* **308,** 69–77.

21. Zhou, H. X. (1997) Enhancement of protein-protein association rate by interaction potential: accuracy of prediction based on local Boltzmann factor. *Biophys. J.* **73,** 2441–2445.

22. Camacho, C. J., Kimura, S. R., DeLisi, C., and Vajda, S. (2000) Kinetics of desolvation-mediated protein-protein binding. *Biophys. J.* **78,** 1094–1105.

23. Gabdoulline, R. R. and Wade, R. C. (1999) On the protein-protein diffusional encounter complex. *J. Mol. Recognit.* **12,** 226–234.

24. Schreiber, G. and Fersht, A. R. (1996) Rapid, electrostatic assisted, association of proteins. *Nat. Struct. Biol.* **3,** 427–431.

25. Myles, T., Le Bonniec, B. F., Betz, A., and Stone, S. R. (2001) Electrostatic steering and ionic tethering in the formation of thrombin-hirudin complexes: the role of the thrombin anion-binding exosite-I. *Biochemistry* **40,** 4972–4979.

26. Taylor, M. G., Rajpal, A., and Kirsch, J. F. (1998) Kinetic epitope mapping of the chicken lysozyme. HyHEL-10 Fab complex: delineation of docking trajectories. *Protein Sci.* **7,** 1857–1867.

27. Gabdoulline, R. R. and Wade, R.C. (2002) Biomolecular diffusional association. *Curr. Opin. Struct. Biol.* **12,** 204–213.
28. Gabdoulline, R. R. and Wade, R. C. (2001) Protein-protein association: investigation of factors influencing association rates by brownian dynamics simulations. *J. Mol. Biol.* **306,** 1139–1155.
29. Marvin, J. S. and Lowman, H. B. (2003) Redesigning an antibody fragment for faster association with its antigen. *Biochemistry* **42,** 7077–7083.
30. Selzer, T. and Schreiber, G. (1999) Predicting the rate enhancement of protein complex formation from the electrostatic energy of interaction. *J. Mol. Biol.* **287,** 409–419.
31. Davis, M. E., Madura, J. D., Luty, B. A., and McCammon, J. A. (1991) Electrostatics and diffusion of molecules in solution: simulations with the University of Houston Brownian dynamics program. *Comp. Phys. Com.* **62,** 187–197.
32. Schlosshauer, M. and Baker, D. (2004) Realistic protein-protein association rates from a simple diffusional model neglecting long-range interactions, free energy barriers, and landscape ruggedness. *Protein Sci.* **13,** 1660–1669.
33. Shaul, Y. and Schreiber, G. (2005) Exploring the charge space of protein–protein association: a proteomic study. *Proteins* **60,** 341–352.
34. Eyal, E., Najmanovich, R., McConkey, B.J., Edelman, M., and Sobolev, V. (2004) Importance of solvent accessibility and contact surfaces in modeling side-chain conformations in proteins. *J. Comp. Chem.* **25,** 712–724.

III

Design of Amyloidogenic Polypeptides and Amyloid Inhibitors

12

Peptide Model Systems for Amyloid Fiber Formation
Design Strategies and Validation Methods

Alexandra Esteras-Chopo, María Teresa Pastor, and Manuela López de la Paz

Summary

The rational understanding of the factors involved in the formation of amyloid deposits in tissue is fundamental to the identification of novel therapeutic strategies to prevent or cure pathological conditions such as Alzheimer's and Parkinson's disease or spongiform encephalopathies. Given the complexity of the molecular events driving protein self-association, a frequent strategy in the field has consisted of designing simplified model systems that facilitate the analysis of the elements that predispose polypeptides toward amyloid formation. In fact, these systems have provided very valuable knowledge on the determinants underlying structural transitions to the polymeric β-sheet state present in amyloid fibers and more disordered aggregates. In this chapter, we will describe different approaches to obtain and design model systems for amyloidogenesis, as well as the methodologies that are typically used to validate them. We will also show how some of the general principles obtained from these studies can be applied for *de novo* design purposes and for the sequence-based identification of amyloidogenic stretches in proteins.

Key Words : Amyloid; amyloid fiber formation; peptide design; peptide model systems; β-sheet; β-sheet aggregation; amyloid pattern; proteolysis; combinatorial library; β-gal complementation.

1. Introduction

Amyloid fiber formation refers to the self-association process of proteins into β-sheet–rich fibrillar structures. In recent years, this phenomenon has become prominent in the literature because it is related to a growing list of diseases, including a number of devastating neurodegenerative disorders such as Alzheimer's and Parkinson's disease *(1)*. The identification of novel therapeutic strategies to prevent or cure these diseases implies the rational understanding of the molecular mechanisms leading to this anomalous protein behavior

From: *Methods in Molecular Biology, vol. 340: Protein Design: Methods and Applications*
Edited by: R. Guerois and M. López de la Paz © Humana Press Inc., Totowa, NJ

(2), as well as the clarification of how self-association of a misfolded protein results in organ dysfunction and neurodegeneration *(3)*.

The group of peptides and proteins capable of forming amyloid fibrils is very diverse. This group does not consist only of proteins involved in amyloid deposits in vivo *(4)*, but also of nonpathogenic peptides and proteins that, under some conditions, form fibrils in vitro *(5–7)*. Although the soluble precursors of these proteins do not have any obvious sequence homology or common folding patterns, X-ray fiber diffraction data indicate that all amyloid fibrils share a characteristic cross-β-structure *(8)*. In such a structure, polypeptide chains form β-strands oriented perpendicular to the long axis of the fibril, resulting in self-assembled β-sheets that propagate in the direction of the fibril. This consensus fiber organization suggests that at least the key elements of the process may be common to all proteins and that, therefore, simplified systems that polymerize into β-sheets will offer novel insights into the molecular details of amyloid fibril formation.

In fact, substantial progress has been made by developing model systems in which the effects of the perturbation of single properties on amyloid fiber formation can be pinpointed *(9)*. Both experimental and theoretical studies on these model systems have provided very valuable knowledge on the structural and sequence determinants of amyloid fibril formation. Here, we will present the different approaches that can be used to design such systems and how the knowledge extracted from these studies can be applied to identify amyloidogenic stretches in proteins and for *de novo* design.

Protein self-association into β-sheets can lead to products of different quaternary structure. Therefore, and for the sake of clarity, in this chapter the terms *amyloid* and *aggregate* refer to β-sheet assemblies of fibrillar and amorphous morphology, respectively.

2. Design Strategies

Ideally, a model system for amyloidogenesis must recapitulate what is known about the in vivo process, it must present suitable physical properties such as reversibility and good solubility to allow a biophysical characterization and it must be small enough to permit rationalization and to extract valuable information on the details of the aggregation process. During the past decade, the search for such simple model systems has been mainly based on the experimental identification of the fragments of amyloidogenic proteins that can self-assemble into amyloid fibrils *(10–12)*. Some rational design principles have been also applied on the construction of combinatorial libraries of peptides and proteins *(13)*. Rational design exercises have been very successful at assessing the principles underlying the structural transitions displayed by proteins before self-assembly and at determining the factors underlying β-sheet polymerization into amyloid assemblies or amorphous aggregates *(14–16)*. The experimental validation of these designs has been typically performed in vitro. More recently, however, several groups have undertaken the challenge of studying the sequence determinants of amyloid fibril formation in vivo *(17)*. Approaches such as those outlined in this section represent a relevant step in our understanding of the more general phenomenon of amyloid deposition.

2.1. Empirical Approaches

The experimental identification of the region(s) carrying the amyloidogenic features of a given protein has been a widely used approach to obtain simplified systems that recapitulate or model the amyloidogenic properties of the full-length protein.

Because a protein must be at least partially unfolded to undergo amyloid fibril formation *(18,19)*, the regions of a protein more prone to unfolding or with higher flexibility are those that most likely will interact intermolecularly. Protein engineering exercises also indicate that there is a good correlation between the solvent exposed regions of a protein and those shown to be critical in the rate-determining steps of protein self-association *(20)*. Such exposed or highly flexible regions of a protein can be identified by limited proteolysis *(21,22)*. After these fragments have been isolated and purified, their ability to form fibrils can be tested by using a range of different techniques described somewhere else in this chapter.

The protease resistance of a protein does not only vary from the native to the amyloid-prone state, but it also depends on the stage of the amyloid process. Therefore, limited proteolysis can be also used to monitor the conformational transitions taking place during the fibril formation process and to characterize the different species that appear on the amyloid pathway *(22)*. Although a consensus filament architecture has not been reached yet, most models propose that the so-called *amyloid domains* stack to form a *core* that is surrounded by the rest of the protein. The folding state of the globular appendages is unknown and it might differ from protein to protein. Controlled digestion of preformed fibrils can be used to identify the *amyloid core* and to study the integrity of the globular domain.

Nuclear magnetic resonance studies on amyloid precursor states can be also conducted to identify the key amyloidogenic regions of a protein *(23)*. Analysis of peptides consisting of these regions can further confirm their amyloidogenic properties *(12)*. Minimal amyloid sequences can be searched within these fragments by spotting overlapping peptides on a membrane and by assaying their binding capabilities to the full-length protein *(11)*. Alanine *(24)* and proline *(10)* scannings can be carried out to determine the role of each residue within a given amyloidogenic fragment (*see* **Note 1**).

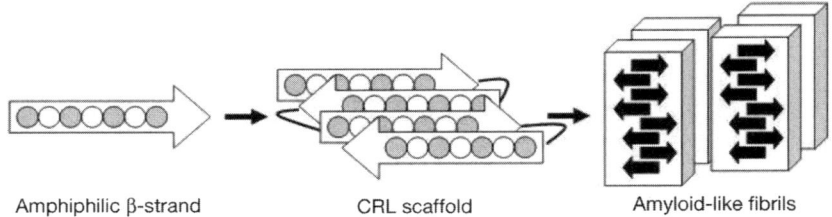

Fig. 1. Schematic representation of the binary code used in the design of amphiphilic β-strands. The alternation of polar (gray beads) and nonpolar (white beads) residues in the strands of a conformationally restricted library drives the assembling of the generated β-sheet proteins into amyloid-like fibers.

2.2. Combinatorial Approaches

In recent years, combinatorial chemistry has been a very popular strategy to generate pools of molecular diversity. The screening of large compound libraries has served to identify a large number of bioactive molecules that have been often used as lead compounds for the development of new drugs.

Combinatorial libraries that consist of random amino acid sequences will rarely yield proteins with desired structural properties. The integration of rational principles in the design reduces the combinatorial diversity and focuses the library on those regions of the sequence space that are most likely to produce well-folded proteins. Libraries generated on theses bases are known as *conformationally restricted libraries* (CRL) *(25)*.

Such design approaches have provided very valuable insights into the sequence factors promoting or preventing β-sheet aggregation. Libraries of proteins intended to adopt β-sheet folds have been designed using a binary polar/nonpolar code *(26)*. This design principle is based on the recognition that hydrophobicity patterns match the periodicity of secondary structures *(27)*. β-sheets show a periodicity of two residues per repeat, and thus a β-strand can be defined by the alternation of polar and nonpolar amino acids. However, proteins designed on these bases do not form monomers but self-assemble into large polymers (**Fig. 1**) *(13,28)*. Thus, alternating binary patterns are prone to forming fibril structures. These highly aggregating β-sheet proteins can be converted into soluble monomers by the application of negative design principles, such as the introduction of charged residues or proline (*see* **Note 2**) *(29,30)*.

2.3. Rational Approaches: De Novo *Design of Peptide-Based Amyloid Fibrils*

Rational protein design implies the creation *de novo* of sequences that adopt defined structures. Rational design is a very useful approach to test general theories experimentally and to assess the limits of our understanding of the sequence-to-structure relationship. Furthermore, if the experiments are devised in a progressive fashion, such that the simplest possible designs are tried first, then it may be possible to identify a minimally sufficient set of elements responsible for a given property. Following this philosophy, in the amyloid field, *de novo* designs have been focused on small peptides that simplify the analysis of the factors underlying the structural transitions displayed by proteins before amyloid assembly *(14)* and the sequence determinants driving β-sheet polymerization into amyloid fibers *(15,16,31)*. Here, we will focus on the peptide-based amyloid fibrils that we have recently designed *(15,16)*.

2.3.1. Design Hypotheses

1. A protein has to be partially or fully unfolded to aggregate *(19)*.
2. There are general rules governing the amyloidogenicity of a polypeptide chain *(15,16,32,33)*.
3. An unfolded polypeptide will undergo a random coil to β-sheet conformational transition if, and only if, its sequence has propensity to self-assemble into a polymeric β-sheet structure *(16,34)*.
4. Propagation and stacking of preformed β-sheets will result in the final assembly of amyloid fibrils *(15)*.
5. Short amyloidogenic stretches are responsible for the amyloidogenic behavior of a given protein *(35)*.

2.3.2. Design Procedure

To test these hypotheses, we have designed a hexapeptide model system that undergoes a random coil to β-sheet conformational transition. We have used a protein design algorithm called *PERLA* (*see* **Note 3**) to search for ideal sequences capable of self-associating into polymeric β-sheet structures (*see* **Subheading 3.5.**). After calculations, a subset from the output sequences has been selected, using as criterion that the set of peptides should be as similar as possible while containing sequences both energetically favorable and unfavorable to adopt a polymeric β-sheet conformation (**Table 1**). To assess the influence of the net charge on the molecule on the self-assembly process, we have studied the aggregation behavior of the selected sequences at different total net charges (0, ±1, and ±2) (*see* **Note 4**).

2.3.3. Experimental Validation of the Design

The synthesized peptides have been checked for β-sheet polymerization and fiber formation by using circular dichroism (CD) and electron microscopy (EM), respectively (*see* **Subheading 3.2.** and **Note 5**). In general, all those sequences predicted to be energetically highly favorable to form polymeric β-sheets show a CD random coil → β-sheet transition at concentrations equal or higher than 0.5 m*M* (**Fig. 2** and **Table 1**). Those sequences predicted to be less

Table 1
Designed Sequences and Self-Assembly Behavior
as a Function of the Net Charge on the Molecule

Sequence[a]		CD signature[b,d,e]			EM[c,d,e]		
		+1	0	−1	+1	0	−1
#1	K TVIIE	β	β	rc + β	fibrils	—	fibrils
#2	STVIIE	β	rc	rc + β	fibrils	—	fibrils
#3	KTVIVE	rc	β	rc	—	—	—
#4	KTVLIE	rc	β	rc	—	—	—
#5	STVIYE	rc + β	rc	rc + β	fibrils	—	fibrils
#6	STVIIT	β	β	rc + β	fibrils	fibrils	fibrils
#7	KTVIYE	rc	rc	rc	—	—	—
#8	KTVIIT	rc	rc	rc	—	—	—

[a]Sequences are given using the one-letter code for amino acid residues. The predicted sequence stability is: #1 > #2 > #4 > #3 > #7 > #5 > #6 > #8.

[b]"rc" indicates random coil CD spectrum; "β" stands for β-sheet CD spectrum and "rc + β" indicates the presence of both β-structure and unfolded conformations in solution.

[c]—, no fibrils are observed by EM.

[d]+1, 0, and −1 are the net charge on the peptide molecule.

[e]Experimental conditions: c = 1 mM; room temperature; incubation time = 1 mo.

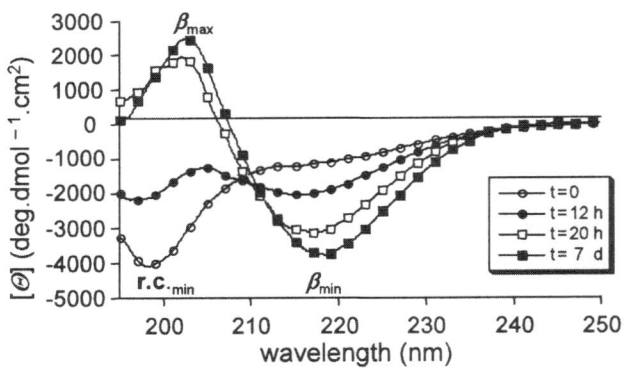

Fig. 2. Far-ultraviolet CD spectra of a solution of STVIIE (c = 0.8 mM, pH = 2.6; net charge = +1) showing the random coil → β-sheet transition experienced by peptide STVIIE on time (t = 0, 12 h, 20 h, and 7 d).

favorable for this polymeric topology display a typical random coil CD spectra under all experimental conditions examined. These results support the usefulness of the computational method for designing mutations that disfavor β-sheet polymerization (e.g., Glu6Thr from #1 to #8) (**Table 1**).

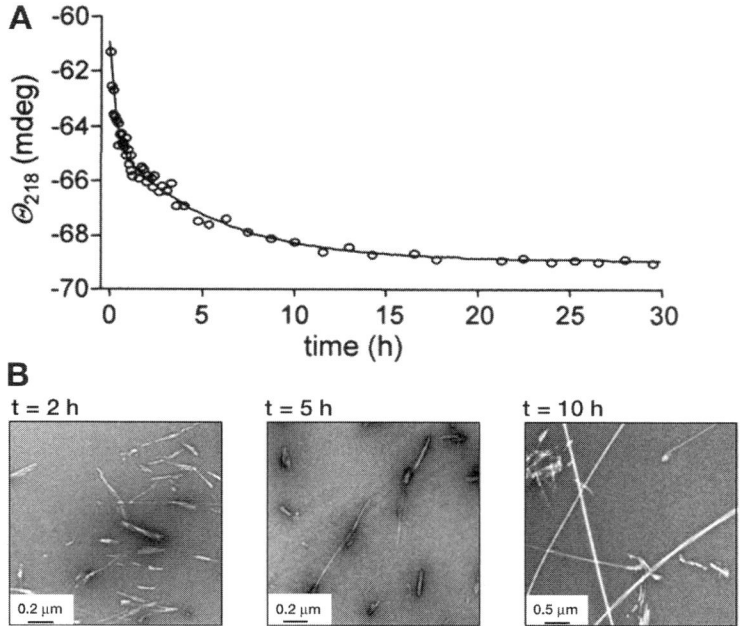

Fig. 3. (**A**) Plot of the ellipticity change at 218 nm (Θ_{218}; β-sheet minimum) of a solution of STVIIE (c = 1.2 m*M*, pH = 2.6) as a function of the time. (**B**) Electron micrographs of aliquots taken at t = 2, 5, and 10 h indicate that although Θ_{218} reaches a plateau at t 4 to 5 h, the degree of maturation and number of fibrils still increases with time. Θ_{218} at t = 2 h indicates a high percentage of β-sheet structure, whereas EM shows the presence of just amorphous species and very short filaments.

2.3.4. Interactions Driving Amyloid Formation From an Unfolded Polypeptide

Kinetic analysis of fibril formation monitored by CD and EM shows that β-sheet polymerization precedes fibril formation (**Fig. 3**). In fact, peptides that do not polymerize into β-sheets do not form amyloid fibrils under any conditions tested. Nevertheless, although β-sheet formation appears to be necessary for fibril formation, it is not a sufficient criterion, as some samples that display a β-sheet CD spectrum do not proceed to form fibrils. Certain point mutations involving substitutions of the Ile residues (e.g., Ile4Leu from #1 to #4; Ile5Val from #1 to #3) completely prevent fibril growth from preformed β-sheets. These results highlight the relevant role of specific side-chain interactions, as some single-point mutations have dramatic effects on the fibril formation process of the peptide.

For amyloidogenic peptides with charged groups, amyloid fibrils are observed in our experiments only when the molecule carries a net charge of ±1 (peptides #1, #2, and #5; **Table 1**). Peptide solutions that display a β-sheet spectrum when the peptide net charge is zero result in the formation of amorphous aggregates or the β-sheet oligomers remain soluble during the longest incubation time checked (i.e., 1 mo). All solutions where the peptide carries a net charge higher than one display a typical random coil CD spectrum and do not form fibrils. As in larger polypeptides the number of charged residues is higher and their distribution more complex, the charge balance that promotes fibril formation must be more complicated. What our findings indicate is that charges play a role of relevance in the supramolecular organization of proteins into fibrils, which suggests that amyloid fibril formation in proteins may take place preferentially for particular charged states of the molecule (*see* **Note 6**) *(15,36)*.

2.3.5. Sequence Determinants of Amyloid Fibril Formation

To investigate the link between sequence and amyloid feature, we have performed a saturation mutagenesis analysis on the *de novo*–designed amyloid peptide STVIIE (peptide #2, **Table 1**) (*see* **Note 7**) *(16)*. The positional scanning mutagenesis has revealed that there is a position dependence upon mutation of amyloid fibril formation and that both very tolerant (*edges*; #1, 2, 6) and restrictive (*core*; #3, 4, 5) positions to mutation can be found within this amyloidogenic sequence (**Table 2**). Position 5 is particularly important for fibril formation because it is restricted to only three hydrophobic amino acids with a high β-sheet propensity (Ile, Phe, and Tyr) *(37)*. That β-branched side chains such as Val and Ile are not permitted at the same place (Val is allowed at #3, but not at #4 or 5, and vice versa for Ile) remarks the high degree of sequence specificity dictated by certain positions. It is also remarkable that Phe is the only amino acid that allows fibril formation at any position. This result is consistent with the relevance that has been put forward for π-stacking interactions in the self-assembly of amyloid fibrils *(24)*.

2.3.6. Applications: Identification of Amyloidogenic Stretches in Protein Sequences and De novo Design of Amyloid Peptides

Amyloid proteins differ widely in sequence and structure. Therefore, the identification of structural and sequence characteristics indicating amyloidogenic propensity in proteins has been very difficult. Nevertheless, investigations carried out in the past have been found to have some success in extracting some common denominators of amyloid polypeptides that can be used for the *de novo* design or for the identification of amyloid-prone fragments *(16,38–40)*.

The existence of some homology amongst amyloid-forming peptide sequences was pointed out a decade ago *(32)*. However, the extraction of *amyloid sequence patterns* that we have reported is probably the first study aimed at identifying potentially amyloidogenic regions in proteins by sequence scanning (*see* **Subheading 3.6.**) *(16)*.

Table 2
Results of the Positional Scanning Mutagenesis of STVIIE[a]

Substitution (X)	Net charge[b]	X₁TVIIE	SX₂VIIE	STX₃IIE	STVX₄IE	STVIX₅E	STVIIX₆
Gly	(+1)	+++	+	—	—	—	+++[c]
Ala	(+1)	+++	++	—	—	—	+++
Val	(+1)	+	++	+++	—	—	+++
Ile	(+1)	++	++	—	+++	+++	++
Leu	(+1)	++	++	+++	+++	—	+
Met	(+1)	+	++	—	—	—	+++
Ser	(+1)	+++	++	+	—	—	++
Thr	(+1)	++	+++	—	++	—	+++
Tyr	(+1)	+	++	—	++	++	++
Trp	(+1)	++	—	++[c]	++	—	++
Phe	(+1)	++	+	++	+++	+++	+++
Asn	(+1)	++	+	++	+++	—	++
Asp⁰	(+1)	++	++	—	—	—	++
Asp⁻	(−1)	++	+	—	—	—	++
Gln	(+1)	+	+++	+	—	—	+++
Glu⁰	(+1)	++	+++	+	+++	—	+++
Glu⁻	(−1)	+++	+	—	—	—	+
Lys⁺	(+1)	+++	—	—	—	—	—
Arg⁺	(+1)	+++	—	—	—	—	—
His⁺	(+1)	+++	—	—	—	—	—
His⁰	(−1)	++	—	—	—	—	—
Pro	(+1)	—	—	—	—	—	—

[a]EM fibril detection at t = 1 month. The amount of fibrils in a sample been quantified by visual inspection of the grids: +++, large amount of fibrils; ++, medium amount of fibrils; +, scarce amount of fibrils; —, no fibrils at all.

[b]Mutants generated by replacement with a non-charged amino acid have been synthesized with free termini (pH 2.6, net charge of +1). If positively charged residues are used, then molecules have both termini protected (pH 2.6, net charge of +1). Asp and Glu substitutions are incorporated into molecules with free termini, if a neutral side chain (Asp⁰ and Glu⁰) is required (pH 2.6, net charge of +1). For negatively charged Asp and Glu side chains (Asp⁻ and Glu⁻), the scanning of positions 1–5 has been performed on the sequence STVIIT with protected termini (pH 7.4, net charge of −1). Asp and Glu mutants at position 6 have protected termini (pH 7.4, net charge of −1).

[c]Fibrils have been detected only at pH 7.4 (net charge = −1).

Shadowed annotations correspond to mutations that lead to amyloid fibril formation.

The use of this pattern for *de novo* design and for the identification of amyloid-prone fragments in protein implies two assumptions: first, that the relative importance of the sequence positions found in peptide #2 will be general in most six-residue amyloidogenic fragments and, second, that most combinations of the allowed residues at a given position will provide an amyloidogenic motif.

Experiments conducted to validate the pattern have shown that the degree of amyloidogenicity of a given six-residue sequence is indeed mainly determined by the amino acid composition at the *core* and that the surrounding residues at the *edges* act just as amyloid modulators. These experiments also indicate that it is likely that most amino acid combinations matching the pattern would generate an amyloidogenic motif, and that amyloid sequences that do not resemble the original peptide **2** at all can be detected or designed *de novo* by using the pattern. *In silico* sequence scanning of amyloid peptides and proteins that form fibrillar aggregates *in vitro* or *in vivo* has shown that the pattern also captures the sequence features of natural amyloidogenic polypeptides *(16)*.

Analysis of protein databases indicates that highly amyloidogenic sequences matching the pattern are less frequent in proteins than innocuous amino acid combinations and that, if present, they are surrounded by amino acids that impair their aggregating capability *(amyloid breakers) (16)*. *Amyloid breakers* may be charged amino acids at particular positions of a given amyloidogenic stretch, a local excess of charges surrounding the sticky motif, or Pro, which is a β-sheet breaker. *Amyloid breaker* residues can be used as negative design elements to abolish amyloid fibril formation from amyloid-prone stretches identified with the pattern.

3. Methods

3.1. Establishment of Fibrillation Conditions

Amyloid fibril formation is a self-association reaction characterized by a nucleation step *(41)*. Therefore, the process is concentration dependent and overcomes the lag phase that precedes the fibril elongation step only above the so-called *critical concentration*. This concentration depends on the amyloidogenic degree and size of the peptide or protein under study. As a consequence, before any biophysical study, a wide enough concentration range should be assayed to assess which is the most suitable concentration to grow fibrils from a given peptide or protein (*see* **Note 8**). For the hexapeptides designed in our group, solutions of 50 μ*M*, 200 μ*M*, 500 μ*M*, and 1000 μ*M* have been prepared by diluting a known volume of stock solution into buffer *(15)*. Typically, these peptides form fibrils at concentrations approx 500 μ*M*. To prevent unspecific binding of the peptide or protein to the test tube, siliconized surfaces or some other special coatings for vials and plates should be used.

For these reasons, rigorous biophysical studies of amyloid fibril formation require starting preparations free of preformed fibrillar material, particulates, or other types of seeds. Preformed aggregates can act as a nucleation point. As a result, the process is accelerated or, eventually, diverted toward the formation of amorphous aggregates. For polypeptides with a moderate amyloidogenic degree, sonication (10 min) followed by centrifugation (5 min at 16,000*g*) is sufficient to dissemble preformed nuclei and to deposit insoluble material *(15)*. If some insoluble material still remains after centrifugation, the sample can be filtered (0.2-μm filters). The concentration of the solutions can be then determined by measuring the absorbance at 280 nm, or at 220 nm for peptides without aromatic residues. In the case of proteins with a highly amyloidogenic tendency, such as the Alzheimer's Aβ peptide, more severe treatments have to be followed to ensure the homogeneity of the starting solution *(42,43)* (*see* **Note 9**).

Because amyloid fibril formation depends on the net charge on the molecule *(15,36)*, the pH of the solution has to be such that the net charge on the peptide favors self-association into amyloid fibrils *vs.* the monomeric state or the formation of amorphous aggregates (*see* **Note 4**). After the most suitable pH for fibril growth has been chosen, the solutions can be incubated at room temperature.

For some model peptides and proteins, amyloid fibril formation takes place only under denaturing conditions; that is, under circumstances in which the native state is destabilized. Low pH, heat, high salt concentration, and additives such as TFE, urea, or GndHCl are external factors typically used to trigger amyloid formation.

3.2. In vitro Characterization of Amyloid Fibril Formation

The identification and characterization of amyloid material is performed by EM and spectroscopic techniques. Fibrillar material detected by EM is considered as amyloid if:

1. Fibrils are straight, unbranched, approx 7 to 12 nm in diameter, and of indeterminate length.
2. Fibrils show a cross-β X-ray diffraction pattern.
3. Fibrils bind dyes, such as Congo Red (CR) and Thioflavine (ThT).
4. Protein solutions display a polymeric β-sheet CD signature.

None of these evidences alone is, however, sufficient to ensure the amyloid nature of the aggregates. Fibrous silk, for example, shows a cross-β diffraction pattern, and amorphous aggregates can also display a polymeric β-sheet CD spectrum and dye binding properties.

3.2.1. Electron Microscopy

EM is typically used for the identification and morphological characterization of the nuclei, filament, and fibrils that appear during amyloid fiber growth (*see* **Fig. 3B**). Aliquots (5 μL) of the fibril preparation are adsorbed to glow-discharged, carbon-coated collodion film on 400-mesh copper grids. The grids are blotted, washed 2 times in droplets of MilliQ water, and stained with 1% (w/v) uranyl acetate.

For cryo-EM, the sample is cooled to liquid nitrogen or helium temperature to reduce the magnitude of ionization damage and the protein is maintained in a hydrated state in ice. High-resolution structural data can be obtained using cryo-EM. For instance, one can estimate the center-to-center spacing of an individual protofilament *(15)* or have a direct visualization of striations across the fibrils that correspond to the hydrogen bonding distance between β-strands *(44)*.

3.2.2. Circular Dichroism

CD is used to monitor the conformational transition of the protein from its native structure to a polymeric β-sheet. For the peptide-based amyloid fibrils designed in our group, there is a transition from a random coil conformation to a polymeric β-sheet (*see* **Fig. 2**). The polymeric β-sheet signature is characterized by a minimum between 215 and 220 nm and a maximum between 195 and 202 nm. The exact wavelengths depend on the sequence of the peptide under study and on the incubation time of the sample. This technique has many advantages to monitor β-sheet self-association because it is fast and it allows recovery of the sample that can be afterwards analyzed by other techniques. The path length of the quartz cuvettes depends on the concentration of monomeric peptide and protein and on whether or not there is salt or an additive in the solution. High salt concentrations and organic solvents, such as DMSO, can prevent CD measurements.

3.2.3. CR Binding

CR is a dye widely used to demonstrate the amyloid nature of protein deposits. CR binding to the cross-β structure of the amyloid fibril induces a change in the ultraviolet absorption spectrum of the dye *(45)*. Because dye binding depends on the amount of amyloid fibrils, this assay can be also used to quantify amyloid formation in vitro *(45)*. The pH of the solution and the CR concentration influence the extent of the binding. Therefore, it is advisable to carry out a titration curve to find the optimal CR concentration and pH to obtain a reliable signal for quantification.

3.2.4. ThT Binding

The formation of polymeric β-sheet structures in solution can be monitored by ThT fluorescence. The binding of ThT to protein aggregates containing extensive β-sheet structure leads to a strong enhancement of the ThT fluorescence emission at 485 nm following excitation at 440 nm (**Fig. 4**). The fluorescence enhancement is proportional to the quantity of aggregates formed during the reaction. As in the case of CR, optimization of the binding condition is indicated to enhance the signal *(46)*.

3.3. Empirical Approaches: Limited Proteolysis Experiments

The first step is the selection of a suitable protease. Because the proteolysis sites are dictated by the structure and dynamics of the protein, one should

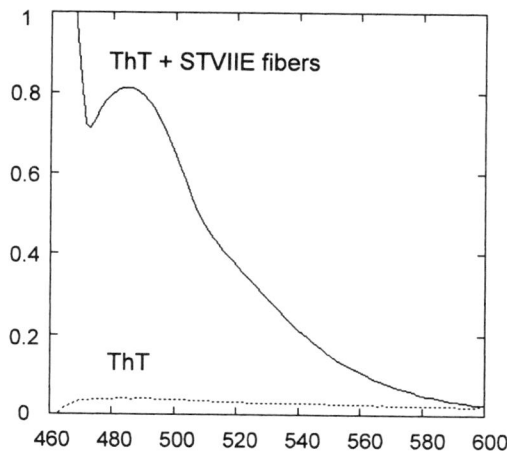

Fig. 4. Fluorescence emission spectra of Thioflavine T (ThT) alone (pH 7.4; 10 μ*M*; dashed line) and in the presence of mature STVIIE fibers (500 μ*M*; solid line). There is a clear enhancement of the emission around 485 nm on fiber binding. $\lambda_{exc} = $ 450 nm band pass = 4 nm; Emission band pass = 8 nm.

choose an enzyme with broad substrate specificity. Because the amyloid and the native state conditions very often differ, at least two enzymes that are active at different pH range have to be used to compare the native and the pro- and amyloidogenic states of the protein. If denaturants or high temperatures are used to promote amyloid formation, it must be also checked that the activity of the selected protease is not dramatically affected by these conditions.

Typically, the final products of the reaction are separated using high-pressure liquid chromatography or sodium dodecyl sulfate-polyacrylamide gel electrophoresis (SDS-PAGE) gels and Western blotting (*see* **Note 10**). If the fragments are too small to be analyzed using SDS-PAGE gels, then electrospray ionization mass spectroscopy can be used to identify them. Sequencing of the N-terminal peptide provides the final proof of the identity of the fragments and it allows for the assignment of the cleavage positions.

The most laborious part of the experiment is the establishment of the optimal ratio enzyme/protein and the appropriate proteolysis time, which varies from minutes to a couple of hours. Too high protease concentrations and too long incubations give rise to a very fast and exhaustive proteolysis of the protein (*see* **Note 11**). The comparison of the fragments obtained under different conditions allows for the selection of the most suitable candidates for further amyloid fibril formation studies. Ideally, one would like to isolate fragments from the proteolytic mixture by high-pressure liquid chromatography in sufficient amount for further conformational and amyloid formation studies (*47*; *see* **Note 12**).

Variations of the optimal conditions to digest the native protein and analysis of the proteolysis products at different incubation times have to be undertaken to characterize the different species that appear during the amyloid process. To identify the regions of the protein that constitute the *amyloid core* of the fibers, samples containing preformed fibrils have to be used as starting material. The *amyloid core* should be resistant to proteolysis, as it forms an extended cross-β structure. The globular appendages should be more susceptible to proteolysis, because they are not involved in the β-sheet network. Lateral association of protofilaments in the fibril structure could, however, provide some protease protection of the globular part of the protein. The presence of soluble oligomers or monomeric protein complicates the interpretation of the results. Therefore, it is advisable to separate the high molecular mass species from the nonfibrillar material by ultracentrifugation. Fibrils can be resuspended later in buffer for the proteolysis experiments (*48; see* **Note 13**). In this case, proteolytic digestion has to be more rigorous, although not complete. Usually, incubation times are longer (16–20 h). The reaction should yield some insoluble material that has resisted proteolysis, and that contains the *amyloid core*. The soluble fraction should contain the proteolysis products of the rest of the protein. These two fractions can be separated by centrifugation for further analysis. Observation by EM of filaments in the pellet provides the final evidence that this fraction contains the *amyloid core* (*see* **Note 14**). The identification of the sequence of the fragment that constitutes the *core* by mass spectroscopy or SDS-PAGE gel can be addressed by dissolving the pellet in a high concentration of denaturant (*see* **Note 15**).

3.4. Combinatorial Approaches: Design Bases of Amyloid Protein Libraries

Peptide and protein combinatorial libraries can be built by using chemical or molecular biology methods. The size, number of random positions, and the variability of the amino acids are dictated by the method used to generate the library. Chemically synthesized libraries can contain both natural and non-natural amino acids. Gene libraries are obtained by using degenerate codons, which restricts the type and number of residues that can be used to generate molecular diversity.

Regardless the combinatorial method of choice, the design of β-sheet libraries that yield amyloid fibrils relies on the generation of amphiphilic β-strands (*13,28*). According to the periodicity of the β-structures, the strands of the scaffold are defined by the alternation of polar and nonpolar residues. The amphiphilic nature of the β-strands drives their self-assembling into amyloid fibers (**Fig. 1**).

In the case of libraries of genes, the use of the binary code as design strategy reduces the combinatorial mixture to six polar (Asn, Asp, Gln, Glu, His, and Lys) and five nonpolar residues (Ile, Leu, Met, Phe, and Val). These polar and nonpolar amino acids are respectively encoded by the degenerated codons VAN and NTN. Validation of the design of the library implies the characterization of the sequences generated by using spectroscopic techniques and EM. For libraries of genes, previous extraction of the proteins generated is required for a characterization *in vitro*.

3.5. Rational Approaches: Design Procedure of Peptide-Based Amyloid Fibrils

The design of the model hexapeptides described in the rational approaches Subheading of this chapter has been carried out using a computer design algorithm called *PERLA*. A detailed description of the algorithm is available elsewhere *(49,50)*. Briefly, *PERLA* enables identification and sorting of amino acid sequences capable of folding into a desired three-dimensional structure. The program performs strict *inverse folding*: a fixed backbone is decorated with selected amino acid side chains from a custom-made side chain rotamer library constructed from the analysis of the protein database, wherein all-atom configurations are given. The ability of a given sequence to fit a protein structure is evaluated through a scoring function that uses an all-atom molecular mechanics force field and a combination of statistical terms including entropy and solvation. *PERLA* finally produces candidate sequences with modeled structures and energy descriptions.

PERLA has been used to search for ideal sequences capable of self-associating into polymeric β-sheet structures. The design procedure has been carried out using a polymeric six-stranded antiparallel β-sheet with six residues per strand as a structural template for calculations (**Fig. 5**). The three-dimensional model for the backbone has been constructed by taking the six central strands of the large single-layer β-sheet exhibited by the outer surface protein A *(51)*. To reduce the sequence space to be explored, we have designed homopolymeric β-sheets (i.e., having the same sequence for all the strands) and selected a simplified amino acid repertoire, containing both polar and hydrophobic residues. Amino acids modeled for the six-residue β-strands of the six-stranded polymeric β-sheet are: at positions #1 and #6 of the β-strand Lys, Glu, Thr, and Ser; at positions #2 and #5 Lys, Glu, Thr, Ser, Val, Ile, Leu, Tyr, and Trp; and at positions #3 and #4 Val, Ile, Leu, Tyr, and Trp (**Fig. 5**). To validate the design hypothesis and the automatic method, we have selected some of the output sequences. The predicted sequence stability is: #1 > #2 > #4 > #3 >#7 >#5 >#6 >#8 (*see* **Table 1**).

3.6. Sequence-Based Identification of Amyloidogenic Stretches in Proteins

Amyloid-prone fragments are identified by scanning protein sequences for six-residue stretches that match the *amyloid pattern* described in the rational approaches Subheading of this chapter. Sequence patterns are based on the

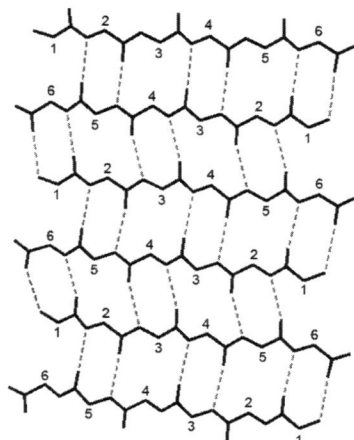

Fig. 5. Backbone template used for the computer-aided design of peptide-based amyloid fibrils. The topology used for the design is a homopolymeric six-stranded β-sheet with six residues per strand. A simplified amino acid repertoire has been modeled: at positions #1 and #6 of the β-strand Lys, Glu, Thr, and Ser; at positions #2 and #5 Lys, Glu, Thr, Ser, Val, Ile, Leu, Tyr, and Trp; and at positions #3 and #4 Val, Ile, Leu, Tyr, and Trp.

allowed amino acids substitutions at each position of the *de novo* designed peptide STVIIE *(16)*. Replacement by charged amino acids at all ionization states has provided us with a different sequence pattern for acidic and neutral pH.

Acidic pH {P}-{PKRHW}-[VLSCWFNQE]-[ILTYWFNE]-[FIY]-{PKRH}

Neutral pH {P}-{PKRHW}-[VLSCWFNQ]-[ILTYWFN]-[FIY]-{PKRH}

Patterns are written using PROSITE syntax (http://www.expasy.org/prosite/). Cys was not used in the mutagenesis experiment. However, because of its similarity to Ser, we assumed that is allowed (or forbidden) at the same positions as Ser.

Protein sequences can be scanned using PATTINPROT (http://npsa-pbil.ibcp.fr) *(52)*. The identified six-residue stretches can be synthesized to confirm their amyloidogenic nature. pH and termini should be such that the net charge on the molecule is ±1 (*see* **Note 4**).

3.7. In vivo Identification of Amyloid Sequences Using Reporter Proteins

Most designs of amyloid sequences are tested in vitro. However, and given that amyloid formation is related to disease, it is desirable to undertake the validation of designs and hypotheses in vivo. During the past several years, several animal models and cell systems have been developed to study the expression, overexpression, and depletion of relevant amyloidogenic proteins *(53–*

56). Unfortunately, the development of such in vivo models is very complicated and time consuming. As an alternative to evaluate the fundamental principles of protein misfolding, amyloid formation, and toxicity in a reasonable
term, several reporter proteins have been used to detect amyloid formation in
prokaryotic and eukaryotic cells. The expression of an amyloid protein fused
to an enzyme or a fluorescent protein enables the detection of amyloid formation by measuring changes in enzymatic activity or in fluorescence,
respectively. Fluorescent proteins, such as the green fluorescent protein (GFP)
(17,57,58), and enzymatic systems, such as β-galactosidase complementation
(β-gal) *(59)* are frequently used in the field to monitor protein aggregation in vivo.

By using these reporter systems, genetic screenings on natural amyloidogenic
sequences have been also tackled to identify nonaggregating variants of the
protein. This procedure has been very successful in identifying the regions of
the molecule promoting Aβ aggregation in vivo (*see* **Note 16**; *17*).

3.7.1. β-Gal as Reporter Protein

β-gal is a homotetramer that catalyzes the hydrolysis of lactose into D-glucose
and D-galactose. Deletions of the N- or C-terminus of the gene that encodes β-gal
(*lacZ*) produces the inactivation of the enzyme. The inactive mutant can be still
complemented with a fragment that contains the missing domain *(60)*. N- and C-
deletion mutants are known as α- and ω-acceptor, respectively *(61)*. The complementary fragments are called α- and ω̄-domains. Intracistronic complementation
of β-gal can be performed both in prokaryotic and eukaryotic systems.

The complementation of β-gal takes place only with the soluble forms of the
domains. Based on this, this reporter system has been used to study the solubility of natural proteins in *Escherichia coli* and to identify aggregating-prone
sequences *(59)*. In the case of α-complementation, the hypothetical aggregating sequence is fused to the α-fragment of β-gal. These constructions have to
be expressed in mutant strains for the α-fragment but that express the α-acceptor. If the protein fused to the α-domain is soluble, the complementation with
the α-acceptor occurs and β-gal recovers its enzymatic activity. However, if
the fused sequence triggers aggregation of the α-domain, the complementation
of β-gal does not take place (**Fig. 6**).

β-gal activity can be monitored using a variety of chromogenic and
fluorogenic substrates. The more commonly used are 5-bromo-4-chloro-3-
indolyl β-D-galactopyranoside (X-gal) for cells growing in solid medium and
o-nitrophenyl-β-D-galactosidase (ONPG) for culture in liquid medium *(62–67)*.

A qualitative monitoring of the reaction can be performed on the basis that
the end-product of X-gal hydrolysis generates blue colonies. Typically, transformed cells are seeded on agar plates that contain X-gal. After incubation
overnight, white colonies indicate the in vivo aggregating capability of the
sequence tested (**Fig. 6**).

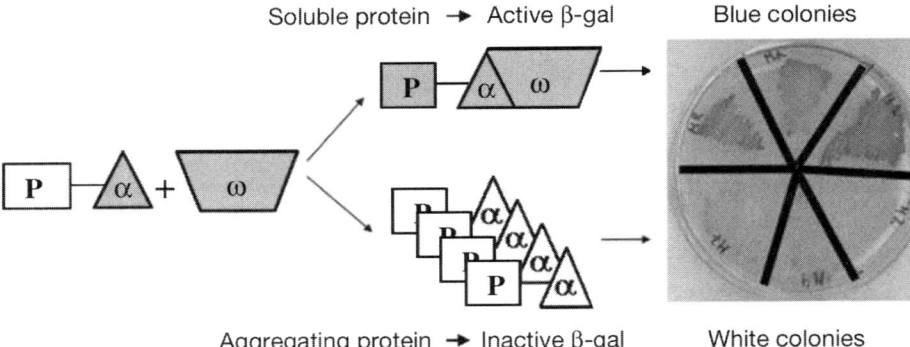

Fig. 6. Blue-white screening of β-galactosidase activity. The fusion of soluble proteins to the α-domain of β-gal allows its complementation with the ω-acceptor, β-gal recovers its enzymatic activity, and blue colonies are obtained. However, the fusion of aggregating sequences drives the aggregation of the α-domain and the complementation is not possible, generating white colonies.

β-gal activity can be quantified using transformed cells grown in liquid medium. In this case, β-gal complementation is measured ex vivo. Depending on the protocol used, the measurements can be carried out in cells extracts or after the permeabilization of the cells. In both cases, β-gal activity is quantified by monitoring production of *o*-nitrophenol, the hydrolysis end-product of ONPG. *o*-Nitrophenol is detected by measuring its absorbance at 420 nm (**62–65**; *see* **Note 17**).

The last step of both methods is always the extraction of the aggregated material to confirm its amyloid nature in vitro (**68**).

4. Notes
1. A mutagenesis scanning experiment consists in the systematic replacement of the residues of a given peptide or protein region by one or several probe amino acids.
2. To avoid polymerization of the designed β-sheet proteins, elements of negative design, such as mutations to proline or charged residues, can be introduced at the first and last β-strands of the β-sheet.
3. A protein design algorithm allows for a fully automatic procedure for predicting the amino acid sequences compatible with a given target structure.
4. Charge variations can be achieved by synthesizing versions of the same sequences with either both termini protected (acetylated at the N-terminus and amidated at the C-terminus), protected at only one terminus, or with both termini free, as well as by varying the pH of the solution (2.6, 6.0, 7.4, and 12.5).
5. Validation of the model system. The small model peptides that we have designed form more regular fibers than proteins do, which increases enormously the quality of the data obtained by cryo-EM and other structural techniques. On the basis

of the structural characterization of the fibrils formed by STVIIE, we have proposed a detailed structural model of these peptide-based amyloid fibrils *(15)*. The designed peptide reproduces the organization of the protofilament core of naturally occurring amyloid fibrils, which validates its use as model system to study amyloid fibril formation and structure.

6. This interpretation of the results is also in agreement with experimental data obtained both from peptides and proteins, in which the propensity to form fibrils is highly dependent on the pH and ionic strength of the solutions from which aggregation or amyloid fibril formation occurs.

7. The saturation mutagenesis analysis of STVIIE has consisted in the systematic replacement of all residues of STVIIE with all natural amino acids, except Cys.

8. The conditions for fibril formation depend on the amyloidogenic propensity of the protein. While the Aβ amyloid peptide forms fibrils at pH 7.4, 230 μ*M* and 37°C in 2 d *(42)*, hen lysozyme has to be incubated for 10 d at pH 2, 1 m*M*, and 65°C *(47)*.

9. If the starting point of the experiments has to be the native state, it should be checked that the treatment has not affected the correct folding of the protein.

10. Different bioinformatics tools are available to determine the expected cleavage sites and the theoretical size of the expected fragments (http://www.expasy.org/tools/#proteome). Although in limited proteolysis experiments, a complete proteolysis of the protein is not expected, this information is useful to guide the choice of the separation technique according to the expected size of the fragments and the resolution of the technique.

11. Often the supplier provides useful information about the optimal working conditions.

12. As an alternative, and depending on their length, they can be synthesized using standard chemical synthesis.

13. EM and ThT binding assays of both the supernatant and the resuspended pellet should be carried out to check for the absence or presence of fibrils.

14. By using cryo-EM, it may be possible to obtain the dimensions of both the digested and undigested fibers. These measurements, in combination with some other structural data, may permit to build a model of the organization of the fiber *(69)*.

15. For proteins containing domains with catalytic or ligand-binding activity, functional assays can be conducted to check whether or not the domain keeps its activity after proteolytic cleavage from the *amyloid core*.

16. We recommend the use of β-gal vs. GFP as reporter protein. For the latter, the selected vector contains the DNA encoding for GFP and a multicloning site where the DNA encoding for the hypothetic amyloidogenic protein is fused. This construction can be expressed in prokaryotic and eukaryotic cells. Aggregation of the fused sequence induces a modification of the fluorescence of GFP. Typically, fluorescence decreases or disappears. However, it has been reported that in some cases GFP-aggregates are still fluorescent. Therefore, it is preferable to work with a non-ambiguous reporter system, such as the complementation of β-gal.

17. In both the qualitative and quantitative methods, it is advisable to use a construction that only expresses the α-domain of β-gal as positive control.

Acknowledgments

We thank Dr. Luis Serrano for support in our investigations. We acknowledge support by a EU grant to AE-C (RTN project 2001-00364: "Protein Folding, Misfolding, and Disease"); by a fellowship of the "Ministerio de Educación y Cultura" (Spain) to MTP; and by a EU grant (APOPIS) to MLP. AE-C and MTP have contributed equally to this chapter.

References

1. Dobson, C. M. (2002) Getting out of shape. *Nature* **418,** 729–730.
2. Thirumalai, D., Klimov, D. K., and Dima, R. I. (2003) Emerging ideas on the molecular basis of protein and peptide aggregation. *Curr. Opin. Struct. Biol.* **13,** 146–159.
3. Soto, C. (2003) Unfolding the role of protein misfolding in neurodegenerative diseases. *Nat. Rev. Neurosci.* **4,** 49–60.
4. Sipe, J. D. and Cohen, A. S. (2000) Review: history of the amyloid fibril. *J. Struct. Biol.* **130,** 88–98.
5. Chiti, F., Webster, P., Taddei, N., Clark, A., Stefani, M., Ramponi, G., and Dobson, C. M. (1999) Designing conditions for *in vitro* formation of amyloid protofilaments and fibrils. *Proc. Natl. Acad. Sci. USA* **96,** 3590–3594.
6. Fandrich, M., Fletcher, M. A., and Dobson, C. M. (2001) Amyloid fibrils from muscle myoglobin. *Nature* **410,** 165–166.
7. Guijarro, J. I., Sunde, M., Jones, J. A., Campbell, I. D., and Dobson, C. M. (1998) Amyloid fibril formation by an SH3 domain. *Proc. Natl. Acad. Sci. USA* **95,** 4224–4228.
8. Sunde, M., Serpell, L. C., Bartlam, M., Fraser, P. E., Pepys, M. B., and Blake, C. C. (1997) Common core structure of amyloid fibrils by synchrotron X-ray diffraction. *J. Mol. Biol.* **273,** 729–739.
9. Pastor, M. T., Esteras-Chopo, A., and López de la Paz, M. (2005) Design of model systems for amyloid formation: lessons for prediction and inhibition. *Curr. Opin. Struct. Biol.* **15,** 57–63.
10. von Bergen, M., Friedhoff, P., Biernat, J., Heberle, J., Mandelkow, E. M., and Mandelkow, E. (2000) Assembly of tau protein into Alzheimer paired helical filaments depends on a local sequence motif ((306)VQIVYK(311)) forming beta structure. *Proc. Natl. Acad. Sci. USA* **97,** 5129–5134.
11. Mazor, Y., Gilead, S., Benhar, I., and Gazit, E. (2002) Identification and characterization of a novel molecular-recognition and self-assembly domain within the islet amyloid polypeptide. *J. Mol. Biol.* **322,** 1013–1024.
12. Jones, S., Manning, J., Kad, N. M., and Radford, S. E. (2003) Amyloid-forming peptides from β(2)-microglobulin-Insights into the mechanism of fibril formation in vitro. *J. Mol. Biol.* **325,** 249–257.
13. West, M. W., Wang, W., Patterson, J., Mancias, J. D., Beasley, J. R., and Hecht, M. H. (1999) *De novo* amyloid proteins from designed combinatorial libraries. *Proc. Natl. Acad. Sci. USA* **96,** 11211–11216.

14. Kammerer, R. A., Kostrewa, D., Zurdo, J., Detken, A., Garcia-Echeverria, C., Green, J. D., et al. (2004) Exploring amyloid formation by a *de novo* design. *Proc. Natl. Acad. Sci. USA* **101**, 4435–4440.

15. López de la Paz, M., Goldie, K., Zurdo, J., Lacroix, E., Dobson, C. M., Hoenger, A., et al. (2002) *De novo* designed peptide-based amyloid fibrils. *Proc. Natl. Acad. Sci. USA* **99**, 16052–16057.

16. López de la Paz, M. and Serrano, L. (2004) Sequence determinants of amyloid fibril formation. *Proc. Natl. Acad. Sci. USA* **101**, 87–92.

17. Wurth, C., Guimard, N. K., and Hecht, M. H. (2002) Mutations that reduce aggregation of the Alzheimer's Aβ42 peptide: an unbiased search for the sequence determinants of Aβ amyloidogenesis. *J. Mol. Biol.* **319**, 1279–1290.

18. Dobson, C. M. (1999) Protein misfolding, evolution and disease. *Trends Biochem. Sci.* **24**, 329–332.

19. Rochet, J. C. and Lansbury, P. T., Jr. (2000) Amyloid fibrillogenesis: themes and variations. *Curr. Opin. Struct. Biol.* **10**, 60–68.

20. Monti, M., Garolla di Bard, B. L., Calloni, G., Chiti, F., Amoresano, A., Ramponi, G., et al. (2004) The regions of the sequence most exposed to the solvent within the amyloidogenic state of a protein initiate the aggregation process. *J. Mol. Biol.* **336**, 253–262.

21. Fontana, A., Zambonin, M., Polverino de Laureto, P., De Filippis, V., Clementi, A., and Scaramella, E. (1997) Probing the conformational state of apomyoglobin by limited proteolysis. *J. Mol. Biol.* **266**, 223–230.

22. Polverino de Laureto, P., Taddei, N., Frare, E., Capanni, C., Costantini, S., Zurdo, J., et al. (2003) Protein aggregation and amyloid fibril formation by an SH3 domain probed by limited proteolysis. *J. Mol. Biol.* **334**, 129–141.

23. McParland, V. J., Kalverda, A. P., Homans, S. W., and Radford, S. E (2002) Structural properties of an amyloid precursor of β(2)-microglobulin. *Nat. Struct. Biol.* **9**, 326–331.

24. Azriel, R. and Gazit, E. (2001) Analysis of the minimal amyloid-forming fragment of the islet amyloid polypeptide. An experimental support for the key role of the phenylalanine residue in amyloid formation. *J. Biol. Chem.* **276**, 34156–34161.

25. Pastor, M. T., Mora, P., Ferrer-Montiel, A., and Pérez-Paya, E. (2004) Design of bioactive and structurally well-defined peptides from conformationally restricted libraries. *Biopolymers* **76**, 357–365.

26. Hecht, M. H., Das, A., Go, A., Bradley, L. H., and Wei, Y. (2004) *De novo* proteins from designed combinatorial libraries. *Protein Sci.* **13**, 1711–1723

27. Kamtekar, S., Schiffer, J. M., Xiong, H., Babik, J. M., and Hecht, M. H. (1993) Protein design by binary patterning of polar and nonpolar amino acids. *Science* **262**, 1680–1685.

28. Matsuura, T., Ernst, A., and Pluckthun, A. (2002) Construction and characterization of protein libraries composed of secondary structure modules. *Protein Sci.* **11**, 2631–2643.

29. Richardson, J. S. and Richardson, D. C. (2002) Natural β-sheet proteins use negative design to avoid edge-to-edge aggregation. *Proc. Natl. Acad. Sci. USA* **99**, 2754–2759.

30. Wang, W. and Hecht, M. H. (2002) Rationally designed mutations convert *de novo* amyloid-like fibrils into monomeric β-sheet proteins. *Proc. Natl. Acad. Sci. USA* **99,** 2760–2765.

31. Tjernberg, L., Hosia, W., Bark, N., Thyberg, J., and Johansson, J. (2002) Charge attraction and beta propensity are necessary for amyloid fibril formation from tetrapeptides. *J. Biol. Chem.* **277,** 43243–43246.

32. Turner, W. G. and Finch, J. T. (1992) Binding of the dye congo red to amyloid protein pig insulin reveals a novel homology amongst amyloid-forming peptide sequences. *J. Mol. Biol.* **277,** 1205–1223.

33. Kallberg, Y., Gustafsson, M., Persson, B., Thyberg, J., and Johansson, J. (2001) Prediction of amyloid fibril-forming proteins. *J. Biol. Chem.* **276,** 12945–12950.

34. Uversky, V. N., Gillespie, J. R., and Fink, A. L. (2000) Why are "natively unfolded" proteins unstructured under physiologic conditions? *Proteins* **41,** 415–427.

35. Ventura, S., Zurdo, J., Narayanan, S., Parreno, M., Mangues, R., Reif, B., et al. (2004) Short amino acid stretches can mediate amyloid formation in globular proteins: the Src homology 3 (SH3) case. *Proc. Natl. Acad. Sci. USA* **101,** 7258–7263.

36. Chiti, F., Calamai, M., Taddei, N., Stefani, M., Ramponi, G., and Dobson, C. M. (2002) Studies of the aggregation of mutant proteins *in vitro* provide insights into the genetics of amyloid diseases. *Proc. Natl. Acad. Sci. USA* **99,** 16419–16426.

37. Minor, D. L., Jr. and Kim, P. S. (1994) Measurement of the β-sheet-forming propensities of amino acids. *Nature* **367,** 660–663.

38. Chiti, F., Stefani, M., Taddei, N., Ramponi, G., and Dobson, C. M. (2003) Rationalization of the effects of mutations on peptide and protein aggregation rates. *Nature* **424,** 805–808.

39. Fernández, A., Kardos, J., Scott, L. R., Goto, Y., and Berry, R. S. (2003) Structural defects and the diagnosis of amyloidogenic propensity. *Proc. Natl. Acad. Sci. USA* **100,** 6446–6451.

40. Fernández-Escamilla, A. M., Rousseau, F., Schymkowitz, J., and Serrano, L. (2004) Prediction of sequence-dependent and mutational effects on the aggregation of peptides and proteins. *Nat. Biotechnol.* **22,** 1302–1306.

41. Harper, J. D. and Lansbury, P. T., Jr. (1997) Models of amyloid seeding in Alzheimer's disease and scrapie: mechanistic truths and physiological consequences of the time-dependent solubility of amyloid proteins. *Annu. Rev. Biochem.* **66,** 385–407.

42. Wood, S. J., Maleeff, B., Hart, T., and Wetzel, R. (1996) Physical, morphological and functional differences between pH 5.8 and 7.4 aggregates of the Alzheimer's amyloid peptide Abeta. *J. Mol. Biol.* **256,** 870–877.

43. Walsh, D. M., Hartley, D. M., Kusumoto, Y., Fezoui, Y., Condron, M. M., Lomakin, A., et al. (1999) Amyloid beta-protein fibrillogenesis. Structure and biological activity of protofibrillar intermediates. *J. Biol. Chem.* **274,** 25945–25952.

44. Serpell, L. C. and Smith, J. M. (2000) Direct visualisation of the beta-sheet structure of synthetic Alzheimer's amyloid. *J. Mol. Biol.* **299,** 225–231.

45. Klunk, W. E., Pettegrew, J. W., and Abraham, D. J. (1989) Quantitative evaluation of congo red binding to amyloid-like proteins with a beta-pleated sheet conformation. *J. Histochem. Cytochem.* **37,** 1273–1281.

46. LeVine, H. 3rd. (1993) Thioflavine T interaction with synthetic Alzheimer's disease beta-amyloid peptides: detection of amyloid aggregation in solution. *Protein Sci.* **2,** 404–410.
47. Frare, E., Polverino De Laureto, P., Zurdo, J., Dobson, C. M., and Fontana, A. (2004) A highly amyloidogenic region of hen lysozyme. *J. Mol. Biol.* **340,** 1153–1165.
48. Kheterpal, I., Williams, A., Murphy, C., Bledsoe, B., and Wetzel, R. (2001) Structural features of the Aβ amyloid fibril elucidated by limited proteolysis. *Biochemistry* **40,** 11757–11767.
49. Finsinger, S., Serrano, L., and Lacroix, E. (2001) Computational estimation of specific side chain interaction energies in α-helices. *Protein Sci.* **10,** 809–818.
50. López de la Paz, M., Lacroix, E., Ramírez-Alvarado, M., and Serrano, L. (2001) Computer-aided design of β-sheet peptides. *J. Mol. Biol.* **312,** 229–246.
51. Pham, T. N., Koide, A., and Koide, S. (1998) A stable single-layer β-sheet without a hydrophobic core. *Nat. Struct. Biol.* **5,** 115–119.
52. Combet, C., Blanchet, C., Geourjon, C., and Deleage, G. (2000) NPS@: network protein sequence analysis. *Trends Biochem. Sci.* **25,** 147–150.
53. Zoghbi, H. Y. and Botas, J. (2002) Mouse and fly models of neurodegeneration. *Trends Genet. TIG* **18,** 463–471.
54. Bonini, N. M. and Fortini, M. E. (2003) Human neurodegenerative disease modeling using *Drosophila. Annu. Rev. Neurosci.* **26,** 627–656.
55. Outeiro, T. F. and Muchowski, P. J. (2004) Molecular genetics approaches in yeast to study amyloid diseases. *J. Mol. Neurosci.* **23,** 49–60.
56. Rochet, J. C., Outeiro, T. F., Conway, K. A., Ding, T. T., Volles, M. J., Lashuel, H. A., et al. (2004) Interactions among alpha-synuclein, dopamine, and biomembranes: some clues for understanding neurodegeneration in Parkinson's disease. *J. Mol. Neurosci.* **23,** 23–34.
57. Ripaud, L., Maillet, L., and Cullin, C. (2003) The mechanisms of [URE3] prion elimination demonstrate that large aggregates of Ure2p are dead-end products. *EMBO J.* **22,** 5251–5259.
58. Burnett, B., Li, F., and Pittman, R. N. (2003) The polyglutamine neurodegenerative protein ataxin-3 binds polyubiquitylated proteins and has ubiquitin protease activity. *Hum. Mol. Genet.* **12,** 3195–3205.
59. Wigley, W. C., Stidham, R. D., Smith, N. M., Hunt, J. F., and Thomas, P. J. (2001) Protein solubility and folding monitored *in vivo* by structural complementation of a genetic marker protein. *Nat. Biotechnol.* **19,** 131–136.
60. Ullmann, A., Jacob, F. and Monod, J. (1967) Characterization by *in vitro* complementation of a peptide corresponding to an operator-proximal segment of the β-galactosidase structural gene of *E. coli. J. Mol. Biol.* **24,** 339–343.
61. Langley, K. E., Villarejo, M. R., Fowler, A. V., Zamenhof, P. J., and Zabin, I. (1975) Molecular basis of β-galactosidase α-complementation. *Proc. Natl. Acad. Sci. USA* **72,** 1254–1257.
62. Miller, J. H. (1972) *Experiments in Molecular Genetics.* Cold Spring Harbor Laboratory Press, Cold Spring Harbor, NY.
63. Arvidson, D. N., Youderian, P., Schneider, T. D., and Stormo, G. D. (1991) Automated kinetic assay of β-galactosidase activity. *Biotechniques* **11,** 733–738.

64. Bianco, P. R. and Weinstock, G. M. (1994) Automated determination of β-galactosidase specific activity. *Biotechniques* **17,** 974–980.
65. Schupp, J. M., Travis, S. E., Price, L. B., Shand, R. F., and Keim, P. (1995) Rapid bacterial permeabilization reagent useful for enzyme assays. *Biotechniques* **19,** 18–20.
66. Menzel, R. (1989) A microtiter plate-based system for the semiautomated growth and assay of bacterial cells for β-galactosidase activity. *Anal. Biochem.* **181,** 40–50.
67. Griffith, K. L. and Wolf, R. E., Jr. (2002) Measuring β-galactosidase activity in bacteria: cell growth, permeabilization, and enzyme assays in 96-well arrays. *Biochem. Biophys. Res. Commun.* **290,** 397–402.
68. Goi, F. and Gallo, G. (2005) Extraction and chemical characterization of tissue-deposited proteins from minute diagnostic biopsy specimens. In: *Amyloid Proteins: Methods and Protocols* (Sigurdsson, E. M., ed.) Humana, Totowa, NJ, pp. 261–266.
69. Baxa, U., Taylor, K. L., Wall, J. S., Simon, M. N., Cheng, N., Wickner, R. B., et al. (2003) Architecture of Ure2p prion filaments: the N-terminal domains form a central core fiber. *J. Biol. Chem.* **278,** 43717–43727.

13

Protein Misfolding Disorders and Rational Design of Antimisfolding Agents

Lisbell D. Estrada, June Yowtak, and Claudio Soto

Summary

Compelling evidence strongly suggests that the conversion of a normal soluble protein into a β-sheet–rich oligomeric structure and further fibril formation is the critical step in the pathogenesis of several human diseases, termed *protein misfolding disorders*. Therefore, a promising therapeutic strategy consists of the design of molecules that prevent the misfolding and aggregation of these proteins. In this chapter, we survey the mechanism of protein misfolding and some strategies to rationally produce inhibitors of this process.

Key Words: Peptide inhibitors; β-sheet breakers; amyloid; conformational disorders; Alzheimer's disease.

1. Introduction

The information required to fold a protein in a functional, specific three-dimensional structure is contained in its amino acid sequence. Proteins usually fold properly into the native conformation and when they do not, the misfolding is usually corrected by chaperone proteins *(1)*. However, in protein misfolding disorders (PMD), accumulation of misfolded proteins results in aggregation and accumulation in diverse tissues as protein deposits referred to as amyloid *(2–5)*. Some diseases included in this group are Alzheimer's disease (AD), Parkinson disease, Huntington's disease, transmissible spongiform encephalopathies, serpin-deficiency disorders, hemolytic anemia, cystic fibrosis, diabetes type II, amyotrophic lateral sclerosis, and dialysis-related amyloidosis. Although proteins involved in those diseases do not present similarities in sequence or structure, all of them are able to adopt at least two different conformations without changes in the amino acid sequence.

Usually in PMD, the misfolded form of the protein contains repetitions of the β-sheet motifs organized in a structure known as cross-β *(6)*. A β-sheet is

From: *Methods in Molecular Biology, vol. 340: Protein Design: Methods and Applications*
Edited by: R. Guerois and M. López de la Paz © Humana Press Inc., Totowa, NJ

composed of several extended β-strands of a polypeptide that run alongside each other, linked by hydrogen bonding between the NH and CO groups of the peptide bond. This forms an extended, flattened surface. Because β-sheets can be stabilized by intermolecular interactions, misfolded proteins have a high tendency to form oligomers, which also stabilize β-sheet formation and promote aggregation.

Several lines of evidence support the involvement of aggregation and amyloid deposition in the pathology of PMD. The presence of abnormal aggregates usually occurs in the tissues with most damage *(7–9)*. In addition, accumulation of these amyloid deposits in diverse organs is the end point in the majority of PMD *(2,5)*. Postmortem studies show amyloid plaques composed by β-amyloid protein (Aβ) and intracellular aggregates of protein tau in AD *(10)*. Patients with Parkinson's disease have neurons in the substantia nigra that contain Lewy bodies consisting of aggregates of α-synuclein *(11)*. A typical feature in Huntington's disease is the presence of intracellular aggregates formed of Huntington protein containing additional polyglutamine repeats *(12)*. Brains of patients with transmissible spongiform encephalopathies contain aggregates of protease-resistant prion protein *(13)*. Another example is the presence of aggregates composed of the pancreatic peptide hormone amylin in 90% of the patients affected by type II diabetes *(14)*. In systemic amyloidosis, diverse proteins misfold, aggregate, and accumulate in different tissues, including heart, kidney, liver, and bones *(15)*.

The possibility that protein aggregation is a cause rather than an effect in PMD is strongly supported by evidence from genetic studies *(5,16)*. Most PMD have both an inherited and sporadic origin and genes linked to familial forms of the disease encode the protein component of fibrillar aggregates. Familial forms usually have an earlier onset and higher severity than sporadic cases, possibly because mutations may destabilize the normal protein folding, favoring the misfolding and aggregation of the protein *(16)*. Studies with animal models additionally support the critical role of misfolding and aggregation in disease pathogenesis *(5,17)*. Transgenic animals overexpressing mutant human genes encoding for the protein component of the aggregates, show a disease-associated phenotype, which includes several of the pathological and clinical characteristics of these disorders.

1.1. Mechanism of Protein Misfolding and Aggregation

At least three hypotheses have been proposed to explain how the native protein, which usually has a mixture of α-helical and random structure, adopts a β-pleated sheet aggregated structure *(5)* (**Fig. 1**). Elucidating the pathway by which a natural protein changes its structure to become pathogenic through aggregate formation will be extremely useful in designing valuable therapeutic strategies.

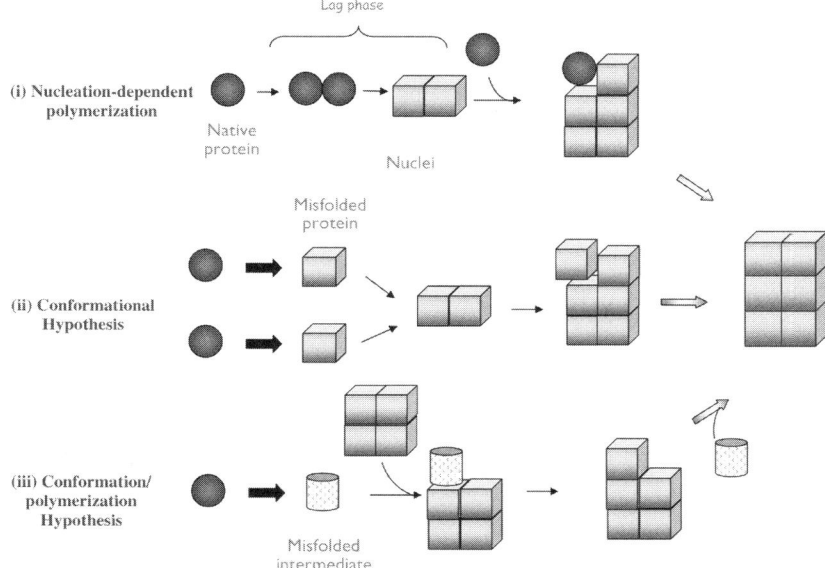

Fig. 1. Models of protein misfolding and aggregation. The nucleation-dependent polymerization model **(i)** proposes that the key step is the formation of nuclei that catalyses protein misfolding and aggregation. In the conformational hypothesis **(ii)**, protein misfolding is independent of aggregation, which is an unnecessary end point of conformational changes. The conformation/polymerization hypothesis **(iii)** implies the formation of an unstable intermediate that is stabilized on protein oligomerization. Square represents the folded native conformation, circles the disease-associated conformer; cylinder corresponds to an unstable conformational intermediate.

Based on the kinetic modeling of protein aggregation, it has been proposed that the critical event in PMD is the formation of protein oligomers that act as seeds to induce protein misfolding *(18)* (**Fig. 1**). In this nucleation-dependent model, it is possible to distinguish three different phases; first, there is formation of an ordered nucleus in a slow lag phase from unfavorable interactions between monomers. Then, polymerization starts, and the nucleus grows, forming a larger polymer. The growth phase continues until the equilibrium between aggregates and monomers is reached (steady-state phase) *(18)*. In a nucleation-dependent polymerization, the aggregation does not occur until the protein concentration exceeds a point known as the critical concentration *(19)*. Below this level, the monomer is the predominant species. If the critical concentration is barely exceeded, there is a long lag time before polymerization. This slow nucleation step can be bypassed by adding preformed nuclei or seeds. Even though amyloid seeding is very selective because it needs a match between the growing face and the monomer, a different seed can accelerate the process, in a process known as heterologous seeding *(19)*.

The second model (the conformational hypothesis) suggests that amyloid formation is just a consequence of conformational instability *(20)* (**Fig. 1**). The central feature in this hypothesis is that protein misfolding is due to a change in protein conformation independent of aggregates formation *(21)*. In this model, the formation of larger order aggregates is an unnecessary consequence of protein misfolding. Genetics and environmental factors have been shown to influence conformational changes, such as mutations that destabilize the folded structure or alterations in the activity of certain proteins known as pathological chaperones *(5)*. Other factors that catalyze misfolding are changes in pH, metal ions, and oxidative stress.

However, the most likely model lies between the previous two and is known as the conformation/polymerization model *(5)*, which implies the initial misfolding of the protein to form an unstable intermediate, which is stabilized by intermolecular interactions and aggregate formation (**Fig. 1**). Because the intermediate form contains hydrophobic segments, it is not stable in an aqueous environment. However, these intermediates are able to form oligomers rich in β-sheet structures, which grow to produce fibrils. In this hypothesis, the generation of the pathological form includes a prior conformational change, but the complete misfolding requires an oligomerization process. This model can explain the presence of some degree of conformational change preceding aggregates appearance. This phenomenon has been reported for diverse proteins, such as transthyretin, serpins, amyloid-β, and prion protein *(2,20,22,23)*.

2. Correcting Protein Misfolding as a Novel Therapeutic Approach

Considering that protein misfolding and aggregation is central in the pathogenesis of PMD, a therapy directed to the cause of the disease should aim to inhibit or reverse the conformational changes that result in the formation of the pathological protein conformer and its posterior aggregation.

Because of the high prevalence of AD in the population and the lack of an efficient treatment to alter the progression of the disease, most of the effort to develop inhibitors of protein misfolding and aggregation has been focused on this disease. However, many of the approaches can be extrapolated to design compounds beneficial for the treatment of other diseases in the PMD group.

One approach that has been used to discover small chemical inhibitors of protein misfolding and aggregation has been the screening of large libraries of compounds using simple in vitro assays. Several unrelated small molecules have been shown to prevent or reverse Aβ polymerization in vitro *(24,25)*. Among these compounds is possible to mention the following: Congo red *(26)*, hexadecyl-*N*-methylpiperidinium bromide *(27)*, small sulfonated anions *(28)*, benzofuran-based compounds *(29)*, rifampicin *(30)*, melatonin *(31)*, nicotine *(32)*, estrogen *(33)*, glycosaminoglycans mimetics *(34)*, nitrophenols *(35)*, tetracycline *(36)*, anthracycline 4'-iodo-4'-deoxydoxorubicin *(37)*, clioquinol *(38)*, ibuprofen *(39)*, and *N,N'*-bis(3-hydroxyphenyl)pyridazine-3,6-diamine *(40)*. The activity of some of these compounds has also been studied in vivo using AD animal models and even some of them are under clinical evaluation in AD patients. However, the usefulness of these small molecules as amyloid inhibitors is compromised by their lack of specificity and their (in most of the cases) unknown mechanism of action, which makes it difficult to improve them. In addition, many of these compounds are highly toxic.

2.1. Inhibiting Misfolding by Endogenous Amyloid-Binding Proteins

Diverse proteins have been identified immunohistochemically to be associated with amyloid deposits *(15)*. Some of these amyloid-associated proteins have been found in many diverse types of amyloid plaques and are known as universal components of amyloid, including apolipoprotein E (apoE), serum amyloid-P (SAP), and heparan sulfate proteoglycans *(41–43)*. It has been hypothesized that some of these proteins and others known to interact with the protein before misfolding may serve as endogenous chaperones to stabilize the native folding and prevent protein aggregation *(44–46)*.

Indeed, several reports have shown that some proteins (e.g., apoE, apolipoprotein J [apoJ], SAP, α-antichymotrypsin, laminin, transthyretin, albumin) can bind Aβ with high affinity and prevent amyloid formation in vitro *(47–53)*. However, other studies have shown that these same proteins can also promote fibrillogenesis in vitro *(54–56)*. These contradictory results have been explained by differences in the peptide concentration, assay sensitivity, solvents, or different peptide sources *(45)*. Nevertheless, the relevance of these proteins in vivo is still unclear.

Transthyretin (TTR), a homotetramer with 127 amino acid residue in each chain, is synthesized in the liver and is found in blood plasma and cerebrospinal fluid *(57)*. TTR is able to form amyloid which accumulate in different peripheral nerves of patients affected by familial amyloid polyneuropathy or in the heart of people affected by familial amyloid cardiomyopathy *(58)*. It has been published that TTR is a major Aβ-binding protein in cerebrospinal fluid *(59)*, and further studies demonstrated that TTR is able to prevent formation of Aβ fibrils in vitro, sequestering Aβ from cerebrospinal fluid by a stable complex formation *(49)*.

Laminin is a four-arm glycoprotein component of the extracellular matrix. This protein is induced by brain injury, and it has been shown to accumulate in senile plaques of patient with AD *(60)*. Thioflavin T-fluorescence assays and electronic microscopy studies shown that laminin inhibits the amyloid formation in a concentration-dependent manner *(53)*. Experiments in rat hippocampal neurons show the ability of laminin-1 and laminin-2 to inhibit fibril formation and reduce Aβ toxicity *(61)*. Also, it was demonstrated that amyloid fibrils are unstable in presence of laminin-1.

apoE is a homotetrameric, highly α-helical protein primarily known for its involvement in cholesterol metabolism *(62)*. Contradictory results have been published for apoE's role in Aβ fibrillogenesis in vitro *(47,54,55)*. In cell cultures, apoE produces the inhibition of Aβ aggregation *(63)*, whereas in canine smooth muscle cells, the addition of apoE3 or apoE4 induces accumulation of Aβ-immunoreactive deposits *(64)*. There are three alleles of ApoE in the human population. ApoE3 is the major allele and is considered the wild type. Considerable evidence demonstrates that the E4 allele of apoE is an important risk factor for AD, being present in 30 to 50% of patients who develop late-onset AD *(65,66)*. By contrast, inheritance of the E2 allele appears to reduce the risk of developing AD *(67)*. Amyloid precursor protein (APP) transgenic mice lacking apoE, develop diffuse, nonfibrillar Aβ deposits, but the amyloid burden is markedly reduced in old mice when compared with the transgenic complemented with murine apoE *(68)*. Expression of apoE4 in the absence of mouse apoE increases hippocampal Aβ levels and amyloid burden *(69)*. One hypothesis to explain the linkage between the E4 allele and late-onset AD is that apoE4 may form a less stable complex with Aβ than does apoE3 or apoE2, thereby rendering Aβ more prompt to aggregate or less sensitive to degradation by endogenous proteases.

Another protein implicated in controlling Aβ aggregation is apoJ, also known as clusterin because of its ability to elicit clustering of Sertoli cells *(70)*. Overexpression of this protein has been observed in many pathological conditions involving injury and chronic inflammation of the brain. Also immunohistochemical studies show apoJ in association with amyloid plaques and vascular amyloid in AD cortex and hippocampus *(71,72)*. Different data sources from in vitro studies demonstrates that apoJ plays an important role as a carrier for Aβ protein *(73)*, forming a stable complex under physiological conditions. As a result of the complex formation, aggregation and polymerization is inhibited, and Aβ remains in the soluble form *(48)*. In addition, resistance to neuronal death in apoJ-immunoreactive neurons suggests a neuroprotective role of clusterin *(74)*. However, in vivo studies in APP transgenic mice show a significant reduction in fibrillar deposits when the apoJ gene is knocked out, arguing for an opposite effect as in vitro *(75)*.

SAP is a calcium-dependent lectin, a normal plasma glycoprotein *(76)*. It consists of five identical 25-kDa subunits of 204 amino acids. Each subunit has multiple antiparallel β-strands arranged in two sheets. It has been reported that SAP binds to DNA, chromatin, and glycosaminoglycans such as heparin, heparan, and dermatan sulfate, which are frequently associated with amyloid deposits *(77)*. Purified SAP inhibits the fibrils formation of Aβ peptide, increasing its solubility in a dose-dependent manner *(50)*. However, other studies provided evidence for an opposite effect of SAP in vitro *(56)*. In addition, SAP has been shown to prevent the proteolysis of the Aβ fibrils when the protein is bound to the aggregates, contributing to their stabilization in the tissue *(78)*. Experiments in SAP knockout animals demonstrated that the induction of reactive amyloidosis is retarded in mice with targeted deletion of the SAP gene *(79)*. These observations indicate that inhibition of SAP binding to amyloid fibrils might be an attractive therapeutic target for drugs to clear amyloidosis *(80)*. The drug CPHPC (R-1-[6-[R-2-carboxy-pyrrolidin-1-yl]-6-oxo-hexanoyl]pyrrolidine-2-carboxylic acid) was developed to prevent the SAP binding to amyloid fibrils *(80)*.Clinical trials with CPHPC show a decrease in circulating SAP, but its efficiency in amyloid clearance and reduction has not been demonstrated.

α-l-Antichymotrypsin (ACT) is a glial-derived serine protease inhibitor of cathepsin and chymotrypsin-like enzymes *(81)*. ACT, a minor protein component of β-amyloid deposits, is able to inhibit Aβ aggregation into fibrils *(82,83)*, but it is incapable of modulating the toxicity in primary rat hippocampal cell cultures *(82)*. However, other in vitro studies have reported the enhancement of amyloid formation by ACT *(55)*. The later conclusion is supported by in vivo experiments in transgenic animals that overexpresses the mutant APP and ACT showed that the protein promotes amyloid deposition *(84)*. Further experiments indicated that depending on the ACT concentration, both inhibitory and promoter effects can be seen, and a molecular explanation was given based on the binding of Aβ to two different β-sheets of ACT *(51)*.

2.2. Rational Design of Short Peptide Inhibitors

An extensively used strategy to rationally develop specific inhibitors of Aβ misfolding and aggregation is to use short peptides based on the identification of the protein region needed for self-recognition. Tjerberg and colleagues showed that $A\beta_{16-20}$ (KLVFF) is the most important region for Aβ protein–protein interaction *(85)*, in agreement with previous reports from several groups using Aβ mutations, which demonstrated that the central hydrophobic domain of Aβ was responsible for protein misfolding and aggregation *(86–88)*. Tjerberg and coworkers also show that this pentapeptide is able to bind to full-length Aβ and inhibit the formation of amyloid fibrils *(85)*. However, $A\beta_{16-20}$ spontaneously aggregates into amyloid-like fibrils, and thus, its use as an inhibitor might be problematic. Therefore, several groups began to modify this sequence to produce peptide derivatives containing the self-recognition motif, but at the same time a disrupting element that might enhance their inhibitory activity (**Fig. 2**).

Our approach using β-sheet breaking amino acids was the first to lead to modified peptides with inhibitory activity *(89)*. In recent years, the β-sheet

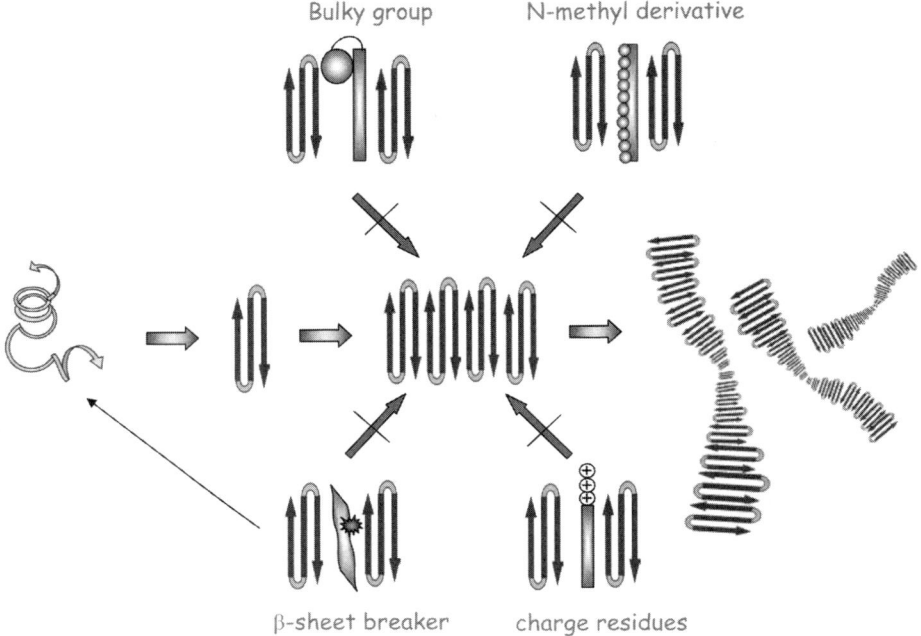

Fig. 2. Diagrammatic representation of several rationally designed peptides for preventing amyloid formation. Based on the structural and sequence determinants of Aβ misfolding and aggregation, several peptide strategies have been designed to prevent and possible reverse amyloid formation, including the use of bulky groups, *N*-methylations, β-sheet breakers, and charge residues.

breaker concept has been used to refer to any compound with amyloid inhibitory properties. However, this name has been created to refer specifically to compounds rationally designed to break β-sheets. Based on the fact that pathological Aβ conformation is stabilized by intermolecular interactions, β-sheet breakers have been designed to bind specifically to Aβ forming a complex that stabilizes the physiological conformation and destabilizes the abnormal conformation of Aβ (**Fig. 2**). To achieve binding and specificity, a sequence homology of Aβ was maintained as a self-recognition motif (residues 17–21), but a key residue for β-sheet formation was replaced by a proline, an amino acid thermodynamically unable to fit inside this structure *(24,89)*. It was shown that valine at position 18 of Aβ plays an important role on stabilizing β-sheet folding in the peptide, but it is not absolutely necessary for self-recognition *(87)*. In light of this, valine was replaced by proline in the inhibitor peptide to block β-sheet formation. In addition, charged residues were added at the ends of the peptide to increase solubility. Several β-sheet–breaker peptides were generated and tested and fragments of 11 and 5 residues (iAβ11 and iAβ5) were reported as active in vitro *(89)*. In particular, iAβ5 binds Aβ with high affinity, inhibiting peptide conformational changes that consequently results in amyloid formation. Even more, quantification of fibrils by thioflavin S, Congo red, sedimentation, and electron microscopy demonstrates that iAβ5 induces the disassembly of preformed fibrils in vitro *(90)*. Also, it had been demonstrated that co-incubation of Aβ with the five-residue peptide can prevent neuronal death in neuroblastoma cultures *(90)*. More importantly, this compound was shown to inhibit and disassembly amyloid deposits in three different animal models of AD, including the most relevant transgenic mice models *(90–92)*. Moreover, we demonstrated that treatment with the compound led to a decrease of the associated cerebral histopathological changes, such as neuronal degeneration and brain inflammation and improvement on behavior *(90–93)*. The pharmacokinetic and toxicological characteristics of this compound have been extensively studied in both animal and human individuals, as well as its ability to cross the blood–brain barrier *(94)*. The proven in vivo potency and favorable pharmacological and toxicological characteristics of the lead β-sheet–breaker peptide has let this compound to be the most advanced rationally designed peptide inhibitor for the treatment of AD.

Another approach pioneered by Murphy and colleagues is based on the fact that the major force driving Aβ aggregation is hydrophobicity, and thus addition of charged residues to the ends of the recognition motif was proposed as a disrupting element *(95,96)* (**Fig. 2**). They showed that at least three charged groups are required as an appropriate disrupting element. The peptides KLVFFKKKK and KLVFFEEEE were shown to alter the fibril morphology and to reduce the cellular toxicity. In contrast, the neutral compound KLVFFSSSS does not show any activity, suggesting the importance of including charged residues.

Findeis and coworkers' *(97,98)* strategy was to retain a peptide sequence that could bind to Aβ and add a bulky group, such as a steroid, at its terminus to hinder Aβ polymerization (**Fig. 2**). Peptides of 15-residue derived from Aβ were tested with and without N-terminal modification. From these studies Aβ $_{16-30}$ was identified as particularly active, especially with the addition of a cholyl modifier *(97)*. The lead compound identified was cholyl-LVFFA-OH, also named PPI-368. This compound potently inhibits the nucleation of Aβ polymerization and also the polymerization in preseeded fibril extension assays. No fibrils were detected by electron micrographic analysis in samples of Aβ incubated with the inhibitor. In addition, samples of Aβ, in which nucleation had been delayed in the presence of PPI-368, were nontoxic in culture cells *(98)*. Unfortunately, the all-D-amino acid peptide cholyl-LVFFA-OH was cleared up almost completely after the hepatic first pass, possibly because the cholyl group was recognized as an endogenous bile component *(98)*.

Several teams are studying the incorporation of N-methyl amino acids into peptides as disrupting elements *(99,100)*. Hughes et al. have hypothesized that N-methylated (NMe) peptide derivatives can act as β-sheet breakers because one side of the peptide presents a hydrogen-bonding "complementary" face to the protein, whereas the other side having N-methyl groups in place of backbone NH groups, represents a "blocking" face (**Fig. 2**). NMe derivatives of $β_{25-35}$ can both prevent the aggregation of fibrils and inhibit the resulting toxicity *(99)*. The effectiveness of these peptides depends on the position of the NMe group. Meredith and coworkers investigated NME peptides on the 16–20 region of Aβ *(100)*. They demonstrate, through two-dimensional NMR and circular dichroic spectroscopy, that a pentapeptide with two N-methyl amino acids, $Aβ_{16-20}$ m or Ac-K(Me)LV(Me)FF-NH2 has the intended structure of an extended β-strand, which is extremely stable to changes in solvent conditions, denaturation by heating, changes in pH, and addition of denaturants. Thioflavin T fluorescence assays demonstrate that $Aβ_{16-20}$ m is an effective fibrillogenesis inhibitor and also disassembles preformed Aβ fibrils *(100)*.

A somehow different strategy has been described by Blanchard and colleagues, who selected peptides from a combinatorial library of hexapeptides containing a tri-glycine spacer with varying degrees of β-sheet–forming potential *(101)*. These peptides were composed of D-amino acids to avoid the proteolytic attack expected when the drugs were given orally or through injection. The selection of these peptides was done by their ability to complex with tagged Aβ. A few of these so-called "decoy peptides" were able to abolish the calcium influx caused by aggregated Aβ. These peptides function as decoys because they have features that make them "attractive" to β-amyloid peptides, which may prefer binding to the decoys rather than to each other *(101,102)*. The mechanism of interaction between the decoy peptide and β-amyloid is not yet clear, but it seems that they coaggregate to form a new fibril that cannot induce calcium influx and cytotoxicity *(101)*. These decoy peptides do not interfere with fibril formation, but they modify and render fibrils harmless *(102)*.

Although some peptides have been used as drugs, the development of peptides as clinically useful compounds is greatly limited by their poor metabolic stability and low bioavailability, which is due in part to their inability to readily cross membrane barriers such as the intestinal and blood–brain barriers *(103)*. However, several strategies are available to minimize the weaknesses of these molecules, including the production of chemically modified peptides, pseudopeptides, and peptidemimetics *(103)*.

3. Conclusions and Perspectives

Although it is not yet formally proven, compelling evidences suggest that protein misfolding, aggregation, and tissue accumulation is the most likely cause of a variety of human degenerative diseases. Therefore, various strategies to correct protein misfolding are under development. Lead compounds can be discovered either by blind screening of large chemical libraries, by identification of endogenous molecules known to bind to amyloid, or by modifying core regions of the amyloidogenic peptide or protein. Hopefully, the coming years will bring important data on the useful of some of these strategies to produce drugs for these devastating diseases.

References

1. Fink, A. L. (1999) Chaperone-mediated protein folding. *Physiol Rev.* **79,** 425–449.
2. Kelly, J. W. (1996) Alternative conformations of amyloidogenic proteins govern their behavior. *Curr. Opin. Struct. Biol.* **6,** 11–17.
3. Dobson, C. M. (1999) Protein misfolding, evolution and disease. *Trends Biochem. Sci.* **24,** 329–332.
4. Carrell, R. W. and Lomas, D. A. (1997) Conformational disease. *Lancet* **350,** 134–138.
5. Soto, C. (2001) Protein misfolding and disease; protein refolding and therapy. *FEBS Lett.* **498,** 204–207.
6. Blake, C. C., Serpell, L. C., Sunde, M., Sandgren, O., and Lundgren, E. (1996) A molecular model of the amyloid fibril. *Ciba Found. Symp.* **199,** 6–15.
7. Glenner, G. G. (1981) Amyloidosis. Its role in Alzheimer's disease and other diseases. *Ann. Pathol.* **1,** 105–108.
8. Sipe, J. D. and Cohen, A. S. (2000) Review: history of the amyloid fibril. *J. Struct. Biol.* **130,** 88–98.
9. Cohen, A. S. and Connors, L. H. (1987) The pathogenesis and biochemistry of amyloidosis. *J. Pathol.* **151,** 1–10.
10. Terry, R. D. (1994) Neuropathological changes in Alzheimer disease. *Prog. Brain Res.* **101,** 383–390.
11. Forno, L. S. (1996) Neuropathology of Parkinson disease. *J. Neuropathol. Exp. Neurol.* **55,** 259–272.
12. DiFiglia, M., Sapp, E., Chase, K. O., Davies, S. W., Bates, G. P., Vonsattel, J. P., et al. (1997) Aggregation of huntingtin in neuronal intranuclear inclusions and dystrophic neurites in brain. *Science* **277,** 1990–1993.
13. Bolton, D. C., McKinley, M. P., and Prusiner, S. B. (1982) Identification of a protein that purifies with the scrapie prion. *Science* **218,** 1309–1311.
14. Hull, R. L., Westermark, G. T., Westermark, P., and Kahn, S. E. (2004) Islet amyloid: a critical entity in the pathogenesis of type 2 diabetes. *J. Clin. Endocrinol. Metab.* **89,** 3629–3643.
15. Ghiso, J., Wisniewski, T., and Frangione, B. (1994) Unifying features of systemic and cerebral amyloidosis. *Mol. Neurobiol.* **8,** 49–64.
16. Buxbaum, J. N. and Tagoe, C. E. (2000) The genetics of the amyloidoses. *Annu. Rev. Med.* **51,** 543–569.
17. Price, D. L., Wong, P. C., Markowska, A. L., Lee, M. K., Thinakaren, G., Cleveland, D. W., et al. (2000) The value of transgenic models for the study of neurodegenerative diseases. *Ann. N.Y. Acad. Sci.* **920,** 179–191.

18. Jarrett, J. T. and Lansbury, P. T., Jr. (1993) Seeding "one-dimensional crystalliza-
    tion" of amyloid: a pathogenic mechanism in Alzheimer's disease and scrapie?
    Cell **73,** 1055–1058.
19. Harper, J. D. and Lansbury, P. T., Jr. (1997) Models of amyloid seeding in
    Alzheimer's disease and scrapie: mechanistic truths and physiological conse-
    quences of the time-dependent solubility of amyloid proteins. *Annu. Rev. Biochem.*
    66, 385–407.
20. Carrell, R. W. and Gooptu, B. (1998) Conformational changes and disease—
    serpins, prions and Alzheimer's. *Curr. Opin. Struct. Biol.* **8,** 799–809.
21. Cohen, F. E. and Prusiner, S. B. (1998) Pathologic conformations of prion pro-
    teins. *Annu. Rev. Biochem.* **67,** 793–819.
22. Soto, C. (1999) Alzheimer's and prion disease as disorders of protein conforma-
    tion: implications for the design of novel therapeutic approaches. *J. Mol. Med.* 77,
    412–418
23. Serpell, L. C., Sunde, M., Fraser, P. E., Luther, P. K., Morris, E. P., Sangren, O.,
    et al. (1995) Examination of the structure of the transthyretin amyloid fibril by
    image reconstruction from electron micrographs. *J. Mol. Biol.* **254,** 113–118.
24. Soto, C. (1999) Plaque busters: strategies to inhibit amyloid formation in
    Alzheimer's disease. *Mol. Med. Today* **5,** 343–350.
25. LeVine, H. III and Scholten, J. D. (1999) Screening for pharmacologic inhibitors
    of amyloid fibril formation. *Meth. Enzymol.* **309,** 467–476.
26. Lorenzo, A. and Yankner, B. A. (1994) Beta-amyloid neurotoxicity requires fibril
    formation and is inhibited by congo red. *Proc. Natl. Acad. Sci. USA* **91,** 12243–12247.
27. Wood, S. J., MacKenzie, L., Maleeff, B., Hurle, M. R., and Wetzel, R. (1996)
    Selective inhibition of Aβ fibril formation. *J. Biol. Chem.* **271,** 4086–4092.
28. Kisilevsky, R., Lemieux, L. J., Fraser, P. E., Kong, X., Hultin, P. G., and Szarek,
    W. A. (1995) Arresting amyloidosis in vivo using small-molecule anionic
    sulphonates or sulphates: implications for Alzheimer's disease. *Nat. Med.* **1,** 143–148.
29. Allsop, D., Gibson, G., Martin, I. K., Moore, S., Turnbull, S., and Twyman, L. J.
    (2001) 3-p-Toluoyl-2-[4'-(3-diethylaminopropoxy)-phenyl]-benzofuran and 2-[4'-
    (3-diethylaminopropoxy)-phenyl]-benzofuran do not act as surfactants or micelles
    when inhibiting the aggregation of beta-amyloid peptide. *Bioorg. Med. Chem.
    Lett.* **11,** 255–257.
30. Tomiyama, T., Asano, S., Suwa, Y., Morita, T., Kataoka, K., Mori, H., and Endo,
    N. (1994) Rifampicin prevents the aggregation and neurotoxicity of amyloid beta
    protein in vitro. *Biochem. Biophys. Res. Commun.* **204,** 76–83.
31. Pappolla, M., Bozner, P., Soto, C., Shao, H., Robakis, N. K., Zagorski, M., et al.
    (1998) Inhibition of Alzheimer beta-fibrillogenesis by melatonin. *J. Biol. Chem.*
    273, 7185–7188.
32. Salomon, A. R., Marcinowski, K. J., Friedland, R. P., and Zagorski, M. G. (1996)
    Nicotine inhibits amyloid formation by the beta-peptide. *Biochemistry* **35,**
    13568–13578.
33. Hosoda, T., Nakajima, H., and Honjo, H. (2001) Estrogen protects neuronal cells
    from amyloid beta-induced apoptotic cell death. *Neuroreport* **12,** 1965–1970.

34. Gervais, F., Chalifour, R., Garceau, D., Kong, X., Laurin, J., Mclaughlin. R., et al. (2001) Glycosaminoglycan mimetics: a therapeutic approach to cerebral amyloid angiopathy. *Amyloid* **8,** 28–35.
35. De Felice, F. G., Houzel, J. C., Garcia-Abreu, J., Louzada, P. R., Jr., Afonso, R. C., Meirelles, M. N., et al. (2001) Inhibition of Alzheimer's disease beta-amyloid aggregation, neurotoxicity, and in vivo deposition by nitrophenols: implications for Alzheimer's therapy. *FASEB J.* **15,** 1297–1299.
36. Forloni, G., Colombo, L., Girola, L., Tagliavini, F., and Salmona, M. (2001) Anti-amyloidogenic activity of tetracyclines: studies in vitro. *FEBS Lett.* **487,** 404–407.
37. Merlini, G., Ascari, E., Amboldi, N., Bellotti, V., Arbustini, E., Perfetti, V., et al. (1995) Interaction of the anthracycline 4'-iodo-4'-deoxydoxorubicin with amyloid fibrils: inhibition of amyloidogenesis. *Proc. Natl. Acad. Sci. USA* **92,** 2959–2963.
38. Cherny, R. A., Atwood, C. S., Xilinas, M. E., Gray, D. N., Jones, W. D., McLean, C. A., et al. (2001) Treatment with a copper-zinc chelator markedly and rapidly inhibits beta-amyloid accumulation in Alzheimer's disease transgenic mice. *Neuron* **30,** 665–676.
39. Lim, G. P., Yang, F., Chu, T., Chen, P., Beech, W., Teter, B., et al. (2000) Ibuprofen suppresses plaque pathology and inflammation in a mouse model for Alzheimer's disease. *J. Neurosci.* 20, 5709–5714
40. Nakagami, Y., Nishimura, S., Murasugi, T., Kaneko, I., Meguro, M., Marumoto, S., et al. (2002) A novel beta-sheet breaker, RS-0406, reverses amyloid beta-induced cytotoxicity and impairment of long-term potentiation in vitro. *Br. J. Pharmacol.* **137,** 676–682.
41. Wisniewski, T. and Frangione, B. (1992) Apolipoprotein E: a pathological chaperone protein in patients with cerebral and systemic amyloid. *Neurosci. Lett.* **135,** 235–238.
42. Pepys, M. B., Dyck, R. F., De Beer, F. C., Skinner, M., and Cohen, A. S. (1979) Binding of serum amyloid P-component (SAP) by amyloid fibrils. *Clin. Exp. Immunol.* **38,** 284–293.
43. Snow, A. D., Willmer, J., and Kisilevsky, R. (1987) Sulfated glycosaminogly-cans: a common constituent of all amyloids? *Lab. Invest.* **56,** 120–123.
44. Frangione, B., Wisniewski, T., and Ghiso, J. (1994) Chaperoning Alzheimer's amyloids. *Neurobiol. Aging* 15, S97–S99.
45. Soto, C., Ghiso, J., and Frangione, B. (1997) Alzheimer's amyloid-β aggregation is modulated by the interaction of multiple factors. *Alzheimer's Res.* **3,** 215–222.
46. Holtzman, D. M. (2003) Potential role of endogenous and exogenous amyloid-beta binding molecules in the pathogenesis, diagnosis, and treatment of Alzheimer disease. *Alzheimer Dis. Assoc. Disord.* **17,** 151–153.
47. Evans, K. C., Berger, E. P., Cho, C. G., Weisgraber, K. H., and Lansbury, P. T., Jr. (1995) Apolipoprotein E is a kinetic but not a thermodynamic inhibitor of amyloid formation: implications for the pathogenesis and treatment of Alzheimer disease. *Proc. Natl. Acad. Sci. USA* **92,** 763–767.
48. Matsubara, E., Soto, C., Governale, S., Frangione, B., and Ghiso, J. (1996) Apolipoprotein J and Alzheimer's amyloid beta solubility. *Biochem. J.* **316,** 671–679.

49. Schwarzman, A. L., Gregori, L., Vitek, M. P., Lyubski, S., Strittmatter, W. J., Enghilde, J. J., et al. (1994) Transthyretin sequesters amyloid beta protein and prevents amyloid formation. *Proc. Natl. Acad. Sci. USA* **91**, 8368–8372.

50. Janciauskiene, S., Garcia, d. F., Carlemalm, E., Dahlback, B., and Eriksson, S. (1995) Inhibition of Alzheimer beta-peptide fibril formation by serum amyloid P component. *J. Biol. Chem.* **270**, 26041–26044.

51. Janciauskiene, S., Rubin, H., Lukacs, C. M., and Wright, H. T. (1998) Alzheimer's peptide Abeta1-42 binds to two beta-sheets of alpha1-antichymotrypsin and transforms it from inhibitor to substrate. *J. Biol. Chem.* **273**, 28360–28364.

52. Castillo, G. M., Lukito, W., Peskind, E., Raskind, M., Kirschner, D. A., Yee, A. G., et al. (2000) Laminin inhibition of beta-amyloid protein (Abeta) fibrillogenesis and identification of an Abeta binding site localized to the globular domain repeats on the laminin a chain. *J. Neurosci. Res.* **62**, 451–462.

53. Bronfman, F. C., Garrido, J., Alvarez, A., Morgan, C., and Inestrosa, N. C. (1996) Laminin inhibits amyloid-beta-peptide fibrillation. *Neurosci. Lett.* **218**, 201–203.

54. Castano, E. M., Prelli, F., Wisniewski, T., Golabek, A., Kumar, R. A., Soto, C., et al. (1995) Fibrillogenesis in Alzheimer's disease of amyloid beta peptides and apolipoprotein E. *Biochem. J.* **306**, 599–604.

55. Ma, J., Yee, A., Brewer, H. B., Jr., Das, S., and Potter, H. (1994) Amyloid-associated proteins alpha 1-antichymotrypsin and apolipoprotein E promote assembly of Alzheimer beta-protein into filaments. *Nature* **372**, 92–94.

56. Hamazaki, H. (1995) Amyloid P component promotes aggregation of Alzheimer's beta-amyloid peptide. *Biochem. Biophys. Res. Commun.* **211**, 349–353.

57. Hamilton, J. A. and Benson, M. D. (2001) Transthyretin: a review from a structural perspective. *Cell Mol. Life Sci.* **58**, 1491–1521.

58. Saraiva, M. J. (2001) Transthyretin amyloidosis: a tale of weak interactions. *FEBS Lett.* **498**, 201–203.

59. Golabek, A., Marques, M. A., Lalowski, M., and Wisniewski, T. (1995) Amyloid beta binding proteins in vitro and in normal human cerebrospinal fluid. *Neurosci. Lett.* **191**, 79–82.

60. Narindrasorasak, S., Altman, R. A., Gonzalez-DeWhitt, P., Greenberg, B. D., and Kisilevsky, R. (1995) An interaction between basement membrane and Alzheimer amyloid precursor proteins suggests a role in the pathogenesis of Alzheimer's disease. *Lab. Invest.* **72**, 272–282.

61. Morgan, C., Bugueno, M. P., Garrido, J., and Inestrosa, N. C. (2002) Laminin affects polymerization, depolymerization and neurotoxicity of Abeta peptide. *Peptides* **23**, 1229–1240.

62. Weisgraber, K. H., Roses, A. D., and Strittmatter, W. J. (1994) The role of apolipoprotein E in the nervous system. *Curr. Opin. Lipidol.* **5**, 110–116.

63. Ohman, T., Dang, N., LeBoeuf, R. C., Furlong, C. E., and Fukuchi, K. (1996) Expression of apolipoprotein E inhibits aggregation of the C-terminal fragments of beta-amyloid precursor protein. *Neurosci. Lett.* **210**, 65–68.

64. Mazur-Kolecka, B., Frackowiak, J., and Wisniewski, H. M. (1995) Apolipoproteins E3 and E4 induce, and transthyretin prevents accumulation of the Alzheimer's beta-amyloid peptide in cultured vascular smooth muscle cells. *Brain Res.* **698**, 217–222.

65. Roses, A. D. (1996) Apolipoprotein E alleles as risk factors in Alzheimer's disease. *Annu. Rev. Med.* **47,** 387–400.
66. Ohm, T. G., Scharnagl, H., Marz, W., and Bohl, J. (1999) Apolipoprotein E isoforms and the development of low and high Braak stages of Alzheimer's disease-related lesions. *Acta Neuropathol. (Berl)* **98,** 273–280.
67. Bales, K. R., Dodart, J. C., DeMattos, R. B., Holtzman, D. M., and Paul, S. M. (2002) Apolipoprotein E, amyloid, and Alzheimer disease. *Mol. Interv.* **2,** 363–375, 339.
68. Holtzman, D. M., Fagan, A. M., Mackey, B., Tenkova, T., Sartorius, L., Paul, S. M., et al. (2000) Apolipoprotein E facilitates neuritic and cerebrovascular plaque formation in an Alzheimer's disease model. *Ann. Neurol.* **47,** 739–747.
69. Hartman, R. E., Laurer, H., Longhi, L., Bales, K. R., Paul, S. M., McIntosh, T. K., et al. (2002) Apolipoprotein E4 influences amyloid deposition but not cell loss after traumatic brain injury in a mouse model of Alzheimer's disease. *J. Neurosci.* **22,** 10083–10087.
70. Calero, M., Rostagno, A., Frangione, B., and Ghiso, J. (2005) Clusterin and Alzheimer's disease. *Subcell. Biochem.* **38,** 273–298.
71. Choi-Miura, N. H., Ihara, Y., Fukuchi, K., Takeda, M., Nakano, Y., Tobe, T., and Tomita, M. (1992) SP-40,40 is a constituent of Alzheimer's amyloid. *Acta Neuropathol. (Berl)* **83,** 260–264.
72. Kida, E., Pluta, R., Lossinsky, A. S., Golabek, A. A., Choi-Miura, N. H., Wisniewski, H. M., et al. (1995) Complete cerebral ischemia with short-term survival in rat induced by cardiac arrest. II. Extracellular and intracellular accumulation of apolipoproteins E and J in the brain. *Brain Res.* **674,** 341–346.
73. Matsubara, E., Frangione, B., and Ghiso, J. (1995) Characterization of apolipoprotein J-Alzheimer's A beta interaction. *J. Biol. Chem.* **270,** 7563–7567.
74. Giannakopoulos, P., Kovari, E., French, L. E., Viard, I., Hof, P. R., and Bouras, C. (1998) Possible neuroprotective role of clusterin in Alzheimer's disease: a quantitative immunocytochemical study. *Acta Neuropathol. (Berl)* **95,** 387–394.
75. DeMattos, R. B., Cirrito, J. R., Parsadanian, M., May, P. C., O'Dell, M. A., Taylor, J. W., et al. (2004) ApoE and Clusterin Cooperatively Suppress Abeta Levels and Deposition. Evidence that ApoE regulates extracellular abeta metabolism in vivo. *Neuron* **41,** 193–202.
76. Pepys, M. B., Baltz, M. L., De Beer, F. C., Dyck, R. F., Holford, S., Breathnach, S. M., et al. (1982) Biology of serum amyloid P component. *Ann. N.Y. Acad. Sci.* **389,** 286–298.
77. Pepys, M. B., Rademacher, T. W., Amatayakul-Chantler, S., Williams, P., Noble, G. E., Hutchinson, W. L., et al. (1994) Human serum amyloid P component is an invariant constituent of amyloid deposits and has a uniquely homogeneous glycostructure. *Proc. Natl. Acad. Sci. USA* **91,** 5602–5606.
78. Tennent, G. A., Lovat, L. B., and Pepys, M. B. (1995) Serum amyloid P component prevents proteolysis of the amyloid fibrils of Alzheimer disease and systemic amyloidosis. *Proc. Natl. Acad. Sci. USA* **92,** 4299–4303.
79. Botto, M., Hawkins, P. N., Bickerstaff, M. C., Herbert, J., Bygrave, A. E., McBride, A., et al. (1997) Amyloid deposition is delayed in mice with targeted deletion of the serum amyloid P component gene. *Nat. Med.* **3,** 855–859

80. Pepys, M. B., Herbert, J., Hutchinson, W. L., Tennent, G. A., Lachmann, H. J., Gallimore, J. R., et al. (2002) Targeted pharmacological depletion of serum amyloid P component for treatment of human amyloidosis. *Nature* **417**, 254–259.
81. Abraham, C. R. and Potter, H. (1989) Alpha 1-antichymotrypsin in brain aging and disease. *Prog. Clin. Biol. Res.* **317**, 1037–1048.
82. Aksenova, M. V., Aksenov, M. Y., Butterfield, D. A., and Carney, J. M. (1996) alpha-1-antichymotrypsin interaction with A beta (1–40) inhibits fibril formation but does not affect the peptide toxicity. *Neurosci. Lett.* **211**, 45–48.
83. Janciauskiene, S., Eriksson, S., and Wright, H. T. (1996) A specific structural interaction of Alzheimer's peptide A beta 1–42 with alpha 1-antichymotrypsin. *Nat. Struct. Biol.* **3**, 668–671.
84. Nilsson, L. N., Bales, K. R., DiCarlo, G., Gordon, M. N., Morgan, D., Paul, S. M., et al. (2001) Alpha-1-antichymotrypsin promotes beta-sheet amyloid plaque deposition in a transgenic mouse model of Alzheimer's disease. *J. Neurosci.* **21**, 1444–1451.
85. Tjernberg, L. O., Naslund, J., Lindqvist, F., Johansson, J., Karlstrom, A. R., Thyberg, J., et al. (1996) Arrest of beta-amyloid fibril formation by a pentapeptide ligand. *J. Biol. Chem.* **271**, 8545–8548.
86. Hilbich, C., Kisters-Woike, B., Reed, J., Masters, C. L., and Beyreuther, K. (1992) Substitutions of hydrophobic amino acids reduce the amyloidogenicity of Alzheimer's disease beta A4 peptides. *J. Mol. Biol.* **228**, 460–473.
87. Soto, C., Castano, E. M., Frangione, B., and Inestrosa, N. C. (1995) The alpha-helical to beta-strand transition in the amino-terminal fragment of the amyloid beta-peptide modulates amyloid formation. *J. Biol. Chem.* **270**, 3063–3067.
88. Wood, S. J., Wetzel, R., Martin, J. D., and Hurle, M. R. (1995) Prolines and amyloidogenicity in fragments of the Alzheimer's peptide beta/A4. *Biochemistry* **34**, 724–730.
89. Soto, C., Kindy, M. S., Baumann, M., and Frangione, B. (1996) Inhibition of Alzheimer's amyloidosis by peptides that prevent beta-sheet conformation. *Biochem. Biophys. Res. Commun.* 226, 672–680.
90. Soto, C., Sigurdsson, E. M., Morelli, L., Kumar, R. A., Castano, E. M., and Frangione, B. (1998) Beta-sheet breaker peptides inhibit fibrillogenesis in a rat brain model of amyloidosis: implications for Alzheimer's therapy. *Nat. Med.* **4**, 822–826.
91. Sigurdsson, E. M., Permanne, B., Soto, C., Wisniewski, T., and Frangione, B. (2000) In vivo reversal of amyloid-beta lesions in rat brain. *J. Neuropathol. Exp. Neurol.* **59**, 11–17.
92. Permanne, B., Adessi, C., Saborio, G. P., Fraga, S., Frossard, M. J., Van Dorpe, J., et al. (2002) Reduction of amyloid load and cerebral damage in a transgenic mouse model of Alzheimer's disease by treatment with a beta-sheet breaker peptide. *FASEB J.* **16**, 860–862.
93. Chacon, M. A., Barria, M. I., Soto, C., and Inestrosa, N. C. (2004) Beta-sheet breaker peptide prevents Abeta-induced spatial memory impairments with partial reduction of amyloid deposits. *Mol. Psychiatry* **9**, 953–961.

94. Adessi, C., Frossard, M. J., Boissard, C., Fraga, S., Bieler, S., Ruckle, T., et al. (2003) Pharmacological profiles of peptide drug candidates for the treatment of Alzheimer's disease. *J. Biol. Chem.* **278,** 13905–13911.
95. Ghanta, J., Shen, C. L., Kiessling, L. L., and Murphy, R. M. (1996) A strategy for designing inhibitors of beta-amyloid toxicity. *J. Biol. Chem.* **271,** 29525– 29528.
96. Pallitto, M. M., Ghanta, J., Heinzelman, P., Kiessling, L. L., and Murphy, R. M. (1999) Recognition sequence design for peptidyl modulators of beta-amyloid aggregation and toxicity. *Biochemistry* **38,** 3570–3578.
97. Findeis, M. A., Musso, G. M., Arico-Muendel, C. C., Benjamin, H. W., Hundal, A. M., Lee, J. J., et al. (1999) Modified-peptide inhibitors of amyloid beta-peptide polymerization. *Biochemistry* **38,** 6791–6800.
98. Findeis, M. A., Lee, J. J., Kelley, M., Wakefield, J. D., Zhang, M. H., Chin, J., et al. (2001) Characterization of cholyl-leu-val-phe-phe-ala-OH as an inhibitor of amyloid beta-peptide polymerization. *Amyloid* **8,** 231–241.
99. Hughes, E., Burke, R. M., and Doig, A. J. (2000) Inhibition of toxicity in the beta-amyloid peptide fragment beta -(25-35) using N-methylated derivatives: a general strategy to prevent amyloid formation. *J. Biol. Chem.* **275,** 25109–25115.
100. Gordon, D. J., Tappe, R., and Meredith, S. C. (2002) Design and characterization of a membrane permeable N-methyl amino acid-containing peptide that inhibits Abeta1-40 fibrillogenesis. *J. Pept. Res.* **60,** 37–55.
101. Blanchard, B. J., Konopka, G., Russell, M., and Ingram, V. M. (1997) Mechanism and prevention of neurotoxicity caused by beta-amyloid peptides: relation to Alzheimer's disease. *Brain Res.* **776,** 40–50.
102. Blanchard, B. J., Hiniker, A. E., Lu, C. C., Margolin, Y., Yu, A. S., and Ingram, V. M. (2000) Elimination of amyloid beta neurotoxicity. *J. Alzheimers. Dis.* **2,** 137–149.
103. Adessi, C. and Soto, C. (2002) Converting a peptide into a drug: strategies to improve stability and bioavailability. *Curr. Med. Chem.* **9,** 963–978.

Index